Multisensory Control
of Posture

Multisensory Control of Posture

Edited by

T. Mergner
University of Freiburg
Freiburg, Germany

and

F. Hlavačka
Slovak Academy of Sciences
Bratislava, Slovakia

Springer Science+Business Media, LLC

Library of Congress Cataloging in Publication Data

Multisensory control of posture / edited by T. Mergner and F. Hlavačka.
    p.   cm.
    Proceedings of an international symposium on sensory interaction in posture and movement
control, held September 9–11, 1994 in Smolenice, Slovakia, as a satellite symposium to the
European Neuroscience Association meeting of 1994—T.p. verso.
    Includes bibliographical references and index.
    ISBN 978-0-306-45101-0   ISBN 978-1-4615-1931-7 (eBook)
    DOI 10.1007/978-1-4615-1931-7
    1. Posture—Congresses. 2. Vestibular apparatus—Congresses. I. Mergner, Th. (Thomas) II.
Hlavačka, F. III. European Neuroscience Association.
    [DNLM: 1. Posture—physiology—congresses. 2. Sensation—physiology—congresses. 3. Ves-
tibule—physiology—congresses. 4. Movement—physiology—congresses. WE 103 M9615 1995]
QP310.3.M85   1995
612′.04—dc20
DNLM/DLC   95-30547
for Library of Congress   CIP

Proceedings of an international symposium on Sensory Interaction in Posture and Movement Control,
held September 9–11, 1994, in Smolenice, Slovakia, as a satellite symposium to the European
Neuroscience Association meeting of 1994

© 1995 Springer Science+Business Media New York
Originally published by Plenum Press, New York in 1995

10 9 8 7 6 5 4 3 2 1

PREFACE

From recent developments in the rapidly growing area of neuroscience it has become increasingly clear that a simplistic description of brain function as a broad collection of simple input-output relations is quite inadequate. Introspection already tells us that our motor behavior is guided by a complex interplay between many inputs from the outside world and from our internal "milieu," internal models of ourselves and the outside world, memory content, directed attention, volition, and so forth. Also, our motor activity normally involves more than a circumscribed group of muscles, even if we intend to move only one effector organ. For example, a reaching movement or a reorientation of a sensory organ almost invariably requires a pattern of preparatory or assisting activities in other parts of the body, like the ones that maintain the body's equilibrium.

The present volume is a summary of the papers presented at the symposium "Sensory Interaction in Posture and Movement Control" that was held at Smolenice Castle near Bratislava, Slovakia, as a Satellite Symposium to the ENA Meeting 1994 in Vienna. The focus of this meeting was not only restricted to the "classical" sensory interactions such as between vestibular and visual signals, or between otolith and semicircular canal inputs. Rather, the symposium tried to consider also the interplay between perception and action, between reflexive and volitional motor acts as well as between sensory driven or self-initiated motor acts and reafferent inputs. Furthermore, it tried to consider the complexity of three-dimensional space in which our activities take place, a fact that poses us with the problem of coordinate transformations of sensory inputs that are picked up by one part of the body (e.g., the head) and are used for the control of other parts (e.g., trunk posture).

The aim of the symposium was to bring together researchers working on different topics, who share an interest in how the brain puts the different pieces of information together. Each symposium member could benefit from the expertise of the other member, so as to be alerted to new vistas and to the large variety of conceptual and methodological approaches used in the different laboratories.

This proceedings volume has been organized according to the following topics: Basic Aspects, Coordination and Motor Plan, Role of Visual, Vestibular and Somatosensory Afferents for Posture and Movement Control, Orientation and Perception, Reafferents, Modeling, Clinical Insights, Aviation and Space Flight. These topics do not form separate chapters, since each of the articles may touch upon several of these topics. Also, from the variety of the issues addressed we had to choose a simple title for the book, which, hopefully, is not too misleading.

We wish to express our gratitude to ENA (European Neuroscience Association) for giving us the opportunity to organize the symposium. Furthermore, we acknowledge the financial support provided by the DARA (Deutsche Agentur für Raumfahrtangelegenheiten) both for the symposium and the publishing of the proceedings. In addition, we thank the

Slovak Academy of Sciences and the DFG (Deutsche Forschungsgemeinschaft), which supported the collaboration between our laboratories in Bratislava and Freiburg as well as the symposium. Last but not least, we remember with pleasure and gratitude the excellent organization of the symposium by the staff of the Institute of Normal and Pathological Physiology in Bratislava.

Th. Mergner, Freiburg
F. Hlavačka, Bratislava

CONTENTS

MULTISENSORY CONTROL OF MOVEMENTS AT SPINAL LEVELS AND ITS SUPRASPINAL MODULATION

E. D. Schomburg

Institute of Physiology
University of Göttingen
Humboldtallee 23
D-37073 Göttingen
Germany

INTRODUCTION

The main domain of spinal functions in movement control is often considered to be the evocation of more or less stereotyped reflex responses to different stimuli as the stretch reflex induced by Ia spindle afferents, autogenetic inhibition induced by afferents from Golgi tendon organs and retraction reflexes evoked by nociceptive afferents. However, low threshold cutaneous afferents, joint afferents and group II and III-IV muscle afferents can be assumed to be at least not of a minor importance in the control of complex normal movements. These groups of afferents, together with nociceptive afferents, have been comprised as flexor reflex afferents (FRA) since they all may evoke the flexion reflex under particular conditions. Characteristic features of these afferents are that they may use different short- and long-latency reflex pathways in common, beside so called "private" pathways, and that the interneurones of the common pathways show a wide multisensorial convergence from afferents of a great variety of receptors and a wide convergent input from descending tracts (for reviews see Lundberg, 1979; Baldissera et al., 1981; Schomburg, 1990; Jankowska, 1992).

TECHNICAL COMMENTS

All investigations were performed in anaemically decapitated high spinal cats, thus excluding supraspinal interference (for technical details see Kniffki et al., 1981). The transmission in segmental reflex pathways was investigated with intracellular recording from alpha-motoneurones or by monosynaptic reflex testing. Afferent input was activated by graded electrical nerve stimulation, weak mechanical stimuli (low threshold mechanosensitive afferents) or noxious radiant heat (nociceptive afferents). Convergence onto common interneurones was investigated by testing for spatial facilitation. Opioids were either intra-venously injected (0.5-3.6 mg/kg) or suffused over the spinal cord (10^{-3} - 10^{-5} M).

RESULTS

Multisensorial Convergence in FRA Pathways

As mentioned above a characteristic feature of segmental FRA reflex pathways is the wide multisensorial convergence onto their interneurones. Fig. 1a summarizes the different patterns of convergence from segmental afferents which have been demonstrated in different experimental series. It has to be assumed that there is a great variation of interneuronal subsets with different patterns of convergence from different afferents and different patterns of projection to different motoneurone pools. Each interneuronal subset may thus integrate and convey an individual combination of information. As can be seen from Fig. 1a, the convergence in segmental FRA pathways does not only reflect convergence patterns already known from ascending sensory pathways, as e.g. convergence between low threshold cutaneous afferents and nociceptive afferents, but also between afferents which at the first glance do not have features in common, as e.g. nociceptive cutaneous afferents and group II muscle spindle afferents. As shown in Fig. 1b the facilitation of the excitation evoked by group II muscle afferents by conditioning activation of nociceptive afferents may be quite considerable.

The functional advantage of the spatial facilitation achieved by the multisensorial convergence consists in the opportunity to keep the gain in the unisensorial ways at a quite low level, thus preventing reflex reactions to weak inputs from a single modality (cf. Lundberg et al., 1987). For example an input from low threshold cutaneous afferents will normally not evoke any reflex actions on its own, but it would facilitate the reaction to a concomitant noxious component of a skin stimulus, thus accelerating and enhancing motor reactions if mechanical skin stimuli exceed the noxious threshold (Schomburg, 1991; Steffens and Schomburg, 1993). Under pathologically changed supraspinal control the situation may be different. For example light stroking of the skin in spastic patients may easily induce strong motor reactions.

A similarly wide multisensorial convergence as observed in FRA pathways has been demonstrated in the disynaptic inhibitory group Ia pathway (Hultborn, 1972) and in path-ways used by Ib afferents from Golgi tendon organs (Jankowska, 1984; Harrison and Jankowska, 1985).

Alternative Inhibitory and Excitatory FRA Pathways

If FRA, as there misleading name partly suggests, would in a stereotype way evoke flexion reflexes, the function in the control of movements would be quite restricted. Indeed the flexion reflex is just one type of reflex action evoked by these afferents under particular conditions (e.g., spinal animal). As already suggested in 1959 by Eccles and Lundberg there are parallel inhibitory and excitatory pathways from the different FRA to each motoneurone pool (Fig. 2A). It depends on the spinal and supraspinal constellation in which of the alternative pathways the transmission is prevailing. Fig. 2 B presents an example of the spinal influence onto the transmission in the alternative pathways in different phases of the step cycle during fictive locomotion. Stimulation of the sural nerve in the extension (B1) and flexion (B4) phase evoked a clear excitation, as expected in a flexor motoneurone of a spinal animal. However, given at the transition of the extension to the flexion phase (B3), the same stimulus evokes an IPSP in the same motoneurone. In a similar way the extensive supraspinal

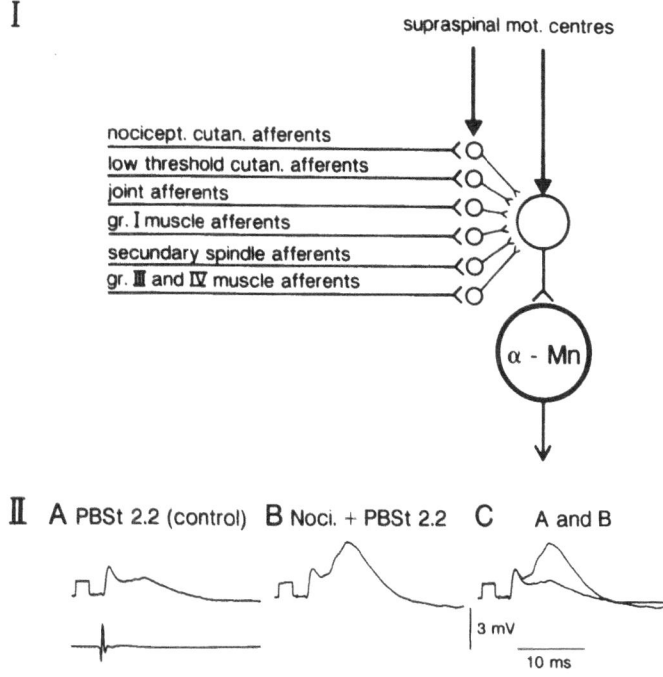

Figure 1. (a) Segmental interneurones as integrators for the performance of centrally commanded movements with a multisensorial convergence from afferents of different receptor systems serving as a feed back for the ongoing movement. (b) Example of the high degree of spatial facilitation in reflex pathways from group II muscle afferents by nociceptive afferents. Intracellular recording of an anterior biceps semimembranosus motoneurone in a high spinal cat. Stimulation of the nerve from posterior biceps semitendinosus (PBSt) with 2.2 times threshold i.e. in the low group II range. (A) Control (lower beam incoming volley recorded from the root entry zone), (B) with conditioning activation of an asynchronous long lasting nociceptive input by noxious radiant heat applied to the skin area innervated by the sural nerve, (C) recordings of A and B superimposed. There is no change of the monosynaptic Ia EPSP, but distinct facilitation of the oligosynaptic group II EPSP (modified from Schomburg, 1990; Schomburg and Steffens, 1993).

control of the segmental FRA interneuronal system may switch the transmission between the alternative excitatory and inhibitory FRA pathways, e.g. in the decerebrate cat all FRA responses are suppressed. If an additional midpontine lesion is performed, FRA evoke inhibition which is also prevalent to flexors. Only after spinalization the "characteristic" flexor reflex pattern with predominant flexor excitation and extensor inhibition occurs.

Descending Monoaminergic and Enkephalinergic Control

What is the functional meaning of the FRA system and its extensive supraspinal control if FRA may evoke such variable reflex actions. As first proposed by Lundberg (1975, 1979; Lundberg et al., 1987) the segmental reflexes of the FRA *per se* are not the main point, but the intercalation of the FRA interneuronal system in the transmission of descending commands for the performance of movements. The supraspinal command for the performance of a movement is mediated via the same segmental interneurones, onto which the multisensorial feed back information from the FRA is channelled back, thus adapting the performance of the movement to the peripheral conditions. As mentioned above the transmission in these feed-back reflex pathways is controlled by different descending pathways.

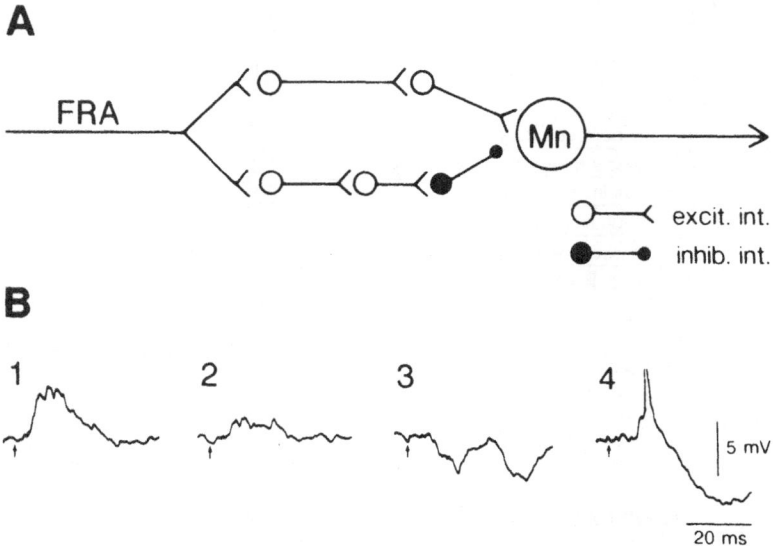

Figure 2. (a) Alternative excitatory and inhibitory pathways (int., interneurone) from FRA to alpha-motoneurones (Mn). The transmission through the alternative pathways is controlled by spinal (see b) and supraspinal mechanisms. (b) Intracellular recordings of a deep peroneal (flexor) motoneurone in a high spinal cat during DOPA induced fictive locomotion. Stimulation of the sural nerve with 8 times threshold (time of stimulus marked by an arrow below the records) during different phases of the step cycle; excitation evoked during the mid extension (B1) and flexion (B4) phase, (B3) inhibition evoked by the same stimulus given at the transition from extension to flexion phase. In the late extension phase (B2) only a small EPSP occurred. (Results from Schomburg and Behrends 1978).

An exhaustive consideration of this descending control would exceed the frame of this presentation (for a review see Baldissera et al., 1981). However, two control systems which gained some particular interest during the recent years should be mentioned: the descending monoaminergic and the enkephalinergic systems which partly show coinciding influences. It was in the sixties that Lundberg and co-workers (Andén et al., 1966) first showed that the transmission in the segmental reflex pathways from FRA is heavily depressed by injection of L-DOPA which activates the turnover and release of transmitter at the spinal terminals of descending monoaminergic pathways. This inhibition of FRA pathways is quite selective and has partly been used as a criterion for the attribution of a pathway to the FRA system. In this context it is interesting that the transmission from group II muscle afferents to α-motoneurones is also depressed by L-DOPA. As shown in Fig. 3 the transmission in the group II pathways is completely suppressed by DOPA, i.e. the depression is even more severe than the transmission in cutaneous pathways. These findings underline the attribution of group II muscle afferents to the FRA (Schomburg and Steffens, 1988; Bras et al., 1988).

It turned out that the selective monoaminergic action onto short latency FRA reflex pathways is not unique. Opioids with a δ- or μ-morphine receptor specificity exert a quite similar specific effect onto these pathways (Schmidt et al., 1991). A particular result underlines that the opioid action is selective with respect to FRA pathways and not to nociceptive pathways and just also to FRA. There is a nociceptive excitatory pathway from the pad of the foot to some ankle and intrinsic foot extensors which is not belonging to the FRA. As shown in Fig. 4 this nociceptive pathway is not depressed by the opioids, or at least not to a similar extent as the FRA pathways, which are distinctly depressed. The transmission in Ib pathways from Golgi tendon organs to alpha-motoneurones also remains unaffected.

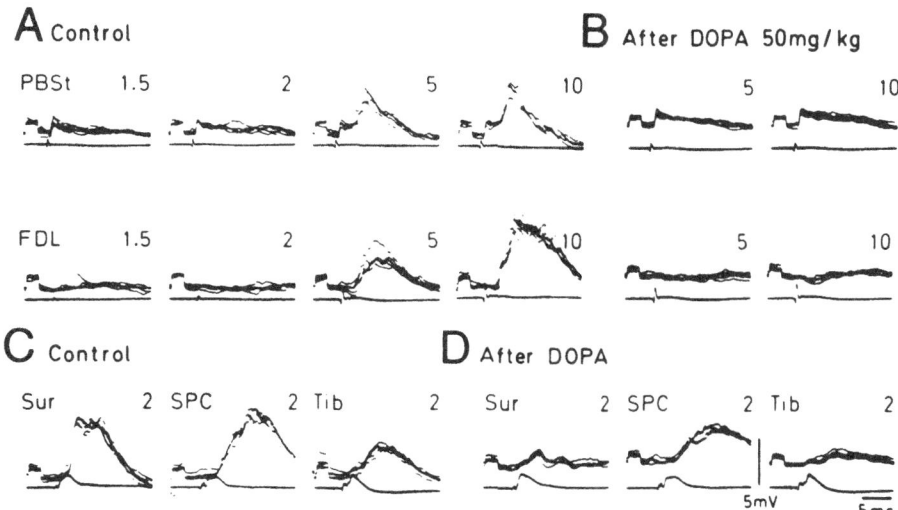

Figure 3. Depression of reflex responses from group II muscle afferents (A and B) and cutaneous afferents (C and D) by L-DOPA. Intracellular recordings from a posterior biceps semitendinosus (PBSt) motoneurone (several traces superimposed) in a high spinal cat. Lower traces of each pair of recordings show the incoming volleys recorded from the dorsal root entry zone. The nerves were stimulated with the strength indicated in multiplies of the threshold strength. Muscle nerves (PBSt and flexor digitorum and hallucis longus, FDL) were stimulated with graded strength, EPSPs occurring in A with a strength between 2 and 5 times threshold were of group II origin; those occurring after an increase of the stimulus strength to 10 times threshold were partly of a group II and partly of a group III origin. Cutaneous nerves (sural, Sur, and superficial peroneal, SPC) and the mixed tibial nerve (Tib) were stimulated with a strength of 2 times threshold. Note that the depression of group II EPSPs (from A to B) is more pronounced than that of the EPSPs evoked by cutaneous afferents (C to D; modified from Schomburg and Steffens, 1988).

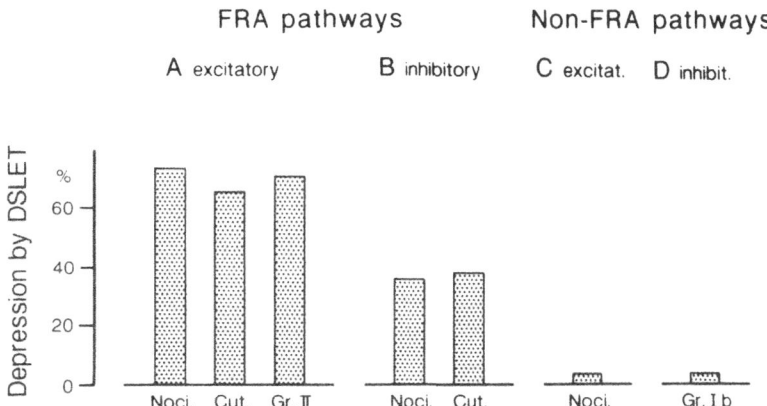

Figure 4. Depression of the transmission in FRA pathways (A and B) versus non-FRA (C and D) pathways by the δ-opioid agonist DSLET. Amount of depression given in percent, 100 percent would reflect a complete depression. (A) strong depression of the transmission in excitatory FRA pathways from nociceptive (Noci.), low threshold cutaneous (Cut.) and group II muscle (Gr. II) afferents to posterior biceps semitendinosus motoneurones. (B) somewhat smaller but still distinct depression of corresponding inhibitory pathways to gastrocnemius-soleus motoneurones. (C) and (D), no depression of the transmission in the excitatory non-FRA pathway from nociceptive afferents of the central foot pad to plantaris motoneurones (C) and in the "autogenetic" inhibitory pathway from Ib afferents to gastrocnemius-soleus motoneurones (D; according to data from Schmidt et al., 1991).

These findings suggest that the function of enkephalins in segmental motor control is not a specifically antinociceptive one but has a more complex character (see below).

SUMMARIZING COMMENTS

Spinal interneuronal systems receive a wide convergent input from different descending pathways and from afferents of a great variety of receptors. Different subsets of interneurones may receive a different convergent information and project to different combinations of motoneurones. Depending on the motor task, particular interneuronal subsets with the appropriate descending and afferent sensory convergence and a projection to the appropriate group of motoneurones may be selected for the performance of the intended movement (Lundberg et al., 1987; Schomburg, 1990). According to Lundberg (1979) the reflexes evoked by the afferents during an active movement (muscle, joint and skin afferents) then "serve to give feed back control of the movements commanded from higher centers". Partly in cooperation with monoaminergic pathways the enkephalinergic systems are obviously able to suppress the relatively unspecific feed back of the FRA. This may be of particular interest during movements which require a predominant supraspinal drive without a disturbance by segmental FRA afferents (Schmidt et al., 1991; Schomburg, 1991).

ACKNOWLEDGMENTS

Supported by the Deutsche Forschungsgemeinschaft (Scho 37/3-3).

REFERENCES

Andén, N.-E., Jukes, M.G.M., Lundberg, A., and Vyklicky, L., 1966, The effect of DOPA on the spinal cord. 1. Influence on transmission from primary afferents, *Acta Physiol. Scand.* 67:373-386.

Baldissera, F., Hultborn, H., and Illert, M., 1981, Integration in spinal neuronal systems. In: V.B Brooks (Ed.), Handbook of Physiology, Vol. 2, Sect. I, Nervous System, Motor Control, Part 1, Am. Physiol. Soc., Bethesda, MD, pp. 509-595.

Bras, H., Cavallari, P., and Jankowska, E., 1988, An investigation of local actions of ionophoretically applied DOPA in the spinal cord, *Exp. Brain Res.* 71:447-449.

Eccles, R.M., and Lundberg, A., 1959, Synaptic actions in motoneurones by afferents which may evoke the flexion reflex, *Arch. Ital. Biol.* 97:199-221.

Harrison, P.J. and Jankowska, E., 1985, Sources of input to interneurones mediating group I non-reciprocal inhibition of motoneurones in the cat, *J. Physiol.* (Lond.) 361:379-401.

Hultborn, H., 1972, Convergence on interneurones in the reciprocal Ia inhibitory pathway to motoneurones, *Acta Physiol. Scand. Suppl.* (375) 85:1-42.

Jankowska, E., 1984, Interneuronal organization in reflex pathways from proprioceptors. In: D.G. Garlick and P.I. Korner (Eds.), Frontiers in Physiological Research, Australian Academy of Science, Canberra, pp. 228-237.

Jankowska, E., 1992, Interneuronal relay in spinal pathways from proprioceptors, *Progress in Neurobiology* 38:335-378.

Kniffki, K.-D., Schomburg, E.D., and Steffens, H., 1981, Effects from fine muscle and cutaneous afferents on spinal locomotion in cats, *J. Physiol.* (Lond.) 319:543-554.

Lundberg, A., 1975, Control of spinal mechanism from the brain. In: D.B. Tower (Ed.), Nervous System. The Basic Neurosciences, Raven, New York, pp. 253-265.

Lundberg, A., 1979, Multisensorial control of spinal reflex pathways, In: R. Granit and D. Pompeiano (Eds.), Reflex Control of Posture and Movement, Progress in Brain Research 50, Elsevier, Amsterdam, pp. 11-28.

Lundberg, A., Malmgren, K., and Schomburg, E.D., 1987, Reflex pathways from group II muscle afferents. 3. Secondary spindle afferents and the FRA; a hypothesis, *Exp. Brain Res.*, 65:294-306.

Schmidt, P.F., Schomburg, E.D., and Steffens, H., 1991, Limitedly selective action of a δ-agonistic leuenkephalin on the transmission in spinal motor reflex pathways in cats, *J. Physiol.* (Lond.) 442:103-126.

Schomburg, E.D., 1990, Spinal sensorimotor systems and their supraspinal control, *Neurosci. Res.* 7:265-340.

Schomburg, E.D., 1991, The role of nociceptive afferents and enkephalins in spinal motor control. In: Restorative Neurology, vol. 5, ed. Wernig, A., pp. 345-353, Elsevier, Amsterdam.

Schomburg, E.D. and Behrends, H.B., 1978, Phasic control of the transmission in the excitatory and inhibitory reflex pathways from cutaneous afferents to α-motoneurones during fictive locomotion in cats. *Neurosci. Lett.* 8:277-282.

Schomburg, E.D., and Steffens, H., 1988, The effect of DOPA and clonidine on reflex pathways from group II muscle afferents to alpha-motoneurones in the cat, *Exp. Brain Res.*, 71:442-446.

Schomburg, E.D., and Steffens, H., 1993, Convergence in segmental reflex pathways from nociceptive and non-nociceptive afferents to α-motoneurones in the cat, *J. Physiol.*(Lond.) 466:191-211.

2

INVOLVEMENT OF DEEP CEREBELLAR NUCLEI IN ATTENTIVE AND ORIENTING MOTOR RESPONSES

J. M. Delgado-García and A. Gruart

Universidad de Sevilla, Facultad de Biología
Laboratorio de Neurociencia
Avda. Reina Mercedes, 6
Sevilla-41012
Spain

INTRODUCTION

Cerebellar nuclear cells are not only under the inhibitory control of overlying Purkinje cells (Ito et al., 1970), but also receive modulatory influences from both mossy and climbing fiber collaterals. Thus, there is convincing morphological evidence indicating that nuclear cells receive collaterals from mossy (Gonzalo-Ruiz and Leichnetz, 1990; Shinoda et al., 1987, 1992) and climbing (Shinoda et al., 1987; van der Want et al., 1989) fiber afferents projecting to the corresponding cerebellar cortex. These results have been confirmed electrophysiologically in the isolated brain stem-cerebellar preparation in vitro (Llinás and Mühlethaler, 1988). The peculiar electroresponsive properties of the membranes of nuclear cells endow them with a background firing (Llinás and Mühlethaler, 1988) that may be further sculpted by inhibitory (Purkinje cell terminals) and excitatory (mossy and climbing fiber collaterals) synaptic inputs.

Although each deep cerebellar nucleus appears to be particularly involved in a specific task related to the control of posture and movement (see Thach et al., 1992 for a detailed account), all nuclear cells seem to be subjected to the same morphofunctional pattern of synaptic inputs, suggesting that they share some common functional mechanisms. For example, fastigial neurons are mostly involved in the control of stance and gait, while interpositus neurons contribute to stabilizing positions of fixation of the extremities and improving the performance of learned movements triggered by natural sensory cues (Thach et al., 1992). In both cases, the animal's level of alertness, as well as the timing of input signals arriving to cerebellar structures, will be crucial for the proper generation of a correct motor-signal response. Theoretical models (Gilbert, 1975), electrophysiological experiments (McDevitt et al., 1987) and behavioral studies in humans (Akshoomoff and Courchesne, 1992) indicate a possible involvement of cerebellar structures in attentive and orienting motor responses. The necessary short term interactions supposedly taking place at the membrane of nuclear cells during those movements have recently been simulated in awake cats by the appropriate stimulation of pontine nuclei (PN) and the inferior olive (IO) while recording the field potentials evoked in deep cerebellar nuclei (Gruart et al., 1994). Results from those experiments indicate that field potentials induced in nuclear zones by IO stimulation can be modulated by conditioning stimuli applied to selected areas of PN. Present experiments were undertaken to substitute natural sensory cues in alert behaving cats for the conditioning electrical stimulation of PN.

EXPERIMENTAL METHODS

Three adult female cats obtained from an official supplier (Iffa-Credo) were used in these experiments. All the experimental procedures were carried out in agreement with the guidelines of the European Union (86/609/EU) and with Spanish Regulations (BOE 67/8509-12, 1988) for the use of laboratory animals. A detailed account of the surgical procedures used here has been reported elsewhere (Gruart and Delgado-García, 1994a; Gruart et al., 1994) and will be explained only summarily below.

Under general anesthesia (35 mg/kg of sodium pentobarbital plus a protective injection of 0.5 mg/kg of atropine sulfate) animals were implanted with 3 mm diameter, stainless-steel eyelid coils and with stimulating electrodes in PN, the IO, the red nucleus (RN) and the restiform body (RB). A recording chamber (5 x 5 mm) was opened in the occipital bone to approach the fastigial and interpositus nuclei contralateral to the implanted electrodes. The chamber was sealed with sterile material between recording sessions. Finally, animals were implanted with a head-holding device to stabilize cerebellar structures during recording sessions.

Two weeks after surgery the animal was introduced into an elastic bag, placed on the recording table and its head immobilized with the holding device. The animal was presented with i.) 90 dB, 600 or 6,000 Hz tones applied for 1-10 s through a loudspeaker located 50 cm over its head; or ii) bright full-field flashes, lasting for ≈ 1 ms, delivered by a xenon lamp located 1 m in front of the animal's eyes. Electrical stimulation consisted of 50 µs, square, cathodal pulses of ≤ 0.5 mA. Electrical stimulation was presented as follows: i.) in pairs at a rate of ≤ 1 Hz, with a conditioning stimulus applied to PN followed 1-200 ms later by a stimulus applied to the IO; ii) a single stimulus applied to the IO at different time intervals before, during and after tone or flash presentations; and iii) electrical stimulation of the RN and the RB was used for the functional location of recording sites -that is, the interpositus and the fastigial nucleus, respectively (see Gruart et al., 1994).

At the end of the recording sessions, the animals were deeply reanesthetized and electrolytic marks were made at selected recording sites. The animals were perfused transcardially with saline and 10% formalin, and their brains removed and histologically processed to locate stimulating and recording sites (Gruart and Delgado-García, 1994a).

RESULTS

As explained in detail elsewhere (Gruart et al., 1994), the interpositus nucleus was located with the help of the positive-negative-positive antidromic field potential evoked in its electrophysiological limits by RN stimulation. Antidromic field potentials of a similar positive-negative-positive profile were evoked in the fastigial nucleus by RB stimulation. The antidromic nature of these field potentials was confirmed by random intracellular recordings of nuclear cells and by the use of the collision test during extracellular single-unit recordings (not illustrated).

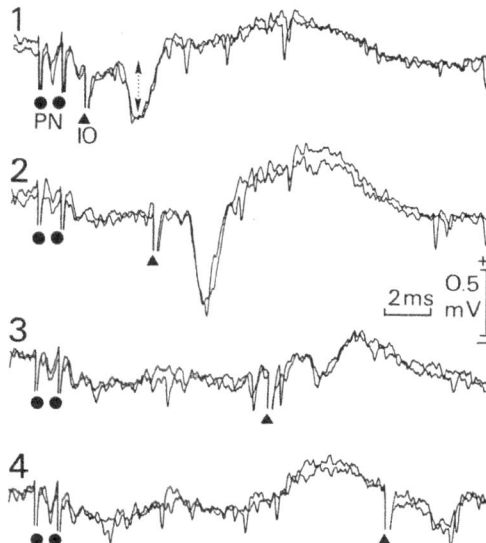

Figure 1. Field potentials recorded in the interpositus nucleus following electrical stimulation of the inferior olive (IO, triangles), preceded by double-pulse conditioning stimulation of the pontine nuclei (PN, dots) at different time intervals.

As described previously (Gruart et al., 1994) and illustrated in Fig. 1, the electrical stimulation of PN produced in nuclear zones a small field potential composed of two negative waves at latencies of 0.5 and 1.5 ms respectively. The stimulation of the IO induced in nuclear zones a field potential composed of a negative wave at 1.5-2.5 ms of latency, followed by a delayed positivity at 3-5 ms (Fig. 1). The stimulation of the IO also evoked an early (0.5 ms) small negative wave restricted to the ventral aspect of the interpositus nucleus (not illustrated). According to results obtained with whole-brain in vitro studies (Llinás and Mühlethaler, 1988) and to extracellular single-unit recordings (Gruart and Delgado-García, 1994a), the early negativity corresponded to the antidromic activation of nuclear cells projecting to the IO. The second negative wave represented the extracellular profile of the synaptic activation of nuclear cells by climbing fiber collaterals. Finally, the delayed positivity corresponded to inhibitory post-synaptic potentials induced by Purkinje cells in deep cerebellar nuclei neurons.

It was observed that for a given recording site the field potential evoked by RN, RB and PN stimulation did not change in amplitude and/or latency in relation to changes in the animal's level of alertness or by the presentation of natural stimuli of different sensory modalities (sounds, body rotation, silhouettes, etc). In contrast, the amplitude of the negative-positive synaptic field potentials evoked by IO stimulation was easily modified by the presentation of stimuli able to draw the animal's attention. As already described (Gruart et al., 1994) and illustrated in Fig. 1, stimuli applied to PN were also able to modify the amplitude of the negative-positive synaptic field potential evoked in the interpositus nucleus by IO stimulation. The time window for this facilitatory effect was very short (maximum of 30-40 ms). Interestingly, different sensory cues successfully substituted for the conditioning stimuli applied to the PN. As illustrated in Fig. 2, the sudden presentation of a loud tone produced an increase in the amplitude of the negative-positive field potential evoked by IO stimulation, as compared with controls. However, this facilitatory effect disappeared in a few seconds if the sensory stimulus was maintained unchanged; i.e., when it lost its novelty for the animal.

It was also shown that the variability in the amplitude of IO-evoked synaptic field potentials was correlated with orienting movements made by the animal during stimulus

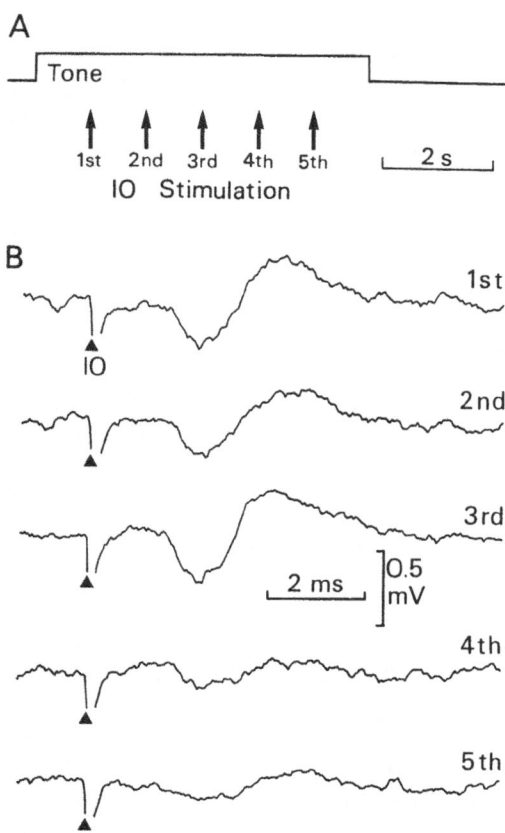

Figure 2. Change in the amplitude of synaptic field potentials recorded in the fastigial nucleus following inferior olive (IO, triangles) electrical stimulation during the presentation of a 90 dB, 6,000 Hz tone for the time indicated in A. Calibrations for the extracellular records are shown below the 3rd record in B.

presentation. As illustrated in Fig. 3, there was a linear relationship (slope = 0.1 mV/deg; coefficient of correlation, r = 0.92) between the amplitude of the reflex eyelid blink evoked by a flash of light and the amplitude of the second negativity evoked in the caudal interpositus by IO stimulation. As known (Gruart and Delgado-García, 1994a), this reflex response is easily modified in amplitude depending upon the animal's level of alertness and on the novelty of the presented stimulus. For data shown in Fig. 3, IO stimulation was presented in coincidence with the flash-evoked blink. When electrical stimulation of the IO was presented outside the time-window for which the eyelid response was reflexively induced, no significant modulation of the evoked field potential was observed (not illustrated).

DISCUSSION AND CONCLUSION

According to the present results, natural sensory stimulation used as a conditioning stimulus produced an increase in the amplitude of the synaptic field potentials evoked by IO stimulation on selected cerebellar nuclear zones. This facilitatory effect seemed to be i.) related to attentive processes; ii) very short lasting (ms); and iii) linearly related to the amplitude of movements evoked by the sensory stimulation. Moreover, the membrane of nuclear cells has to be endowed with electroresponsive properties able to detect the orderly arrival of inputs from mossy and climbing fiber sources. As sensory stimuli of different modalities could successfully substitute for the electrical stimulation of PN it may be proposed that sensory cues are carried out by mossy fiber afferent collaterals, while climbing

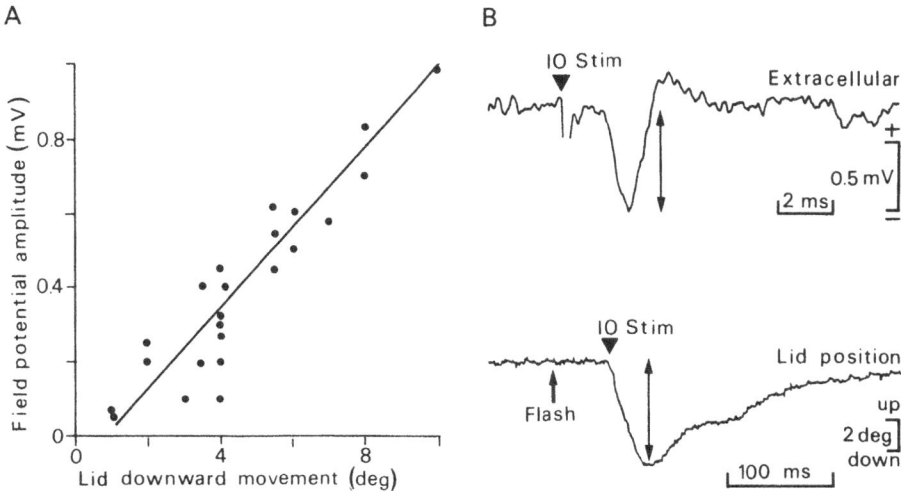

Figure 3. In A is shown a plot illustrating the linear relationship between the amplitude of flash-evoked reflex blinks and the amplitude of the negative field potential evoked in the interpositus nucleus by inferior olive (IO) stimulation. B. Example of lid movement evoked by a flash (bottom) and the field potential (top) evoked by IO stimulation. The moment of IO stimulation is indicated by a triangle.

fiber collaterals seem to act as a coincidence detector device, i.e., producing a strong excitation-inhibition sequence on nuclear cells when arriving (in a narrow time window) after mossy fiber input signals. The variability of the facilitatory effects of IO stimulation on cerebellar nuclear cells was also correlated with the amplitude of the motor responses induced by stimulus presentation.

Present results indicate that the facilitatory effects on cerebellar output produced by the timed arrival of IO signals could be related to the generation of the necessary motor commands to produce fast and accurate movements during attentive and orienting motor responses. Similar facilitatory mechanisms have been described for other cerebral circuits. For example, the activation of cholinergic neurons located in the basal forebrain facilitates cortical processing in awake animals (Hars et al., 1993). On the other hand, many neuronal types described in deep cerebellar nuclei (see Gruart and Delgado-Garcia, 1994a, b and Thach et al., 1992 for references) seem to be involved in attentive processes, as they modify their firing rate accompanying eye and arm movements in the periphery of the visual field in response to conditioning sensory cues and during changes in the animal's level of alertness.

Serotoninergic (Takeuchi et al., 1982) and/or noradrenergic (Hökfelt and Fuxe, 1969) inputs to cerebellar structures could also be involved in the potentiation of the discharge rate of nuclear cells by their timed activation by mossy and climbing fiber collaterals. The presence of low threshold spikes (Llinás and Mühlethaler, 1988) in cerebellar nuclear cells allows the suggestion that this could be one of the membrane phenomena triggered by the orderly arrival of inhibitory (Purkinje cell) and excitatory (mossy and climbing fiber afferent collaterals) inputs to deep cerebellar nuclei neurons.

ACKNOWLEDGMENTS

This work was supported by the Spanish D.G.I.C.Y.T. under grant #PB93-1175.

REFERENCES

Akshoomoff, N.A., and Courchesne, E., 1992, A new role for the cerebellum in cognitive operations, *Behav. Neurosci.* 106:731-738.

Gilbert, P., 1975, How the cerebellum could memorize movements, *Nature* 254:688-689.

Gonzalo-Ruiz, A., and Leichnetz, G.R., 1990, Afferents of the caudal fastigial nucleus in a New World monkey (Cebus apella), *Exp. Brain Res.* 80:600-608.

Gruart, A., and Delgado-García, J.M., 1994a, Discharge of identified deep cerebellar nuclei neurons related to eye blinks in the alert cat, *Neuroscience* 61:665-681.

Gruart, A., and Delgado-García, J.M., 1994b, Signaling properties of identified deep cerebellar nuclear neurons related to eye and head movements in the alert cat, *J. Physiol. (Lond.)* 478:37-54.

Gruart, A., Blázquez, P., Pastor, A., and Delgado-García, J.M., 1994, Very short-term potentiation of climbing fiber effects on deep cerebellar nuclei neurons by conditioning stimulation of mossy fiber afferents, *Exp. Brain Res.* 101:173-177.

Hars, B., Maho, C., Edeline, J.-M., and Hennevin, E., 1993, Basal forebrain stimulation facilitates tone-evoked responses in the auditory cortex of awake rat, *Neuroscience* 56:61-74.

Hökfelt, T., and Fuxe, K., 1969, Cerebellar monoamine nerve terminals: a new type of afferent fiber to the cortex cerebelli, *Exp. Brain Res.* 9:63-72.

Ito, M., Yoshida, M., Obata, K., Kawai, N., and Udo, M., 1970, Inhibitory control of the intracerebellar nuclei by the Purkinje cell axons, *Exp. Brain Res.* 10:64-80.

Llinás, R., and Mühlethaler, M., 1988, Electrophysiology of guinea-pig cerebellar nuclear cells in the in vitro brain stem-cerebellar preparation, *J. Physiol. (Lond.)* 404:241-258.

McDevitt, C. J., Ebner, T., and Bloedel, J.R., 1987, Changes in the responses of cerebellar nuclear neurons associated with the climbing fiber response of Purkinje cells, *Brain Res.* 425:14-24.

Shinoda, Y., Sugiuchi, Y., and Futami, T., 1987, Excitatory inputs to cerebellar dentate nucleus from the cerebral cortex in the cat, *Exp. Brain Res.* 67:299-315.

Shinoda, Y., Sugiuchi, Y., Futami, T., and Izawa, R., 1992, Axon collaterals of mossy fibers from the pontine nucleus in the cerebellar dentate nucleus, *J. Neurophysiol.* 67:547-560.

Takeuchi, Y., Kimura, H., and Sano, Y., 1982, Immunohistochemical demonstration of serotonin-containing nerve fibers in the cerebellum. Cell Tissue Res, 226:1-12.

Thach, W.T., Kane, S.A., Mink, J.W., and Goodkin, H.P., 1992, Cerebellar output: multiple maps and modes of control in movement coordination, in The Cerebellum Revisited, R. Llinás and C. Sotelo, eds., Springer-Verlag, New York, pp 301-334.

van der Want, J.J.L., Wiklund, L., Guegan, M., Ruigrok, T., and Voogd, J., 1989, Anterograde tracing of the rat olivocerebellar system with phaseolus vulgaris leucoagglutinin (PHA-L). Demonstration of climbing fiber collateral innervation of the cerebellar nuclei. *J. Comp. Neurol.* 288:1-18.

ADAPTABILITY OF ADULT MAMMALIAN MOTONEURONS TO NEW MOTOR TASKS

A. Gruart,[1] A. Gunkel,[2] W.F. Neiss[3], E. Stennert,[4] and J.M. Delgado-García[1]

[1] Laboratorio de Neurociencia, Facultad de Biología
Universidad de Sevilla
Avda. Reina Mercedes, 6
Sevilla-41012, Spain
[2] Klinik für Hals-, Nasen- und Ohrenheilkunde
Universität Innsbruck
A-6020 Innsbruck, Austria
[3] Institut I für Anatomie and
[4] Klinik für Hals-, Nasen- und Ohrenheilkunde
Universität zu Köln, Lindenthal
D-50924 Köln, Germany

INTRODUCTION

It is commonly accepted that neural structures of most living beings are endowed with the capacity of modifying their command signals in order to adapt observable behavior to new environmental constraints. Where in the nervous system, according to which timed process, and how these modifications are achieved are the subject matter of one of the central debates of current Neuroscience. As motoneurons represent the final common pathway in which motor commands are elaborated, it seems important to clarify whether some adaptive phenomena reported in the behavior of adult mammals can take place at the level of these motor cells. Although for Sperry (1945) post-embryonic motoneurons lack the capacity to reprogram their firing properties to undergo the functional needs of new motor targets when reinnervating foreign muscles, clinical reports and animal experimentation reopen from time to time the issue (Stennert, 1972; Willer et al., 1992). Available data for the extraocular motor system of adult cats indicate that, following central cut of IIIrd, IVth and VIth cranial nerves, surviving motoneurons are unable (first) to find their parent muscles and (second) to respecify their firing patterns in the amount and sequence required by their new motor tasks (Baker, 1985; Baker et al., 1985). The successful reproduction in rats (Angelov et al., 1993; Neiss et al., 1992) of the classical hypoglossal-facial anastomosis used in humans for the treatment of facial paralysis prompted us to carry out a detailed physiological study on the adaptability of adult cat hypoglossal motoneurons when confronted with the specific tasks of upper lid motor performance.

Multisensory Control of Posture, Edited by T. Mergner and F. Hlavačka
Plenum Press, New York, 1995

EXPERIMENTAL PROCEDURES

Experiments were carried out in five adult female cats weighing from 2-2.5 kg. Cats were provided with electrodes for the recording of the electromyographic (EMG) activity of the orbicularis oculi (OO) muscle and with upper eyelid coils to record the position of the lid with the magnetic search-coil technique (Fig. 1). All experimental procedures were performed according to the guidelines of the European Union (86/609/EU) and the Spanish Government Directives (BOE 67/8509-12, 1988) for the use of laboratory animals in chronic experiments.

Experimental procedures for this chronic preparation have been described in detail elsewhere (Gruart and Delgado-García, 1994) and will only be summarized here. Under general anesthesia (sodium pentobarbital, 35 mg/kg plus atropine sulfate, 0.5 mg/kg) animals were implanted with a 3 mm diameter, 3-turn, stainless-steel coil into the center of the left upper lid near the margin edge. Animals were also implanted with bipolar, hook, stainless-steel electrodes in the lateral and upper aspects of the left OO muscle and with a head-holding system consisting of three bolts affixed with acrylic cement to the skull. In a subsequent surgical step, the left hypoglossal nerve of three of the animals was dissected out, transversely cut and its proximal stump sutured with 3 atraumatic sutures (11-0, Ethicon EH 7438G) to the distal end of the facial nerve (Angelov et al., 1993; Neiss et al., 1992). The other two animals were left without hypoglossal-facial anastomosis for control purposes.

Recording sessions started one week after surgery and were repeated every 15 days for one year. Animals were lightly restrained with an elastic bandage, placed on the recording table and their heads adjusted at the geometrical center of the magnetic field frame with the help of the head-holding system. Eyelid responses were evoked by the presentation of 100 ms, 3 Kg/cm^2 air puffs directed to the cornea and eyelids from a valve located 1 cm away from the eye. Licking by the animal was induced by milk drops delivered through a computed-controlled delivering pipette.

For the animal's conditioning, the air puff was presented to the left eye followed 350 ms later by the delivery of a few drops of milk to the animal's mouth. Both air puff and milk delivery ended simultaneously. This pair of stimuli was presented 120 times/day at a random interval of 60±30 s, for a maximum of 15 days. Conditioning was carried out from 7-9 months

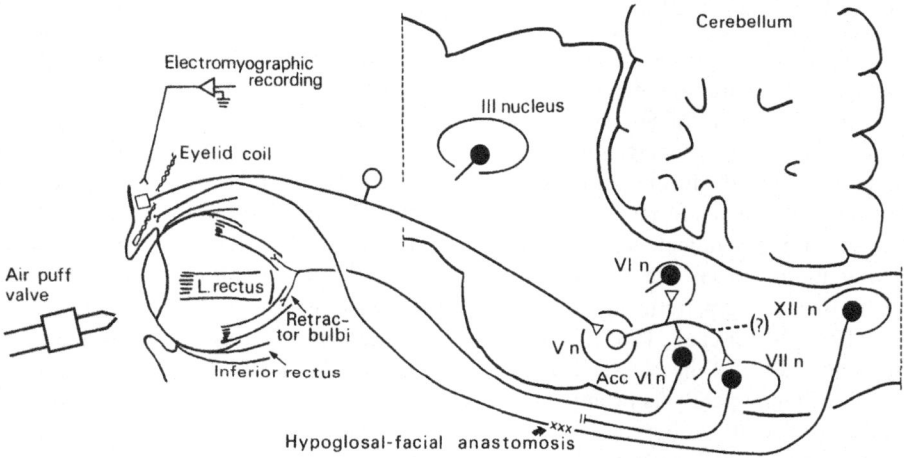

Figure 1. A diagram of the basic neural circuit involved in the blink reflex and the changes introduced by the hypoglossal-facial anastomosis.

after surgery, i.e., well after OO muscle reinnervation by hypoglossal motoneurons (see below). Data was recorded and stored with conventional procedures, and represented and analyzed with available commercial programs (Gruart and Delgado-García, 1994; Gruart et al., 1994).

RESULTS

As illustrated in Fig. 2, air puff-evoked blink responses in control animals consisted of a fast (up to 1,500 deg/s) downward lid movement, of about 20-22 deg in amplitude, that started 10-12 ms after stimulus presentation. The downward movement was preceded, by 4-5 ms, of a sharp activation of the OO muscle. Further bursts of activity of the OO muscle produced late downward sags in the position of the lid, of a smaller amplitude than the initial component. The return movement of the lid was very much slower (peak velocity = 300 deg/s) and showed a time constant of ≈ 35 ms. These control values seemed very stable and showed no significant changes throughout the study. Besides this reflexively-evoked eyelid response, the OO muscle was not active during upward or downward lid movements accompanying spontaneous eye saccades or during experimentally-induced licking. Lid position during normal deglutition was not noticeably modified (Fig. 2, C1).

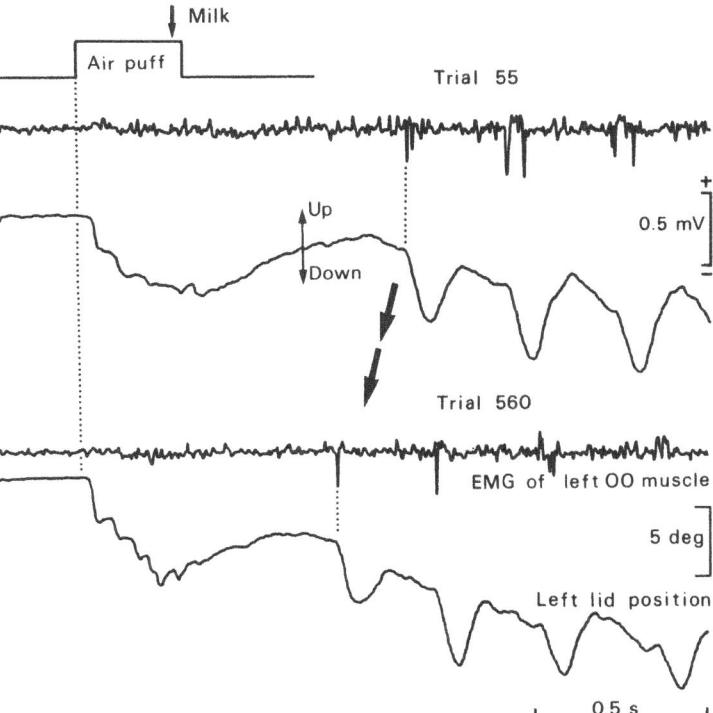

Figure 2. Electromyographic (EMG) activity of the left orbicularis oculi (OO) muscle and left lid position of control (1) animals, and of experimental animals during the denervation period (2) and following reinnervation of the OO muscle by hypoglossal motoneurons (3). Examples are illustrated in the absence of any noticeable stimulus (A), during air puff presentation to the left cornea (B) and during experimentally-induced licking (C). Calibrations in C are also for A-B.

The EMG of the left OO muscle in anastomosed animals was not recovered until 6-7 weeks following hypoglossal-facial anastomosis. During the period of denervation (Fig. 2, A2-C2), left lid movements in response to air puff presentation were very much slower (peak velocity of 350 deg/s) and of very much smaller amplitude (5-7 deg), when recorded 15 days after surgery. However, the presence of these reflexively-evoked lid movements before reinnervation by the hypoglossal nerve indicated that they were produced by the combined action of the extraocular recti and retractor bulbi muscles (Delgado-García et al., 1988; Gruart and Delgado-García, 1994; Gruart et al., 1994), a fact confirmed by the complete absence of EMG activity in the denervated OO muscle. Fast and slow lid movements during the denervation period evoked by visual stimulus were similar (no significant differences) to those observed in control animals, confirming their passive nature.

The recovery in background EMG activity of the OO muscle following its reinnervation by hypoglossal motoneurons was not accompanied for the period tested (one year) by any noticeable electrical response of the reinnervated muscle to air puff stimulation of the ipsilateral cornea (Fig. 2, B3). Nevertheless, a progressive increase in lid responses to air puff stimuli was observed. For example, values of amplitude (18-22 deg) and peak velocity (800-850 deg/s) of air puff-evoked blinks eight months after surgery were similar to those obtained for controls. These adaptive changes have to be ascribed to an increased gain in corneal reflex circuits involving extraocular recti and retractor bulbi motor systems (Fig. 2, B3, stars). Obviously, this increased reflex response produced by a motor system not directly involved in the performed hypoglossal-facial anastomosis is a very interesting plastic phenomenon produced to compensate the lack of activity of blink responses originated in the OO muscle.

The reinnervation of the OO muscle by hypoglossal motoneurons posed some evident constraints in lid motor performance. For example, during attempted ocular following of ramp (1-5 deg/s) optokinetic stimulation in the vertical plane, the lid reinnervated by hypoglossal motoneurons was frequently disturbed in its slow displacement by spontaneous tongue movements or by experimentally-evoked licking, never observed in controls. This situation showed no improvement during the period tested. As illustrated in Fig. 2, C3, licking produced the same oscillatory movements of the left lid as those produced in the tongue. Because of the peculiar structure of the OO muscle, licking always produced an active downward pulling of the lid followed by its slower and passive return in the upward direction.

An attempt was made to train the animal to relate the timed arrival of an air puff applied to the cornea and lids of the anastomosed (left) side with the delivery of a few drops of milk.

As illustrated in Fig. 3, although the animal showed a certain degree of improvement during the trials in the sense of a decrease in the latency to initiate tongue (i.e., left lid) movements, hypoglossal-evoked lid movements were never observed during the time at which the air puff was presented. As already indicated, one reason for this could be the compensatory lid closing produced by the extraocular recti and retractor bulbi systems. More importantly, Fig. 3 illustrated quite nicely the different kinetic properties of extraocular and hypoglossal motor systems. When the lid was moved by recti and retractor bulbi muscles, it presented fast oscillations at a mean rate of 25 Hz. In contrast, when moved by the evoked activity of hypoglossal motoneurons, the dominant oscillatory frequency ranged from 3 to 4 Hz, depending on the animal's central pattern generator. As illustrated in Fig. 3, only these latter movements were accompanied by a noticeable EMG activity of the hypoglossal-reinnervated left OO muscle.

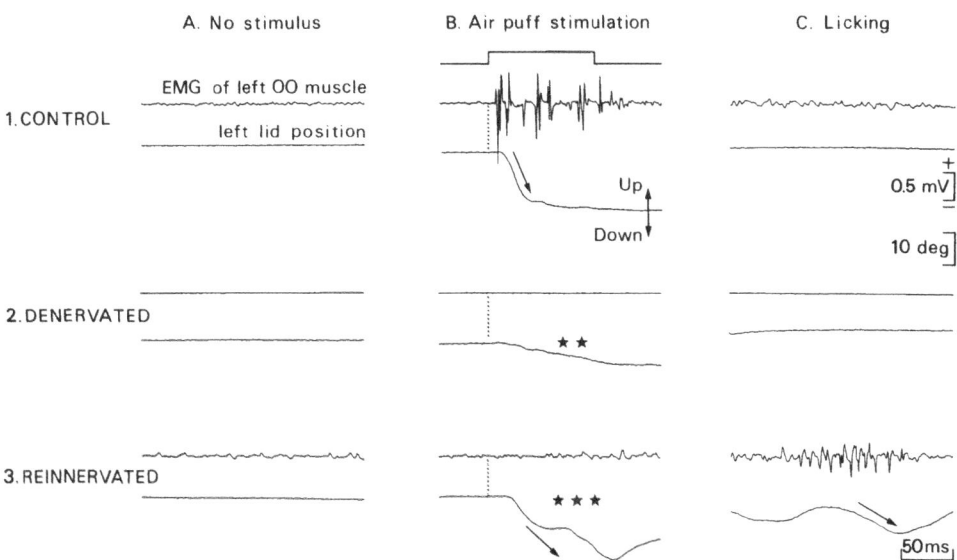

Figure 3. A classical-conditioning paradigm during which animals were presented with an air puff applied to the cornea of the left (anastomosed) side followed by the delivery of a few drops of milk to the animal's mouth.

DISCUSSION AND CONCLUDING REMARKS

According to the present results, the reinnervation of the OO muscle by hypoglossal motoneurons in adult cats is not followed by any noticeable change in the expected behavior of these tongue-related motor cells. At the same time, these results indicate that motoneurons reinnervating a new motor target are unable to de-differentiate in order to deal with the motor properties of the reinnervated target. On the contrary, they seemed to maintain their functional properties, motor commands and reflex responses. In opposition to a recent report (Willer et al., 1992), no evidence was obtained from present experiments suggesting a rewiring of brain stem connections able, for example, to convey newly-formed trigeminal inputs to the hypoglossal motor nucleus. It should be pointed out that such a connection has been reported in control humans (Stennert and Limberg, 1982). No evident modification was observed in the central pattern generator controlling tongue movements, in the sense of increasing its dominant oscillation frequency (3-4 Hz) in order to adapt it to the new motor task f that is, moving the lid at a main frequency of 25 Hz. In this line of thought, adaptive changes in passive and/or active firing properties of chromatolysed motoneurons reported in acute experiments (Eccles et al., 1958; Gustafsson and Pinter, 1984; Kuno and Llinás, 1970), particularly in those related to axotomized hypoglossal motoneurons (Takata 1993), are not in the range needed to cope with the problems posed by a new motor task.

In conclusion, the supposed 'adaptive plasticity' in the physiological responses of axotomized motoneurons following reinnervation of foreign muscles (Baker, 1985) has to be ascribed to compensatory phenomena, such at the one reported here, or to the participation of higher neural centers able to restitute some motor disarrangements through reeducative learning - a process surely involving higher neural centers (Sperry, 1945, 1947), but apparently absent in motoneuronal pools.

ACKNOWLEDGMENTS

This work was supported by the Spanish D.G.I.C.Y.T. under grant #PB93-1175.

REFERENCES

Angelov, D.N., Gunkel, A., Stennert, E., and Neiss, W.F., 1993, Recovery of original nerve supply after hypoglossal-facial anastomosis causes permanent motor hyperinnervation of the whisker-pad muscles in the rat, *J. Comp. Neurol.* 338:214-224.

Baker, R., 1985, Neural mechanisms of adaptation: a viewpoint, in: Keller, E.L., and Zee, D.S., (eds.), Adaptive Processes in Visual and Oculomotor Systems, Pergamon Press, Oxford, pp. 419-427.

Baker, R., Peck, C., Spencer, R.F., Delgado-García, J.M., and Winterkorn, J., 1985, Structural and functional assessment of the reinnervation pattern of cat extraocular muscles following central cut of the IIIrd, IVth and VIth cranial nerves, *Soc. Neurosci. Abstr.* 11:973.

Eccles, J.C., Libet, B., and Young, R.R., 1958, The behaviour of chromatolysed motoneurones studied by intracellular recording, *J. Physiol. (Lond.)* 143:11-40.

Delgado-García, J.M., Del Pozo, F., Spencer, R.F., and Baker, R., 1988, Behavior of neurons in the abducens nucleus of the alert cat - III. Axotomized motoneurons, *Neuroscience* 24(1):143-160.

Gruart, A., and Delgado-García, J.M., 1994, Discharge of identified deep cerebellar nuclei neurons related to eye blinks in the alert cat. *Neuroscience* 61:665-681.

Gruart, A., Gunkel, A., Neiss, W.F., Stennert, E. and Delgado-García, J.M., 1994, A long-term study of eye blink responses after hypoglossal-facial anastomosis, *Soc. Neurosci. Abstr.* 20 (2):1403.

Gustafsson, B., and Pinter, M.J., 1984, Effects of axotomy on the distribution of passive electrical properties of cat motoneurones, *J. Physiol. (Lond.)* 356: 33-442.

Kuno, M., and Llinás, R., 1970, Enhancement of synaptic transmission by dendritic potentials in chromatolysed motoneurones of the cat, *J. Physiol. (Lond.)* 210:807-821.

Neiss, W.F., Guntinas-Lichius, O., Angelov, D.N., Gunkel, A., and Stennert, E., 1992, The hypoglossal-facial anastomosis as a model of neuronal plasticity in the rat, *Ann. Anat.* 174:419-433.

Sperry, R.W., 1945, The problem of central nervous reorganization after nerve regeneration and muscle transposition, *Q. Rev. Biol.* 20(4):311-369.

Sperry, R.W., 1947, Effects of crossing nerves to antagonistic limb muscles in the monkey, *Arch. Neurol. Psych.* 58:452-473.

Stennert, E., 1979, I. Hypoglossal facial anastomosis: Its significance for modern facial surgery. II. Combined approach in extratemporal facial nerve reconstruction, *Clin. Plast. Surg.* 6:471-486.

Stennert, E., and Limberg, C.H., 1982, Central connections between fifth, seventh, and twelfth cranial nerves and their clinical significance, in: Graham, M.D., and House, W.F., eds., Disorders of the Facial NerveRaven, Press, Oxford, pp. 419-427.

Takata, M. , 1993, Two types of inhibitory postsynaptic potentials in the hypoglossal motoneurons, *Progr. in Neurobiol.* 40:385-411.

Willer, J.C., Lamas, G., Poignonec, S., Fligny, I., and Soudant, J., 1992, Redirection of the hypoglossal nerve to facial muscles alters central connectivity in human brainstem, *Brain Res.* 594:301-306.

4

TYPE I MEDIAL VESTIBULAR NEURONS DURING ALERTNESS, FOLLOWING ADAPTATION, AND DURING REM SLEEP EPISODES IN THE HEAD-FIXED GUINEA-PIG

M. Serafin,[1] M. Mühlethaler,[2] and P. P. Vidal[1]

[1] Laboratoire de Physiologie de la Perception et de l'Action
CNRS-Collège de France
UMR C-9950
15 rue de l' Ecole de Médecine
75270 Paris Cedex 06, France
[2] Département de Physiologie
CMU
1 rue Michel-Servet
1211 Genève 4, Switzerland

INTRODUCTION

The horizontal vestibulo-ocular reflex (HVOR) and vestibulo-collic reflex (HVCR) contribute to gaze stabilization during head movements in the horizontal plane (Baker et al., 1981; Berthoz, 1989). During head rotations, sensory inputs from the horizontal semicircular canals of the labyrinth modulate the discharge of the first-order vestibular neurons in proportion to head velocity (Curthoys, 1982; Fernandez and Goldberg, 1971). These primary afferents contact monosynaptically the second-order vestibular neurons located in the central vestibular nuclei, which in turn project to the prepositus hypoglossi neurons, and to the appropriate extraocular and spinal motoneuron pools in order to implement the vestibulo-ocular and vestibulo-spinal reflexes. Whereas part of the second-order vestibular neurons are inhibitory and project through ipsilateral pathways (reviewed in Uchino and Isu, 1992a), another subgroup of second-order vestibular neurons are excitatory and project through contralateral pathways (Berthoz et al., 1989; Iwamoto et al., 1990; McCrea et al., 1987; Uchino et al., 1981,1982; Uchino and Isu, 1992b). According to their axonal projections to the extraocular and neck motor nuclei, the excitatory second-order vestibular neurons have been classified into three distinct groups (reviewed in Uchino and Isu, 1992b): the vestibulo-ocular neurons, which project exclusively towards cells within the contralateral abducens nucleus, the vestibulo-collic neurons, which project exclusively towards contralateral neck and spinal motoneurons and, finally, the vestibulo-oculo-collic neurons, which project contralaterally towards both the abducens and neck motor nuclei (Isu and Yokota, 1983;

McCrea et al., 1987; Uchino et al., 1982). The second-order vestibular neurons, as the first-order ones, can be broadly segregated into tonic (regular) and kinetic (irregular) cells (Boyle et al., 1992; Highstein et al. 1987; Iwamoto et al., 1990 a,b; Sato et al., 1993; Shimazu and Precht, 1965). Interestingly, Iwamoto et al. (1990 a,b) have recently reported in the alert cat the existence of differences in the firing regularity between vestibulo-ocular and vestibulo-oculo-collic neurons: whereas vestibulo-ocular cells were characterized by a regular discharge, vestibulo-oculo-collic neurons displayed an irregular firing rate. The dynamic response of the type I (Duensing and Schaeffer, 1958) medial vestibular nuclei neurons to sinusoidal head rotations, has been studied in details in the cat (Berthoz et al. 1989; Escudero et al., 1992; McCrea et al., 1980; Melvill Jones and Milsum, 1970; Shinoda and Yoshida, 1974) and the monkey (Fuchs and Kimm, 1975; Lisberger and Miles, 1980; Scudder and Fuchs, 1992). As a rule these neurons code head velocity over a wide frequency range, pause and burst during rapid eye movements, and part of them display an eye position sensitivity (Berthoz et al. 1989; Escudero et al. 1992; Fuchs and Kimm, 1975; Iwamoto et al. 1990 a,b; McCrea et al. 1987; Scudder and Fuchs, 1992).

In order to investigate the respective contribution of the intrinsic membrane properties of the vestibular neurons and of the emerging properties of the vestibular network in the processing of vestibular information, our groups have undertaken combined in vivo and in vitro studies of the medial vestibular nuclei neurons (MVNn) in the guinea-pig. Indeed, in contrast with the cat and the monkey, this animal allows to perform both extracellular recordings in vivo (see bellow) and intracellular recordings in various in vitro preparations (slice, isolated and perfused whole brain). This last preparation (Mühlethaler et al., 1993) in particular is interesting because it could potentially bridge the gap between the functional properties of identified neurons recorded in vivo, and their membrane properties determined in vitro (Serafin et al., 1991 a,b; Serafin et al., 1992a). A problem related to the choice of the guinea-pig as an animal model, however, is the relatively few studies devoted to investigate in vivo the characteristics of response of identified second-order MVNn in that species. Hence, in the first part of that chapter, we report our preliminary results concerning the activity of identified second-order type I MVNn recorded in the alert head-fixed guinea-pig during the HVOR.

The vestibular system is also a valuable model to study the plasticity of the central nervous system. In particular, the VOR is well known to undergo adaptive changes aimed at minimizing the retinal slip during head movements (Melvill Jones, 1985). Various models of VOR adaptation have been proposed (Du Lac and Lisberger, 1992; Ito, 1982; Kawato and Gomi, 1992; Lisberger, 1994; Lisberger and Sejnowski, 1992), and several electrophysiological studies have focused on determining the sites where plastic changes might occur (see Lisberger and Miles, 1980; Lisberger and Pavelko, 1988; Lisberger et al., 1994 b,c; Miles and Braitman, 1980; Miles et al., 1980). Nevertheless, the neuronal substrate of adaptation remains open to question (Lisberger and Sejnowski, 1993; Ito, 1993). The medial vestibular nuclei neurons, which mediate the VOR in the horizontal plane, appear to be good candidates, amongst other sites, where the molecular mechanism underlying adaptive changes could take place (Lisberger, 1994; Lisberger et al., 1994 a,b,c; Partsalis et al., 1993; Pastor et al., 1992, 1993). As part of our effort to validate the guinea-pig model, we have therefore attempted to record identified second-order type I MVNn during an adaptive change of the HVOR induced by a conflictual visuo-vestibular adaptation. Our results are summarized in the second part of that chapter.

Finally, the MVNn have been demonstrated, in vitro, to be endowed with intrinsic membrane properties which provide them the capability to display a rhythmical bursting activity in different experimental conditions (Serafin et al., 1992b; de Waele et al., 1993). The question is therefore open of the functional relevance of these oscillations in the alert preparation. We have recently studied the qualitative and quantitative characteristics of cortical activity (EEG), ocular motility and muscular activity in head-restrained guinea-pigs during different stages of vigilance (Escudero and Vidal, 1992). In particular we have described a new type of eye movement in this preparation during REM sleep. It consist in episodes of eye oscillations (8-14 Hz), occurring quite regularly (every 1.6 s) at high velocities (up to more than 1000 °/s peak to peak) and having a mean duration of 1.4 s. Therefore, we have tried to elucidate whether the MVNn membrane oscillations observed in vitro could be relevant to explain the in vivo pattern of discharge of the second-order MVNn during such episodes of eye oscillations. The results are summarized in the third part of that chapter.

MATERIAL AND METHODS

Surgical Procedure

Adult guinea-pigs (400-500 g) were prepared for the chronic recording of identified second-order medial vestibular nuclei neurons. In brief, the animals were anesthetized with Nembutal (40 mg/kg) delivered intraperitoneally. As a first step, a bipolar labyrinthine electrode was implanted over the round window in the middle ear cavity and above the horizontal and anterior semicircular canal ampullae respectively, in order to stimulate the vestibular nerve (Azzena et al., 1976; Shimazu and Precht, 1965; deWaele et al., 1990). Secondly, a coil was sutured on the eyeball for eye movement measurement with the magnetic search coil method (Fuchs and Robinson, 1966) and a craniotomy was performed over the ipsilateral hemicerebellum. Then, three stainless steel T-shaped screws were cemented in the parietal and frontal bones to secure the fixation of the head holder, and record the electroencephalogram signal. Finally, a head-holder was stereotaxically fixed onto the skull with dental cement to restrain the animal's head during the recording sessions. The stereotaxic plane was determined by head rotation around the interaural axis, until the calvarium was horizontal between anterior 6 and 14 mm from the stereotaxic zero (Rapisarda and Bacchelli, 1977).

Recording Conditions

In order to record the HVOR, the animal was placed on a servo-controlled turntable for vestibular stimulation. Its head was secured to the head-holder in a 35 degrees nose-down position to orient the horizontal semicircular canals in the earth horizontal plane (Curthoys et al., 1975). Glass microelectrodes (resistance between 1 to 5 M Ω) were lowered towards the vestibular nuclei. The vestibular field potential evoked by orthodromic stimulation of the vestibular nerve and the mass discharge of the neurons were used to map out the location of the medial vestibular nucleus. Once a vestibular neuron was isolated and identified as second-order (orthodromic activation latency inferior to 1.5 ms), its activity was recorded at rest and during the HVOR in the darkness. The HVOR was elicited by submitting the animal's head to horizontal sinusoidal rotations (0.1, 0.2, 0.5, 1, 2 and 3 Hz) about the vertical axis at a peak angular velocity of ±40°/s. In some experiments, the gain of the HVOR was modified by a training protocol that induced a conflictual visuo-vestibular adaptation: the guinea-pig was rotated in the light (0.05 or 0.1 Hz at ±20 °/s) with the animal surrounded by a visual scene stabilized with respect to its head. The HVOR characteristics and the dynamic response of type I second-order MVNn were tested in the darkness at different frequencies (0.05 or 0.1 Hz at ±20°/s and 0.1, 0.5 and 1 Hz at ±40°/s) before and after one hour of this protocol. The horizontal and vertical components of the eye position, the turntable position and velocity signals, and the discharge of a single MVNn were all recorded on an FM magnetic tape, displayed on a paper chart recorder and processed off-line on a PC computer.

RESULTS

About seventy identified second-order type I medial vestibular nuclei neurons (MVNn), were recorded extracellularly in the head-fixed adult guinea-pig during alertness. Part of them were also recorded following adaptation of the HVOR or during transitions in the sleep-waking cycle, in particular during the occurrence of REM sleep episodes. The quantification of all these data is currently in progress and will be published elsewhere.

Response of Identified Type I MVNn to Sinusoidal Head rotations

As previously reported in other species (Highstein et al., 1987; Shimazu and Precht, 1965), guinea-pig's type I MVNn could be broadly segregated into tonic (regular) and kinetic (irregular) units. They coded head velocity over a wide frequency range (0.1 to 3 Hz) and displayed a phase lead (with respect to the head velocity), which decreased when the frequency of rotation increased. They could, moreover, be distinguished into purely vestibular (discharge unaffected by the position of the eye in the orbit) and eye position-sensitive neurons. Figure 1 illustrates the characteristic response of an irregular eye position-sensitive type I second-order MVNn during horizontal sinusoidal head rotations (0.1, 0.2 and 0.5 Hz with a peak angular velocity of $\pm 40°/s$) in the darkness. This MVNn recorded on the right side increased its discharge during rotations of the head towards the side of its soma (i.e. when the head moved to the right) and became silent during contralateral ones. Furthermore, similarly to what had been previously reported in other species (reviewed in Berthoz et al. 1989), part of the second-order type I MVNn displayed pauses during the quick phases oriented towards the side ipsilateral to their soma and bursts during the contralaterally-directed ones (see inset in Fig. 1A for a typical example of such behaviors, which were however better observed in the light).

Adaptive Change in the Velocity Sensitivity of MVNn

The HVOR and the response of identified second-order type I MVNn were tested in the darkness at different frequencies (0.05 or 0.1 Hz at $\pm 20°/s$ and 0.1, 0.5 and 1 Hz at $\pm 40°/s$) before and after one hour of a training protocol that induced a conflictual visuo-vestibular adaptation (see material and methods). Figure 2A illustrates the horizontal component of the eye position and the instantaneous firing rate of an identified second-order type I MVNn recorded in response to sinusoidal head rotations (0.05 Hz at $\pm 20°/s$) in the darkness before and after adaptation. Whilst the gain of the HVOR (G_E) was largely decreased (from 0.3 to 0.15), the velocity sensitivity of the neuron (i.e. its gain, G_N), calculated by a method allowing to measure it independently of the position of the eye in the orbit (adapted from Godaux and Cheron, 1993), was also decreased (from 0.93 Hz/°/s before adaptation to 0.68 Hz/°/s following adaptation, Fig. 2B). A similar adaptive change of the gain was observed in all type I MVNn tested so far. In the cases where we succeeded to test different frequencies of stimulation both before and after adaptation, the adaptive changes were found to be clearly frequency-selective (not shown). Indeed, whereas the gain of both the HVOR and the MVNn were decreased at the training frequency (0.05 Hz at $\pm 20°/s$), they were not significantly modified at the other frequencies tested (0.1, 0.5 and 1 Hz at $\pm 40°/s$).

Discharge of Type I MVNn During REM Sleep Episodes

Several identified type I second-order MVNn were recorded during both alertness and REM sleep episodes characterized by a desynchronized EEG, a muscular atonia interrupted at

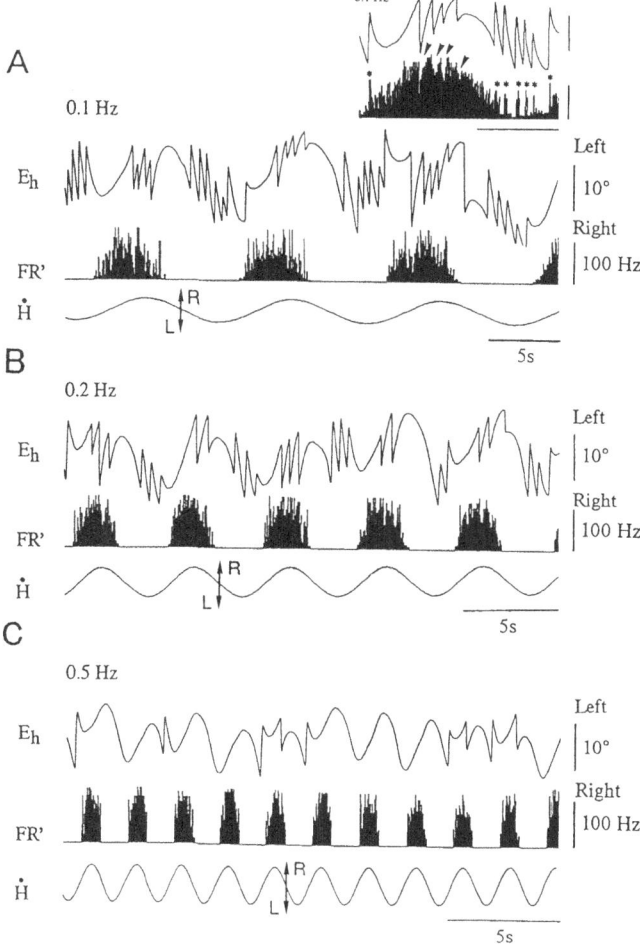

Figure 1. Response of an identified irregular type I second-order MVNn to horizontal sinusoidal head rotations. A-C. Horizontal component of the eye position (E_h), head velocity signal ($\dot{H}$) and instantaneous firing rate (FR') of an identified type I second-order MVNn during horizontal head rotations at 0.1 (A), 0.2 (B) and 0.5 Hz (C) in the darkness. During vestibular stimulation in the light (inset in Fig. 1A), this neuron displayed pauses (arrowheads) during ipsilaterally-oriented quick phases and bursts (asterisks) during contralateral ones. The calibration bars in the inset are the same as in panel A.

times by twitches, and the occurrence of characteristic 10 Hz eye oscillations (see Escudero and Vidal, 1992). During the REM sleep episodes the MVNn still coded head velocity and still exhibited pauses in their discharge during rapid eye movements directed towards the side ipsilateral to their soma. These pauses were strongly reminiscent of those observed in the awake guinea-pig during the ipsilaterally-directed quick phases (see inset in Fig. 1A).

DISCUSSION

Activity of Identified Type I MVNn at Rest and During Sinusoidal Head Rotations

Regularity of Discharge at Rest. Our results demonstrate, not surprisingly, the existence of regular and irregular type I second-order MVNn in the alert head-fixed guinea-pig. Previous electrophysiological studies in the cat and the monkey had also reported the presence of tonic (regular) and kinetic (irregular) second-order vestibular neurons (Highstein et al., 1987; Shimazu and Precht, 1965). The input of regular and irregular first-order

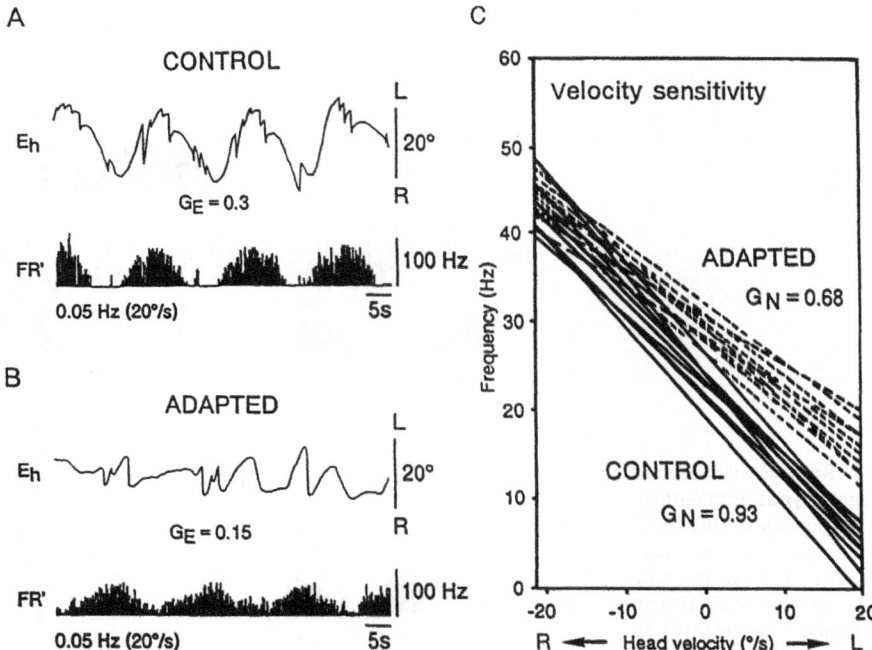

Figure 2. Change in the gain of a second-order MVNn following a conflictual visuo-vestibular adaptation. A, B. Horizontal component of the eye position (E_h) and instantaneous firing rate (FR') of an identified second-order MVNn in response to 0.05 Hz head rotations at ±20°/s before (A, CONTROL) and after (B, ADAPTED) one hour of visuo-vestibular conflict. C. Plots of the velocity sensitivity (G_N) measured before (CONTROL, solid lines) and after adaptation (ADAPTED, dotted lines) for the neuron illustrated in panels A and B (the slope of each line represents the gain of the neuron for a distinct eye position).

vestibular afferents (Curthoys, 1982; Ezure et al., 1978; Fernandez and Goldberg, 1971; Goldberg and Fernandez, 1971; Yagi and Ueno, 1988) to the second-order vestibular neurons has been shown, at least in the monkey, to remain partially segregated (Boyle et al., 1992; Highstein et al., 1987; Iwamoto et al., 1990 a,b; Minor and Goldberg, 1991), which supports the hypothesis of the existence of "frequency-tuned channels" in the vestibular system (Baker et al., 1981; Collewijn and Grootendorst, 1979; Godaux et al., 1983; Lisberger et al., 1983; Paige and Sargent, 1991). According to this hypothesis, the regular units are thought to be involved in the vestibulo-ocular pathways, whereas the irregular ones could be rather involved in vestibulo-spinal pathways (Boyle et al., 1992; Highstein et al., 1987; Iwamoto et al., 1990 a,b; Minor and Goldberg, 1991). An alternative view (Angelaki and Perachio, 1993) would be that the irregular units would in fact be best suited to feed the velocity storage integrator which is responsible for the lengthening of the time constant of the VOR. In view of their intrinsic membrane properties on slice (see Serafin et al., 1991 a,b) and of their mode of firing in the isolated whole brain preparation (Serafin et al., 1992a), we have proposed that type A and type B MVNn might correspond in vivo to the tonic and kinetic second-order MVNn, respectively. Independently of the functional meaning of such segregation, the present demonstration of the existence of regular and irregular second-order vestibular neurons in the medial vestibular nucleus of the alert guinea-pig strengthens this hypothesis.

Eye Position Sensitivity. In the cat and the monkey, part of the second-order type I vestibular neurons display an eye position sensitivity (Berthoz et al., 1989; Escudero et al.,

1992; Fuchs and Kimm, 1975; Lisberger and Miles, 1980; McCrea et al., 1987). Our results indicate that this holds true in the guinea-pig where part of the type I MVNn exhibited such a characteristic.

Discharge of Second-Order MVNn during Quick Phases. Different authors have demonstrated the occurrence of an ipsilateral saccade-related or quick phase-related decrease of discharge in both the cat and the monkey (see discussion in Berthoz et al., 1989). The neuronal mechanism underlying such a decrease in the firing of vestibular neurons has been elucidated: during the quick phases the excitatory burst neurons excite inhibitory type II vestibular neurons, which in turn inhibit type I neurons on the same side (Nakao et al., 1982; Sasaki and Shimazu, 1981). Given the similar results obtained in the guinea-pig, the same mechanism should be true in that species.

Besides, various authors have also reported in vivo the occurrence of contralateral (with respect to the side of the soma) quick phase-related bursts of discharge in both type I and type II vestibular neurons (Baker and Berthoz, 1974; Berthoz et al., 1989; Hikosaka et al., 1977; Maeda et al., 1972; Nakao et al., 1982). The neuronal mechanism leading to these bursts has been described in the cat: during head movements directed to the right, the second-order type I MVNn on the left side are continuously inhibited by the inhibitory type II MVNn on the same side and hence remain silent during the slow phase of the eye (Ishizuka et al., 1980; Shimazu and Precht 1965, 1966). At times however, the slow phase is interrupted by quick phases directed towards the opposite direction. At that time the excitatory burst neurons on the right side become active and excite inhibitory type II MVNn which in turn inhibit type I MVNn on the same side (Hikosaka et al., 1980; Nakao et al., 1982). Becoming silent, these neurons no longer excite inhibitory type II MVNn on the left side (via the commissural pathway), which results in a brisk disinhibition of type I MVNn on the left side. Thus, in the midst of a continuous hyperpolarization, type I MVNn respond to that brisk disinhibition by a burst of spikes. It is noteworthy that in vitro, it is exactly this succession of events at the membrane level which is required to trigger a calcium-dependent low-threshold spike (LTS) that we have shown to be present in about one fourth of type B MVNn (Serafin et al., 1991a) in the guinea-pig. The present results, since they demonstrate that the guinea-pig's MVNn have the capacity to burst in vivo, support the hypothesis that the LTS could possibly be involved in the bursting activity of these neurons during contralaterally-directed quick phases. In addition the fact that, as shown in the cat (Berthoz et al., 1989; Escudero et al., 1992; Yoshida et al., 1981), the burst parameters (duration, amplitude and time course) do not appear to code any aspect of the ongoing quick phase, further supports the idea that an intrinsic mechanism such as an LTS might play a role in the burst generation. Finally, it is noteworthy that the duration of the vestibular quick phases (50 to 100 ms) in the alert guinea-pig (Escudero et al., 1993) is similar to the duration of the LTS recorded in vitro.

Adaptive Change in the Response of Type I MVNn

In the head-fixed guinea-pig, similarly to what had been previously described in other species (see Melvill Jones, 1985 for a review), the gain of the HVOR could be adaptively decreased using a training protocol that induced a visual-vestibular conflict. Here we report that the gain of the response of all the identified second-order type I MVNn tested so far was also decreased following HVOR adaptation. In addition, the decrease of the neuronal gain was found to be frequency-selective, which is in good agreement with what had been previously demonstrated in a number of species for the adaptation of the VOR (Collewijn and Grooten-dorst, 1979; Godaux et al., 1983; Lisberger et al., 1983; Paige and Sargent, 1991). Our results should however be taken with caution due to the small number of neurons that could be recorded for a sufficient time to allow vestibular adaptation. Nevertheless, it is noteworthy that they are in agreement with recent studies in the goldfish, which strongly suggest that VOR learning appears to take place, amongst other sites, at the level of the second-order vestibular neurons (Baker et al., 1992; Pastor et al., 1992,1993), and with the results of a parallel study in the monkey, which demonstrate large changes in the gain of putative vertical second-order vestibular neurons in association with learning (Partsalis et al. 1993).

Discharge during REM Sleep Episodes

During REM sleep, at least three neuronal groups could be involved in the generation of the horizontal component of the eye oscillations: the MVNn, the saccadic generator, and the prepositus hypoglossi neurons. Our present results do not favor a predominant partici-pation of the second-order type I MVNn: they do not appear to oscillate during REM sleep. In fact, as previously described in the cat (Bizzi et al., 1964 a,b), they pause and burst during paradoxical sleep as they do during the saccades and the quick phases occurring in the awake state. Moreover, during REM sleep, the MVNn continue to code head velocity during natural vestibular stimulation while the eyes are oscillating. Hence we are left with our previous suggestion concerning the in vitro oscillatory behavior of the MVNn: it might play a functional role during locomotion since vestibular neurons were demonstrated to oscillate during fictive locomotion in both the cerebellectomized cat (Orlovski, 1972), the decere-brated guinea-pig (Marlinsky, 1992), and in vitro in the lamprey (Bussières and Dubuc, 1993). As far as the REM sleep is concerned, our results tend to indicate that it is the saccadic generator and/or the prepositus hypoglossi neurons which could participate in the generation of the eye oscillations characterizing the REM sleep episodes.

CONCLUSIONS

In the alert head-fixed guinea-pig second-order type I MVNn behave like to what has been previously described in other species. They can be broadly segregated according to the regularity of their discharge and code head velocity over the frequency range of 0.1 to 3 Hz. Moreover, some neurons pause during ipsilaterally-directed quick phases (i.e. the phases directed towards the side of their soma) and some exhibit a burst of activity during contralateral ones.

As reported in other species, the gain of the guinea-pig's HVOR can be adaptively decreased by a training protocol that induces a visual-vestibular conflict. We report here that the gain of identified second-order type I MVNn was also decreased following HVOR adaptation and that this adaptive change was frequency-selective.

Finally, during REM sleep, the type I MVNn still code head velocity. In addition, the apparent periodicity of their phasic discharge during the 10 Hz eye oscillations appears to be the result of the pauses occurring during the ipsilaterally-directed rapid eye movements (with respect to the side of their soma) rather than to an original oscillatory mode of discharge based on their intrinsic membrane properties.

Altogether these results appear to confirm that the guinea-pig is a valuable model to undertake a combined in vivo and in vitro study of the neuronal network underlying gaze control.

ACKNOWLEDGMENTS

This work was supported by grants from the Swiss NSF (No. 823A-030710), the exchange program between France and Switzerland of the French Ministère des Affaires Etrangères, and the CNRS.

REFERENCES

Angelaki, D.E., and Perachio A.A., 1993, Contribution of semicircular canal afferents to the horizontal vestibuloocular response during constant velocity rotation, *J. Neurophysiol.* 69:996-999.

Azzena, G.B., Mameli, O., and Tolu, E., 1976, Vestibular nuclei of hemilabyrinthectomized guinea pigs during decompensation, *Arch. Ital. Biol.* 114:389-398.

Baker, R., and Berthoz, A., 1974, Organization of vestibular nystagmus in the oblique oculomotor system, *J. Neurophysiol.* 37:195-217.

Baker, R., Evinger, C., and McCrea, R.A., 1981, Some thoughts about the three neurons in the vestibulo-ocular reflex, *Ann. NY Acad. Sci.* 374:171-188.

Baker, R.A., Pastor, A.M., De La Cruz, R.R., and Simpson, J.I., 1992, Purkinje cell eye and head velocity sensitivity are not altered during VOR adaptation, *Soc. Neurosci. Abstracts* 18, 178.5, 407.

Berthoz, A., 1989, Coopération et substitution entre le système saccadique et les "réflexes" d'origine vestibulaire: faut-il réviser la notion de réflexe?, *Rev. Neurol. (Paris)* 145:513-526.

Berthoz, A., Droulez, J., Vidal, P.P., and Yoshida, K.,1989, Neural correlates of horizontal vestibulo-ocular reflex cancellation during rapid eye movements in the cat, *J. Physiol. (London)* 419:717-751.

Bizzi, E., Pompeiano, O., and Somogyi, I., 1964a, Vestibular nuclei: activity of single neurons during natural sleep and wakefulness, *Science* 145:414-415.

Bizzi, E., Pompeiano, O., and Somogyi, I., 1964b, Spontaneous activity of single vestibular neurons of unrestrained cats during sleep and wakefulness, *Arch. Ital. Biol.* 102:308-330.

Boyle, R., Goldberg, J.M., and Highstein, S.M., 1992, Inputs from regularly and irregularly discharging vestibular nerve afferents to secondary neurons in squirrel monkey vestibular nuclei. III. Correlation with vestibulospinal and vestibuloocular output pathways, *J. Neurophysiol.* 68:471-484.

Bussières, N., and Dubuc, R., 1992, Phasic modulation of vestibulospinal neuron activity during fictive locomotion in lampreys, *Brain Res.* 575:174-179.

Collewijn, H., and Grootendorst, A.F., 1979, Adaptation of optokinetic and vestibulo-ocular reflexes to modified visual input in the rabbit. In: Reflex control of posture and movement, Granit, R., and Pompeiano, O. (eds.). Amsterdam, Elsevier, pp 771-781.

Curthoys, I.S., 1982, The response of primary horizontal semicircular canal neurones in the rat and guinea-pig to angular acceleration, *Exp. Brain Res.* 47:286-294.

Curthoys, I.S., Curthoys, E.J., Blanks, R.H.I., and Markham, C.H., 1975, The orientation of the semi-circular canals in the guinea-pig, *Acta Otolaryngol. (Stockholm)* 80:197-205.

Duensing, F., and Schaefer, K.P., 1958, Die aktivität einzelner neuron im bereich der vestibulariskerne bei horizontal-beschleunigungen unter besonderer berücksichtigung des vestibulären nystagmus, *Arch. Psychiatr. Nervenkr.* 198:225-252.

du Lac, S., and Lisberger, S.G., 1992, Eye movements and brainstem neuronal responses evoked by cerebellar and vestibular stimulation in chicks, *J. Comp. Physiol.* 171:629-638.

Escudero, M., De la Cruz, R.R., and Delgado-Garcia, J.M., 1992, A physiological study of vestibular and prepositus hypoglossi neurones projecting to the abducens nucleus in the alert cat, *J. Physiol. (London)* 458:538-560.

Escudero, M., de Waele, C., Vibert, N., Berthoz, A., and Vidal P.P., 1993, Saccadic eye movements and horizontal vestibulo-ocular and vestibulo-collic reflexes in the intact guinea-pig, *Exp. Brain Res.* 97:254-262.

Escudero, M., and Vidal, P.P., 1992, Eye movements during paradoxical sleep in guinea pig, Soc. Neurosci. Abstracts 18, 92.8, 196.

Ezure, K., Schor, R.H., and Yoshida, K., 1978, The response of horizontal semicircular canal afferents to sinusoidal rotation in the cat, *Exp. Brain Res.* 33:27-39.

Fernandez, C., and Goldberg, J.M., 1971, Physiology of peripheral neurons innervating semicircular canals of the squirrel monkey. II. Response to sinusoidal stimulation and dynamics of peripheral vestibular system, *J. Neurophysiol.* 34:661-675.

Fuchs, A.F., and Kimm, J., 1975, Unit activity in vestibular nucleus of the alert monkey during horizontal angular acceleration and eye movement, *J. Neurophysiol.* 38:1140-1161.

Fuchs, A., and Robinson, D.A., 1966, A method for measuring horizontal and vertical eye movements chronically in the monkey, *J. Appl. Physiol.* 21:1068-1070.

Godaux, E., and Cheron, G., 1993, Testing the common neural integrator hypothesis at the level of the individual abducens motoneurones in the alert cat, *J. Physiol. (London)* 469:549-570.

Godaux, E., Halleux, J., and Gobert, C., 1983, Adaptative change of the vestibulo-ocular reflex in the cat: the effects of a long-term frequency-selective procedure, *Exp. Brain Res.* 49:28-34.

Goldberg, J.M., and Fernandez, C., 1971, Physiology of peripheral neurons innervating semicircular canals of the squirrel monkey. III. Variations among units in their discharge properties, *J. Neurophysiol.* 34:676-684.

Goldberg, J.M., Highstein, S.M., Moschovakis, A.K., and Fernandez, C., 1987, Inputs from regularly and irregularly discharging vestibular nerve afferents to secondary neurons in the vestibular nuclei of the squirrel monkey. I. An electrophysiological analysis, *J. Neurophysiol.* 58:700-718.

Highstein, S.M., Goldberg, J.M., Moschovakis, A.K., and Fernandez, C., 1987, Inputs from regularly and irregularly discharging vestibular nerve afferents to secondary neurons in the vestibular nuclei of the squirrel monkey. II. Correlation with output pathways of secondary neurons, *J. Neurophysiol.* 58:719-738.

Hikosaka, O., Maeda, M., Nakao, S., Shimazu, H., and Shinoda, Y., 1977, Presynaptic impulses in the abducens nucleus and their relation to postsynaptic potentials in motoneurons during vestibular nystagmus, *Exp. Brain Res.* 27:355-376.

Hikosaka, O., Nakao, S., and Shimazu, H., 1980, Postsynaptic inhibition underlying spike suppression of secondary vestibular neurons during quick phases of vestibular nystagmus, *Neurosci. Letters* 16:21-26.

Ishizuka, N., Mannen, H., Sasaki, S., and Shimazu, H., 1980, Axonal branches and terminations in the cat abducens nucleus of secondary vestibular neurons in the horizontal canal system, *Neurosci. Letters* 16:143-148.

Isu, N., and Yokota, J., 1983, Morphophysiological study of the divergent projection of axon collaterals of medial vestibular nucleus neurons in the cat, *Exp. Brain Res.* 53:151-162.

Ito, M., 1982, Cerebellar control of the vestibulo-ocular reflex - around the flocculus hypothesis, *Annu. Rev. Neurosci.* 5:275-296.

Ito, M., 1993, Cerebellar flocculus hypothesis, *Nature* 363:24-25.

Iwamoto, Y., Kitama, T., and Yoshida, K., 1990a, Vertical eye movement-related secondary vestibular neurons ascending in medial longitudinal fasciculus in cat. I. Firing properties and projection pathways, *J. Neurophysiol.* 63:902-917.

Iwamoto, Y., Kitama, T., and Yoshida, K., 1990b, Vertical eye movement-related secondary vestibular neurons ascending in medial longitudinal fasciculus in cat. II. Direct connections with extraocular motoneurons, *J. Neurophysiol.* 63:918-935.

Kawato, M., and Gomi, H., 1992, The cerebellum and VOR/OKR learning models, *TINS* 15:445-453.

Lisberger, S.G., 1988, The neural basis for the learning of simple motor skills, *Science* 242:728-735.

Lisberger, S.G., 1994, Neural basis for motor learning in the vestibuloocular reflex of primates. III. Computational and behavioral analysis of sites of learning, *J. Neurophysiol.* 72:974-998.

Lisberger, S.G., and Miles, F.A., 1980, Role of primate medial vestibular nucleus in long-term adaptive plasticity of vestibuloocular reflex, *J. Neurophysiol.* 43:1725-1745.

Lisberger, S.G., Miles, F.A., and Optican, L.M., 1983, Frequency-selective adaptation: evidence for channels in the vestibulo ocular reflex ?, *J. Neurosci.* 3:1234-1244.

Lisberger, S.G., and Pavelko T.A., 1988, Brain stem neurons in modified pathways for motor learning in the primate vestibulo-ocular reflex, *Science* 242:771-773.

Lisberger, S.G., Pavelko, T.A., Bronte-Stewart, H.M., and Stone, L.S., 1994c, Neural basis for motor learning in the vestibuloocular reflex of primates. II. Changes in the responses of horizontal gaze velocity Purkinje cells in the cerebellar flocculus and the ventral paraflocculus, *J. Neurophysiol.* 72:954-973.

Lisberger, S.G., Pavelko, T.A., and Broussard, D.M., 1994a, Responses during eye movements of brain stem neurons that receive monosynaptic inhibition from the flocculus and the ventral paraflocculus in monkeys, *J. Neurophysiol.* 72:909-927.

Lisberger, S.G., Pavelko, T.A., and Broussard, D.M., 1994b, Neural basis for motor learning in the vestibuloocular reflex of primates. I. Changes in the responses of brain stem neurons, *J. Neurophysiol.* 72:928-953.

Lisberger, S.G., and Sejnowski, T.J., 1992, Motor learning in a recurrent network model based on the vestibulo-ocular reflex, *Nature* 360:159-161.

Lisberger, S.G., and Sejnowski, T.J., 1993, Cerebellar flocculus hypothesis, *Nature* 363:25.

Maeda, M., Shimazu, H., and Shinoda, Y., 1972, Nature of synaptic events in cat abducens motoneurons at slow and quick phase of vestibular nystagmus, *J. Neurophysiol.* 35:279-296.

Marlinsky, V.V., 1992, Activity of lateral vestibular nucleus neurons during locomotion in the decerebrate guinea pig, *Exp. Brain Res.* 90:583-588.

McCrea, R.A., Strassman, A., May, E., and Highstein, S.M., 1987, Anatomical and physiological characteristics of vestibular neurones mediating the horizontal vestibulo-ocular reflex of the squirrel monkey, *J. Comp. Neurol.* 264:547-570.

McCrea, R.A., Yoshida, K., Berthoz, A., and Baker, R., 1980, Eye movement related activity and morphology of second-order vestibular neurons terminating in the cat abducens nucleus, *Exp. Brain Res.* 40:468-473.

Melvill Jones, G., 1985, Adaptive modulation of VOR parameters by vision. In: Adaptive mechanisms in gaze control. Facts and theories. Berthoz, A., and Melvill Jones, G. (eds). Elsevier, New York, pp 21-50.

Melvill Jones, G., and Milsum, J.H., 1970, Characteristics of neural transmission from the semicircular canal to the vestibular nuclei of cats, *J. Physiol. (London)* 209:295-316.

Miles, F.A., and Braitman, D.J., 1980, Long-term adaptive changes in primate vestibuloocular reflex. II. Electrophysiological observations on semicircular canal primary afferents, *J. Neurophysiol.* 43:1426-1436.

Miles, F.A., Fuller, J.H., Braitman, D.J., and Dow, B.M., 1980, Long-term adaptive changes in primate vestibuloocular reflex. III. Electrophysiological observations in flocculus of normal monkey, *J. Neurophysiol.* 43:1437-1476.

Miles, F.A., and Lisberger, S.G., 1981, Plasticity in the vestibuloocular reflex: a new hypothesis, *Annu. Rev. Neurosci.* 4:273-299.

Minor, L.B., and Goldberg, J.M., 1991, Vestibular-nerve inputs to the vestibulo-ocular reflex: a functional-ablation study in the squirrel monkey, *J. Neurosci.* 11:1636-1648.

Mühlethaler, M., de Curtis, M., Walton, K., and Llinás, R., 1993, The in vitro isolated and perfused guinea-pig brain, *Eur.J. Neurosci.* 5:915-926.

Nakao, S., Sasaki, S., Schor, R.H., and Shimazu, H., 1982, Functional organization of premotor neurons in the cat medial vestibular nucleus related to slow and fast phases of nystagmus, *Exp. Brain Res.* 45:371-385.

Orlovsky, G.N., 1972, Activity of vestibulospinal neurons during locomotion, *Brain Res.* 46:85-98.

Paige, G.D., and Sargent, E.W., 1991, Visually-induced adaptive plasticity in the human vestibulo-ocular reflex, *Exp. Brain Res.* 84:25-34.

Partsalis, A.M., Zhang, Y., and Highstein, S.M., 1993, The Y group in vertical visual-vestibular interactions and VOR adaptation in the squirrel monkey, *Soc. Neurosci. Abstracts* 19, 60.2, 138.

Pastor, A.M., De La Cruz, R., and Baker, R., 1992, Characterization and adaptive modification of the goldfish vestibuloocular reflex by sinusoidal and velocity step vestibular stimulation, *J. Neurophysiol.* 68:2003-2015.

Pastor, A.M., De La Cruz, R.R., and Baker, R., 1993, Cerebellectomy reveals that storage and expression of vestibulo-ocular reflex adaptation occurs in the brainstem, *Soc. Neurosci. Abstracts* 19, 401.4, 982.

Rapisarda, C., and Bacchelli, B., 1977, The brain of the guinea-pig in stereotaxic coordinates, *Arch. Sci. Biol.* 61:1-37.

Sasaki, S., and Shimazu, H., 1981, Reticulovestibular organization participating in generation of horizontal fast eye movements, *Ann. NY Acad. Sci.* 374:130-143.

Sato, F., and Sasaki, H., 1993, Morphological correlation between spontaneously discharging primary vestibular afferents and vestibular nucleus neurons in the cat, *J. Comp. Neurol.* 333:554-566.

Scudder, C.A., and Fuchs, A.F., 1992, Physiological and behavioral identification of vestibular nucleus neurons mediating the horizontal vestibuloocular reflex in trained rhesus monkeys, *J. Neurophysiol.* 69:244-264.

Serafin, M., de Waele, C., Khateb, A., Vidal, P.P., and Mühlethaler, M., 1991a, Medial vestibular nucleus in the guinea-pig: I. Intrinsic membrane properties in brainstem slices, *Exp. Brain Res.* 84:417-425.

Serafin, M., de Waele, C., Khateb, A., Vidal, P.P., and Mühlethaler, M., 1991b, Vestibular nuclei neurons in the guinea-pig: II. Ionic basis of the intrinsic membrane properties in brainstem slices, *Exp. Brain Res.* 84:426-433.

Serafin, M., Khateb, A., de Waele, C., Vidal, P.P., and Mühlethaler, M., 1992a, Electrophysiology and pharmacology of 2 types of neurons in the medial vestibular nucleus and in the nucleus gigantocellularis of the guinea-pig in vitro, In: The head-neck sensory motor system, Berthoz, A., Graf, W., and Vidal, P.P. (eds). Oxford University Press, New-York, pp 244-250.

Serafin, M., de Waele, C., Khateb, A., Vidal, P.P., and Mühlethaler, M., 1992b, Medial vestibular nucleus in the guinea-pig. NMDA-induced oscillations, *Exp. Brain Res.* 88:187-192.

Shimazu, H., and Precht, W., 1965, Tonic and kinetic responses of cat's vestibular neurons to horizontal angular acceleration, *J. Neurophysiol.* 28:991-1013.

Shimazu, H., and Precht, W., 1966, Inhibition of central vestibular neurons from the contralateral labyrinth and its mediating pathway, *J. Neurophysiol.* 29:467-492.

Shinoda, Y., and Yoshida, K., 1974, Dynamic characteristics of responses to horizontal head angular acceleration in vestibuloocular pathway in the cat, *J. Neurophysiol.* 37:653-673.

Uchino, Y., Hirai, N., and Suzuki, S., 1982, Branching pattern and properties of vertical- and horizontal-related excitatory vestibuloocular neurons in the cat, *J. Neurophysiol.* 48:891-903.

Uchino, Y., Hirai, N., Suzuki, S., and Watanabe, S., 1981, Properties of secondary vestibular neurons fired by stimulation of ampullary nerve of the vertical, anterior or posterior, semicircular canals in the cat, *Brain Res.* 223:273-286.

Uchino, Y., and Isu, N., 1992a, Properties of inhibitory vestibulo-ocular and vestibulo-collic neurons in the cat. In: Vestibular and brainstem control of eye, head and body movements. Shimazu, H., and Shinoda, Y. (eds), Jpn. Sci. Soc. Press, Tokyo, pp 31-43.

Uchino Y, and Isu, N., 1992b, Properties of vestibulo-ocular and/or vestibulo-collic neurons in the cat. In: The head-neck sensory motor system. Berthoz, A., Graf, W., and Vidal, P.P. (eds), Oxford, Oxford, pp 266-272.

de Waele, C., Serafin, M., Khateb, A., Yabe, T., Vidal, P.P., and Mühlethaler, M., 1993, Medial vestibular nucleus in the guinea-pig: apamin-induced rhythmic burst firing. An in vitro and in vivo study, *Exp. Brain Res.* 95:213-222.

Yagi, T., and Ueno, H.,1988, Behavior of primary horizontal canal neurons in alert and anesthetized guinea pigs, *Exp. Neurol.* 101:356-365.

Yoshida, K., Berthoz, A., Vidal, P.P., and McCrea, R., 1981, Eye movement related activity of identified second order vestibular neurons in the cat. In: Progress in Oculomotor Research, Developments in Neuroscience. Fuchs, A. and Becker, W. (eds). Elsevier, Amsterdam, pp 371-378.

5

ADAPTIVE CHANGES IN GAIN OF THE VESTIBULOSPINAL REFLEX DURING SUSTAINED NECK-VESTIBULAR STIMULATION

O. Pompeiano

Dipartimento di Fisiologia e Biochimica
Via S.Zeno 31
56127 Pisa
Italy

INTRODUCTION

Both labyrinth and neck inputs exert a prominent influence on postural mechanisms involving the limb musculature. In particular, rotation of the whole animal on one side, leading to stimulation of labyrinth receptors, induces a contraction not only of the limb extensors of that side, thus stabilizing posture of the limbs during head rotation, but also of the dorsal neck extensors of the opposite side, thus counteracting the head displacement (Schor and Miller, 1981). This righting of the head would then lead to stimulation of neck receptors, which contributes synergistically to the labyrinth-induced contraction of the limb extensors to maintain the support of the body over the limbs during animal tilt (Lindsay, Roberts and Rosenberg, 1976).

In the decerebrate animal, in which the proprioceptive neck input does not occur due to fixation of the head to the stereotaxic equipment, the vestibulospinal reflex (VSR) acting on limb extensors is barely compensatory, but its amplitude can be enhanced by out-of-phase body-to-head rotation, leading to appropriate stimulation of neck receptors (cf. Lindsay, Roberts and Rosenberg, 1976). No attempt was made in the past to find out whether, in addition to these rapid postural adjustments, the neck input could also produce slow adaptive changes in gain of the VSR. To test this hypothesis we tried to investigate whether a sustained roll tilt of the animal, leading to sinusoidal stimulation of labyrinth receptors, associated with out-of-phase body-to-head rotation, leading to sinusoidal stimulation of neck receptors, could produce an adaptive increase in gain of the VSR, thus contributing to adjust the gain of this reflex to match the neck displacement (cf. Andre, d'Ascanio, Manzoni and Pompeiano, 1993). Since Purkinje (P)-cells located in the zone B of the cerebellar anterior vermis (Corvaja and Pompeiano, 1979), which send inhibitory axons to lateral vestibulospinal (VS) neurons (cf. Ito, 1984), respond to roll tilt of the animal as well as to neck rotation (Denoth, Magherini, Pompeiano and Stanojevic, 1979, 1980), we postulated that heterosynaptic interactions between proprioceptive neck input and labyrinth input, leading to adaptation of the VSR, occurred within the vermal cortex of the cerebellar anterior lobe. For this purpose, the method of intravermal microinjection of GABA agonists, which decreased the amplitude of the VSR elicited during intermittent stimulation of labyrinth receptors (Andre, d'Ascanio, Manzoni and Pompeiano, 1994), was used to investigate the role of the spinocerebellum in the adaptive control of the VSR gain during sustained neck-vestibular stimulation (cf. Manzoni, Andre, d'Ascanio and Pompeiano, 1994).

METHODS

Experiments were performed in precollicular decerebrate cats, operated under ether anesthesia. The head of the animal was fixed through the stereotaxic frame in horizontal position and pitched 10° nose-down, while both fore- and hindlimbs were extended and clamped. Sinusoidal rotation about the longitudinal axis of the whole animal (at 0.15 Hz, ±10°) produced selective stimulation of labyrinth receptors, leading to the VSRs (Fig. 1A). On the other hand, sinusoidal rotation of the body in both directions of the coronal plane at the same parameters indicated above, while maintaining the horizontal position of the head, selectively stimulated the neck receptors, thus leading to cervicospinal reflexes (Fig. 1B). As reported in the literature (Lindsay, Roberts and Rosenberg, 1976), the forelimb extensor tonus increased during side-down animal tilt (asterisk in A), but decreased during side-down neck rotation (upper arrow in B). Finally, 10° rotation of the stereotaxic equipment (head) in one direction of the coronal plane, leading to stimulation of labyrinth receptors, was

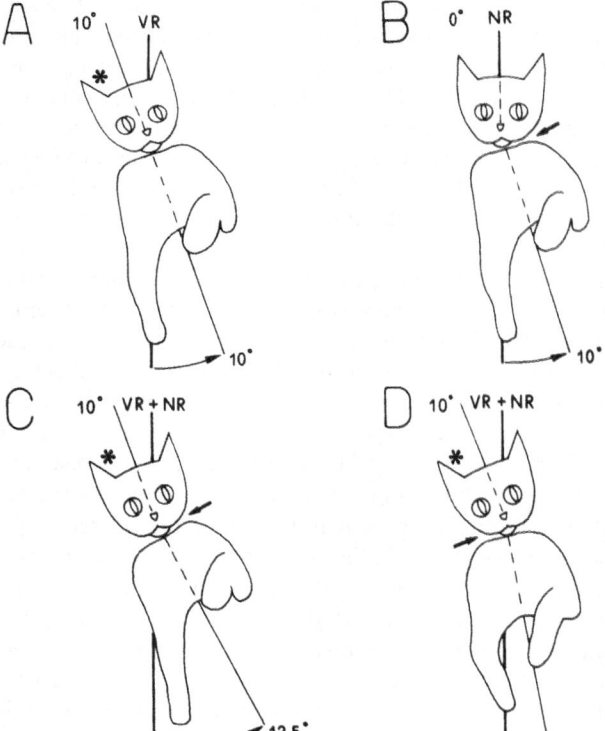

Figure 1. Schematic representation of different head and/or body displacements, leading to individual or combined stimulation of labyrinth or neck receptors. (A) vestibular stimulation; (B) neck stimulation; (C) out-of-phase neck-vestibular stimulation (synergistic strategy); (D) in-phase neck-vestibular stimulation (conflicting strategy).

associated with a synchronous 12.5° or 7.5° rotation of the table (body), which produced a 2.5° out-of-phase (Fig. 1C) or in-phase (Fig. 1D) body-to-head rotation, as indicated by the upper arrow. This led to a neck input, which slightly increased (synergistic strategy, Fig. 1C) or decreased (conflicting strategy, Fig. 1D) the postural changes induced by the labyrinth signal (shown in Fig. 1A). These responses (called "neck-vestibulospinal reflexes" or N-VSRs), were also elicited at the frequency of head and body rotation of 0.15 Hz.

The multiunit EMG activity of the medial head of the triceps brachii of one or both sides was conventionally amplified, selected through a window discriminator and analyzed by a digital signal averager (Correlatron 1024, Laben), which provided sequential pulse density histograms (SPDHs). Synchronizing pulses permitted superimposing and averaging of the muscle discharge during 6 sweeps, each containing the responses to two successive cycles of stimulation (128 bins, 0.1 sec bin width). The digital data of these groups of averaged responses to 12 cycles of stimulation (AR) were processed with a computer system (PC BIT 286), which performed a fast Fourier transform and allowed to evaluate the base frequency (BF, in imp./sec), gain (in imp./sec/deg) and phase angle of the first harmomic responses (in arc degrees with respect to the peak of the side-down displacement of the head; see asterisk in Fig. 1A, C and D). A stainless steel cannula with an outer diameter of 200-300 μm connected to an Hamilton 1 μl syringe, was lowered into the cerebellar anterior vermis (culmen) through the second or third folium rostral to the fissura prima, at the laterality of 1.4-1.8 mm and at the depth of 3-5 mm below the surface. Usually 0.25 μl of the GABA-A agonist muscimol (RBI, Natick, MA, USA) or the GABA-B agonist baclofen (Ciba Geigy, Basel, Switzerland) at the concentration of 2-8 μg/μl saline (0.9% NaCl) were slowly (1-2 min) injected in a given spot. These solutions, with a pH adjusted to 7.4, contained 5% of pontamine sky blue (Gurr, Poole, England) as a marker. The region chosen for injection corresponded to an area that upon previous monopolar stimulation with 3 cathodal pulses at 300/sec, 0.2 msec duration, 0.1-10 V inhibited the EMG activity of the ipsilateral triceps brachii.

Baseline measurements of the VSR were recorded from the triceps brachii during roll tilt of the whole animal at 0.15 Hz, ±10°. Each group of AR was recorded intermittently at regular intervals of 8-10 min for at least 0.5-1 h. Baseline measurements of out-of-phase or in-phase N-VSRs were also recorded. As soon as the control periods were completed, the procedures for adaptation were started. In particular, the animal was submitted to a sustained head rotation (at 0.15 Hz, ±10°) associated with a 2.5° out-of-phase or in-phase body-to-head rotation. These stimuli were applied in different experiments almost continuously for 3 h, during which successive groups of AR of the triceps brachii to neck-vestibular stimulation (the N-VSRs) were recorded. The first measurement of the pure VSR was taken 10 min after the beginning of the adaptation procedure, which was then interrupted every 10-15 min to allow the recording of the VSR, until the 3-h period of sustained neck-vestibular stimulation was over. Thus, the adaptive process was continuous over successive periods of 8.5-13.5 min, and interrupted only for 1.5 min to record the pure VSR. This short interruption did not noticeably affect the adaptive process.

In order to evaluate the role of the cerebellar anterior vermis in the adaptive changes, two groups of experiments were performed. In a *first* group the animals were submitted to microinjection into the cerebellar anterior vermis of muscimol prior to the adaptive stimulation. In particular, the resulting effects on the baseline gain of the ipsilateral and/or the contralateral VSR were tested intermittently every 8-10 min for about 40 min before a 3-h period of sustained out-of-phase head and body rotation at the parameters indicated above was started. The N-VSRs were also recorded from the same muscles during the adaptive stimulation. It was then possible to evaluate whether intravermal microinjection of GABA agonists prevented the occurrence of the adaptive process. In a *second* group of experiments, however, intravermal microinjection of muscimol or baclofen was performed 130 min after the end of a 3-h period of sustained out-of-phase neck-vestibular stimulation and their effects on the baseline gain of the VSR and the N-VSR were tested intermittently at regular intervals of 8-10 min for about 2 h. It was then possible to investigate whether the adapted increase in gain of the VSR and the N-VSR was suppressed by the GABAergic agents. At the end of each experiment, the localization and the extent of the pontamine sky blue-stained tissue were identified on histological serial sections of the cerebellum counterstained with neutral red.

RESULTS

Adaptation of the VSR and the N-VSR During Sustained Neck-Vestibular Stimulation

The EMG responses of the triceps brachii of one or both sides to animal tilt were recorded in 8 experiments as to obtain 11 curves of adaptation (Andre, d'Ascanio, Manzoni and Pompeiano, 1993). Before the adaptation was started, the average gain of the VSR recorded during intermittent roll tilt of the animal was 1.17 ± 0.78, S.D. imp./sec/deg. This baseline value remained constant throughout the whole control period which lasted between 30 and 230 min. The animals were then submitted to a 3-h period of sustained out-of-phase neck-vestibular stimulation (synergistic strategy). All the experiments showed an adaptive increase in gain of the pure VSR in this condition. Fig. 2A (filled triangles) illustrates that the difference between *adapted gain* minus *baseline gain* (ΔG), which was used as an index of adaptation, significantly increased on the average from 0 during the control period to 1.06 ± 0.61, S.D. imp./sec/deg during the third h of stimulation (paired t-test, $p<0.001$). This effect persisted unmodified during the first h of post-adaptation (ΔG=1.02 ± 1.08, S.D. imp./sec/deg).

In all the experiments reported above, before the adaptation was started, we evaluated not only the gain values of the VSR alone, but also those of the reflex responses to combined neck-vestibular stimulation used to elicit the adaptation itself (called "out-of-phase N-VSR"). The baseline gain of the N-VSR evaluated in the 8 experiments reported above, leading to 11 curves of adaptation, corresponded on the average to 1.79 ± 1.08, S.D. imp./sec/deg, thus being about 50% higher than that of the VSR alone, due to the synergistic contribution of the proprioceptive neck input. An adaptive increase in gain, which was more prominent than that displayed in the same experiments by the pure VSR, affected the N-VSR during the 3-h period of adaptive stimulation. Fig. 2B (filled triangles) shows that the ΔG value for the out-of-phase N-VSR increased on the average from 0 during the control period to 1.35 ± 0.74, S.D. imp./sec/deg during the third h of adaptation (paired t-test, $p<0.001$) and persisted unmodified during the first h of post-adaptation (ΔG=1.49 ± 1.80, S.D. imp./sec/deg).

An additional finding was obtained in 2 experiments in which 6 curves of adaptation were obtained during a 3-h period of sustained in-phase neck-vestibular stimulation (conflicting strategy). In these instances, the gain of the pure VSR, which corresponded on the average to 0.84 ± 0.39, S.D. imp./sec/deg during the control period, decreased during the first h of adaptation by an average ΔG of -0.19 ± 0.27, S.D. imp./sec/deg. This change, however, was unable to overcome the tendency to the adaptation of the VSR, which slightly increased to 0.30 ± 0.55, S.D. imp./sec/deg during the last h of adaptation (paired t-test, $p<0.001$). In the same experiments, the gain of the "in-phase N-VSR" remained below or close to the control value throughout the adaptation period.

It is of interest that only a weak increase in gain of the VSR was on the average obtained during a 3-h period of sustained roll tilt of the animal leading to selective stimulation

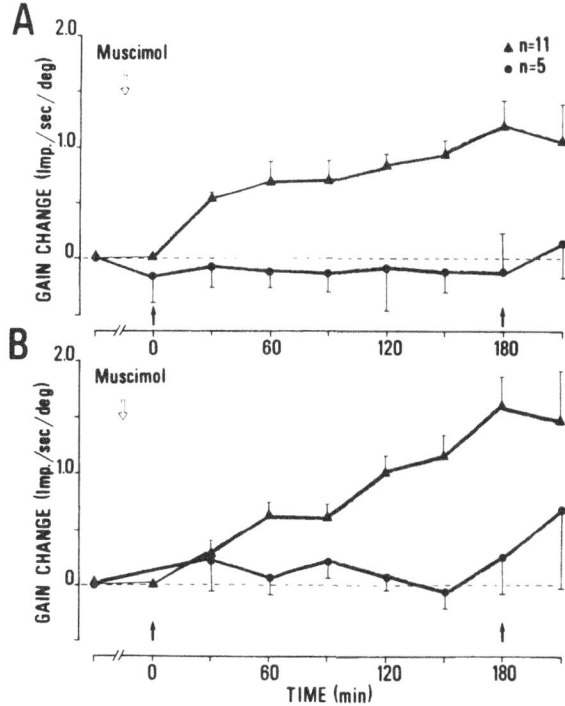

Figure 2. Averaged curves of the adaptive changes in gain of the VSR (A) and N-VSR (B) induced during out-of-phase neck-vestibular stimulation, in untreated animals or after microinjection into the zone B of the cerebellar anterior vermis of the GABA-A agonist muscimol. (A) changes in gain of the AR of the triceps brachii of one or both sides to roll tilt of the whole animal at 0.15 Hz, ±10° during and after a 3-h period of sustained out-of-phase neck-vestibular stimulation (between arrows). The filled triangles indicate the averaged curve obtained in control experiments (8 cats, 11 curves of adaptation), while the filled dots indicate the averaged curve obtained in other experiments (3 cats, 5 curves of adaptation) after intravermal microinjection of muscimol (0.25μl at 8 μg/μl saline). (B) changes in gain of the AR of the triceps brachii to out-of-phase neck-vestibular stimulation during and after the 3-h period of sustained head and body rotation. The averaged curves obtained in the control situation (filled triangles) as well as after microinjection of the GABA-A agonist (filled dots) are illustrated. Each symbol indicates mean ± S.E. value of the AR.

of labyrinth receptors (9 cats, 13 curves of adaptation). On the other hand, no changes in gain of both the VSR and the N-VSR were found in non-adaptation experiments in which these reflexes were tested intermittently (every 8-10 min), for periods which ranged between 90 and 230 min (7 cats, 11 control periods).

In conclusion, the increase in gain of the VSR evoked during the adaptive stimulation depended on the appropriate direction of neck rotation, rather than on aspecific changes in excitability of the neuronal networks continuously activated during combined neck-vestibular stimulation.

Effects of Intravermal Microinjection of a GABA-A Agonist on Adaptation of the VSR and the N-VSR

In a first group of 3 experiments (Manzoni, Andre, d'Ascanio and Pompeiano, 1994), the VSRs were recorded from the triceps brachii of one or both sides as to obtain 5 curves of adaptation, but only after unilateral microinjection into the cerebellar anterior vermis of the GABA-A agonist muscimol (0.25 µl at 8 µg/µl saline). In these selected experiments, the average gain of the VSR, intermittently recorded for about 40 min after muscimol injection, was only slightly but insignificantly depressed with respect to the control value obtained during the pre-injection period (ΔG=-0.17±0.57, S.D. imp./sec/deg). However, administration of the GABA-A agonist prevented the occurrence of any adaptive increase in gain of the VSR during the 3-h period of sustained out-of-phase neck-vestibular stimulation (ΔG at the third h of stimulation =-0.13±0.56, S.D. imp./sec/deg) as well as during the successive 30 min of post-adaptation (ΔG=0.14±0.68, S.D. imp./sec/deg) (Fig. 2A, filled circles). This finding indicates that at this time the muscimol was still sufficiently active to prevent the resumption of the adaptive changes. In contrast to the results obtained in the absence of any intravermal injection, the difference between the baseline value and that obtained at the end of the adaptation period was not statistically significant (paired t-test).

The effects of intravermal injection of muscimol were tested not only on the pure VSR, but also on the out-of-phase N-VSR. Fig. 2B (filled circles) illustrates the average changes in gain (ΔG) of the N-VSR obtained after injection of the GABA-A agonist. Similarly to the VSR, the average ΔG value of the N-VSR corresponded to 0.08±0.49, S.D. imp./sec/deg during the third h of adaptive stimulation, but slightly increased to 0.67±1.19, S.D. imp./sec/deg during the 30 min of post-adaptation. This finding indicates that the effects of muscimol were partially overwhelmed by the resumption of some adaptive changes. However, the difference between the baseline value and that obtained at the end of the adaptation period was not statistically significant. An additional finding was that the average ΔG values obtained for the VSR as well as for the N-VSR were severely depressed after intravermal injection of the GABA-A agonist with respect to those obtained in the normal adaptation experiments (p<0.0001, MANOVA for both the VSR and the N-VSR).

In conclusion, the intravermal injection of muscimol, which produced in selected experiments only slight or negligible changes in gain of the tested reflexes, prevented the occurrence of adaptive changes in gain of both the VSR and the out-of-phase N-VSR, which thus approached the baseline values.

In a second group of 5 experiments (Manzoni, Andre, d'Ascanio and Pompeiano, 1994), the animals were first submitted to a 3-h period of sustained vestibular or neck-vestibular stimulation, leading to an adapted increase in gain of both the VSR and the N-VSR, and then subjected to unilateral microinjection into the cerebellar anterior vermis of a GABA agonist. Administration of muscimol (0.25 µl at 8 µg/µl saline; n=3 experiments) or baclofen (0.25 µl at 2 µg/µl saline; n=2 experiments) decreased within 20 min the gains of the adapted VSR and N-VSR to values comparable to or lower than those obtained in the control records. This effect, which remained unmodified for about 2 h after the injection time, occurred both ipsilaterally and contralaterally to the side of the injection. Results obtained in 2 of these experiments showed that the already adapted gain values of the pure VSR as well as of the N-VSR, which were maximally depressed 25 min after intravermal injection of muscimol, slowly recovered to the previously increased values if the animal was submitted to a second period of sustained (3 h) neck-vestibular stimulation.

The region that suppressed or prevented the occurrence of the VSR adaptation after GABAergic injections corresponded to the zone B of the cerebellar anterior vermis, at the level of the culmen. These effects were not due to irritative phenomena following administration of the fluid, since no changes in the adapted VSR gain were observed after injection into the same corticocerebellar area of an equal volume of saline, stained with a 5% solution of pontamine.

DISCUSSION

Experiments of unit recording performed in decerebrate cats had previously shown that the inhibitory P-cells located in the zone B of the cerebellar anterior vermis and projecting to the lateral vestibular nucleus (LVN), responded to roll tilt of the animal with a predominant response pattern (Denoth, Magherini, Pompeiano and Stanojevic, 1979), which was just opposite in sign to that of the excitatory VS neurons (cf. Marchand, Manzoni, Pompeiano and Stampacchia, 1987 for ref.). The hypothesis that the P-cells of the cerebellar vermis contribute positively to the VSR gain was verified by the fact that intravermal injection of a GABA-A or GABA-B agonist, leading to reversible inactivation of these P-cells, decreased the response gain of the ipsilateral triceps brachii to intermittent labyrinth stimulation (Andre, d'Ascanio, Manzoni and Pompeiano, 1994). In addition to the vestibular input, the P-cells of the cerebellar vermis projecting to the LVN responded also to neck rotation by utilizing in part at least climbing fibers (Denoth, Magherini, Pompeiano and Stanojevic, 1979, 1980).

The main result of our experiments is the demonstration that, in addition to rapid postural adjustments induced by stimulation of neck receptors, the proprioceptive neck input may also lead to slow adaptive changes in gain of the VSR which were recorded from the triceps brachii during a 3-h period of sustained out-of-phase neck-vestibular stimulation and persisted for at least 1 h after the end of the adaptive stimulation (cf. Andre, d'Ascanio, Manzoni and Pompeiano, 1993). We postulated that, in this condition, the proprioceptive neck input, impinging on the zone B of the cerebellar anterior vermis, could not only increase the proportion of P-cells showing an out-of-phase modulatory response to labyrinth stimulation, but also give rise to an efficient process of adaptation, whatever this process might be. This hypothesis was supported by the fact that intravermal microinjection of the GABA-A agonist muscimol at concentrations and/or sites which produced only a slight decrease in gain of the intermittently recorded VSR, fully prevented the occurrence of the adaptive changes in gain of the VSR which always appeared under normal conditions during a sustained out-of-phase neck-vestibular stimulation. Moreover, the same agent injected in other experiments after this sustained stimulation determined a decrease of the already adapted VSR gain to values lower than those obtained prior to the adaptive stimulation. Similar results were also obtained in other experiments after intravermal administration of the GABA-B agonist baclofen (cf. Manzoni, Andre, d'Ascanio and Pompeiano, 1994).

We conclude, therefore, that GABAergic inhibition of the P-cells, which probably suppresses the spontaneous discharge as well as the normal interaction induced at this cellular level by the labyrinth and neck inputs, can also abolish the short-term adaptive changes in gain of the VSR following a sustained neck-vestibular stimulation.

The adaptive increase in gain of the VSR, produced by a 3-h period of sinusoidal out-of-phase head and body rotation leading to a synergistic neck-vestibular stimulation, should be compared with that which affects the vestibulo-ocular reflex (VOR) gain during a sustained sinusoidal rotation of the animal in the horizontal plane associated with an out-of-phase moving visual field, leading to a synergistic visual-vestibular interaction (cf. Ito, 1982, 1984). In this case, however, the corticocerebellar area which integrates the labyrinth signals is represented by the flocculus. There is, in fact, evidence that some of the floccular P-cells fired out-of-phase with respect to the vestibulo-ocular neurons during animal rotation, suggesting a positive role of their discharge to the VOR gain (cf. Ito, 1982, 1984). This hypothesis was supported by the fact that a bilateral local injection of GABA-A or GABA-B agonist in the flocculus decreased the gain of the intermittently recorded VOR in rabbits (van Neerven, Pompeiano and Collewjin, 1989). It appeared also that chemical (by kainic acid) or surgical flocculectomy resulted in an impairment of the VOR adaptation during combined visual-vestibular stimulation (Ito, 1982, 1984; cf. however Ito, Lisberger and Sejnowski, 1993).

Since the P-cells of the flocculus responded not only to the vestibular input, but also to the visual input (cf. Ito, 1982, 1984), it has been postulated that the latter input determines the shift of dominance from the in-phase to the out-of-phase modulatory responses of floccular P-cells during the VOR adaptation (cf. Nagao, 1992) and also contributes to the plastic changes underlying this phenomenon (cf. Ito, 1989).

It is of interest that activation of the visual input conveyed to the flocculus through climbing fibers (Maekawa and Simpson, 1973), produces a long-term depression (LTD) of the mossy fiber responses of the P-cells to labyrinth stimulation, due to desensitization of the excitatory synapses made by parallel fibers on the P-cell dendrites (cf. Ito, 1989), thus leading to an adaptive increase in the VOR gain. Similarly, we propose that in our experimental conditions the neck input, acting through climbing fibers on the P-cells of the cerebellar anterior vermis, provides the interaction signal for the reorganization of the mossy fiber-parallel fiber input to the P-cells requested for VSR adaptation. The suppression of this adaptive process after GABAergic inhibition of the P-cells of the cerebellar anterior vermis suggests that these neurons represent an important site for the plastic changes underlying adaptation of the VSR. Indeed, there is evidence that postsynaptic inhibition of P-cell dendrites during conjunctive parallel fiber-climbing fiber stimulation prevents LTD from taking place (Ekerot and Kano, 1985; cf. also Sakurai, 1987). On the other hand, arguments reported in detail in our original paper (Manzoni, Andre, d'Ascanio and Pompeiano, 1994) make very unlikely the hypothesis that, at least in experiments of short-term adaptation, the P-cells of the cerebellar anterior vermis represent either a relay for adaptive changes occurring before it (i.e., at precerebellar level) or else the generator of error signals that elicits plasticity elsewhere (i.e., at postcerebellar level) (cf. Ito, Lisberger and Sejnowski, 1993).

SUMMARY

In decerebrate cats, a 3-hours (h) period of sustained roll tilt of the head (at 0.15 Hz, $\pm 10°$) leading to selective stimulation of labyrinth receptors, associated with a 2.5° out-of-phase body-to-head rotation leading to stimulation of neck receptors, produced an adaptive increase in gain of the vestibulospinal reflex (VSR) elicited from the triceps brachii during roll tilt of the whole animal (at 0.15 Hz, $\pm 10°$). This increase reached the maximum at the end of the third h of stimulation and persisted unmodified during the first h after stimulation. Microinjection into the zone B of the cerebellar anterior vermis of the GABA-A agonist muscimol (0.25 µl at 8 µg/µl saline), producing only a slight or negligible depression of the VSR gain in non-adaptive conditions, prevented the occurrence of the adapted increase in gain of the VSR following a 3-h period of sustained head and body rotation. Moreover, intravermal injection of the GABA-A agonist muscimol or the GABA-B agonist baclofen (0.25 µl at 2-8 µg/µl saline) suddenly suppressed the already adapted VSR gain. We postulate that the adaptive increase in gain of the VSR following a sustained neck-vestibular stimulation depends on plastic changes which affect the Purkinje cells of the cerebellar anterior vermis.

ACKNOWLEDGMENTS

This work was supported by the N.I.H. Research Grant NS 07685-25A1 and by grants of the MURST and the Agenzia Spaziale Italiana (ASI 1994 RS 124), Rome, Italy.

REFERENCES

Andre, P., d'Ascanio, P., Manzoni, D., and Pompeiano, O., 1993, Adaptive modification of the cat's vestibu-lospinal reflex during sustained vestibular and neck stimulation, *Pflügers Arch.* 425:469-481.

Andre, P., d'Ascanio, P., Manzoni, D., and Pompeiano, O., 1994, Depression of the vestibulospinal reflex by intravermal microinjection of $GABA_A$ and $GABA_B$ agonists in the decerebrate cat, *J. Vest. Res.* 4:251-268.

Corvaja, N., and Pompeiano, O., 1979, Identification of cerebellar corticovestibular neurons retrogradely labeled with horseradish peroxidase, *Neuroscience* 4:507-515.

Denoth, F., Magherini, P.C., Pompeiano, O., and Stanojevic, M., 1979, Responses of Purkinje cells of the cerebellar vermis to neck and macular vestibular inputs, *Pflügers Arch.* 381:87-98.

Denoth, F., Magherini, P.C., Pompeiano, O., and Stanojevic, M., 1980, Responses of Purkinje cells of the cerebellar vermis to sinusoidal rotation of neck, *J. Neurophysiol.* 43:46-59.

Ekerot, C.-F., and Kano,M., 1985, Long-term depression of parallel fibre synapses following stimulation of climbing fibres, *Brain Res.* 342:357-360.

Ito, M., 1982, Cerebellar control of the vestibulo-ocular reflex. Around the flocculus hypothesis, *Ann. Rev. Neurosci.* 5:275-296.

Ito, M., 1984, The Cerebellum and Neural Control, New York, Raven Press, pp. XVII-580.

Ito, M., 1989, Long-term depression, *Ann. Rev. Neurosci.* 12:85-102.

Ito, M., Lisberger, S.G., and Sejnowski, T.J., 1993, Cerebellar flocculus hypothesis, *Nature* 363:24-25.

Lindsay, K.W., Roberts, T.D.M., and Rosenberg, J.R., 1976, Symmetric tonic labyrinth reflexes and their interaction with neck reflexes in the decerebrate cat, *J. Physiol., Lond.* 261:583-601.

Maekawa, K., and Simpson, J.I., 1973, Climbing fiber synapses evoked in vestibulocerebellum of rabbit from visual system, *J. Neurophysiol.* 36:649-666.

Manzoni, D., Andre, P., d'Ascanio, P., and Pompeiano, O., 1994, Depression of the vestibulospinal reflex adaptation by intravermal microinjection of GABA-A and GABA-B agonists in the cat, *Arch. ital. Biol.* 132:243-269.

Marchand, A.R., Manzoni, D., Pompeiano, O., and Stampacchia, G., 1987, Effects of stimulation of vestibular and neck receptors on Deiters neurons projecting to lumbosacral cord, *Pflügers Arch.* 409:13-23.

Nagao, S., 1992, Role of cerebellar flocculus in adaptive gain control of the vestibulo-ocular reflex, In: Shimazu H., and Shinoda Y., (Eds.), Vestibular and Brain Stem Control of Eye, Head and Body Movements, Tokyo, Japan Sci. Soc. Press, Basel, S.Karger, pp. 439-449.

Sakurai, M., 1987, Synaptic modification of parallel fiber-Purkinje cell transmission in *in vitro* guinea pig cerebellar slices, *J. Physiol., Lond.* 394:463-480.

Schor, R.H., and Miller, A.D., 1981, Vestibular reflexes in neck and forelimb muscles evoked by roll tilt, *J. Neurophysiol.* 46:167-178.

van Neerven, J., Pompeiano, O., and Collewjin, H., 1989, Depression of the vestibulo-ocular and optokinetic responses by intrafloccular microinjection of GABA-A and GABA-B agonists in the rabbit, *Arch. ital. Biol.* 127:243-263.

6

NEURAL INTEGRATION OF VISUAL INFORMATION AND DIRECTION OF GRAVITY IN PRESTRIATE CORTEX OF THE ALERT MONKEY

X. M. Sauvan and E. Peterhans

Department of Neurology
University Hospital Zurich
Frauenklinikstr. 26
CH-8091 Zurich
Switzerland

INTRODUCTION

The perception of form requires a generalized representation of contours both with regard to the quality by which they are defined on the retina (e.g. discontinuities of luminance, color, motion, or texture) and with regard to their position and orientation in space. Earlier studies from this laboratory showed that contour generalization with regard to image quality begins early in visual processing, in monkey visual cortex at the level of area V2. While neurons in area V1 preferred continuous contrast lines or edges, we found neurons in area V2 that were orientation selective for contours defined by occlusion cues (illusory contours) as they were for contours defined by luminance contrast (for a review see Peterhans and von der Heydt, 1991). Other neurons were orientation selective for lines defined by coherent motion as they were for lines defined by luminance contrast (Peterhans and Baumann, 1994). These neurons were found in area V2 and in the area V3/V3A complex. Subsequently, contour generalization for simple, geometrical forms has been reported for neurons in monkey inferotemporal cortex (Sáry et al., 1993).

Less is known about the representation of contours with regard to their position and orientation in space. Neurons showing gaze dependent responses were recorded in the thalamic intralaminar nuclei of the cat (Schlag et al., 1980), and in the posterior parietal cortex (area 7a) of the macaque monkey (Andersen and Mountcastle, 1983). In the visual cortex of the macaque similar neurons were found at the level of area V3A (Galletti and Battaglini, 1989). These results suggested a representation of visual space in head centered coordinates. Neurons showing orientation constancy in space centered coordinates have first been found in cat visual cortex. In the anesthetized cat, Horn et al. (1972) found neurons which preferred similar orientations independent of whether the body of the animal was held upright, or tilted about the naso-occipital (roll) axis. A similar result was found by Denney

Multisensory Control of Posture, Edited by T. Mergner and F. Hlavačka
Plenum Press, New York, 1995

and Adorjani (1972) in the paralyzed cat using head tilt. Also in the paralyzed cat, Tomko et al. (1981) found orientation constancy in space centered coordinates for about 25% of the neurons of area 17. On the other hand, a brief report of Schwartzkroin (1972) suggests that there is no effect of body tilt on the direction selectivity of cortical neurons. Similar studies have, to our knowledge, not been done in awake behaving monkeys where the status of the animal and the natural viewing conditions allow a more accurate assessment of eye position and center of gaze.

We here report the effect of body tilt on the orientation selectivity of neurons of areas V1, V2 and V3/V3A in the visual cortex of the awake behaving monkey.

METHODS

The responses of single neurons were studied while the monkey performed a visual fixation task in a static body position, either upright or tilted about the naso-occipital (roll) axis by 25-30 deg, clockwise or counterclockwise (Fig. 1A). The animal was seated in a primate chair in the center of a turntable with its head restrained, but otherwise free to perform the behavioral task. During the periods of active visual fixation moving light or dark

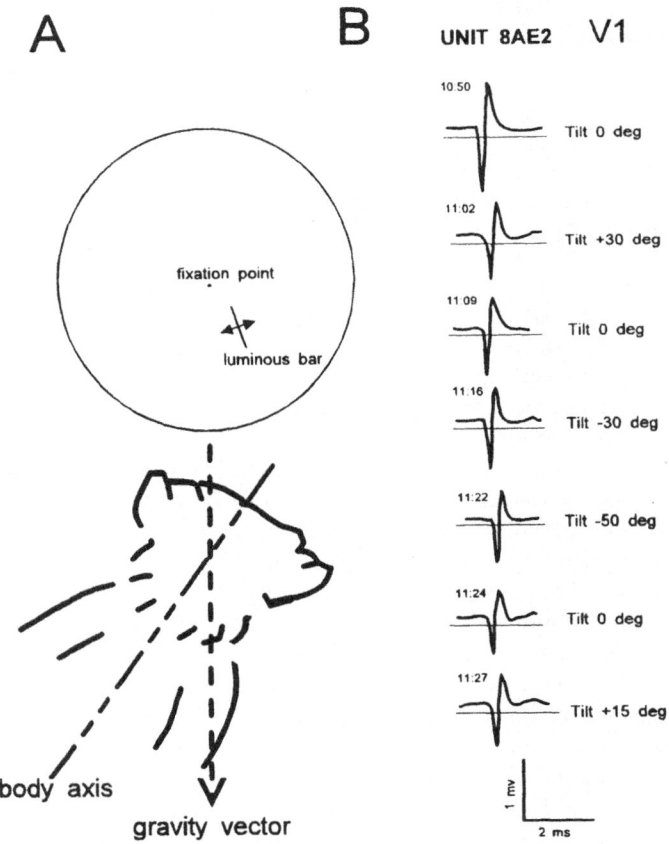

Figure 1. Experimental setup and stability. (A) The monkey was facing a visual display at a distance of 57 cm. It was performing a visual fixation task in a static position either upright or tilted 25-30 deg, clockwise or counterclockwise. (B) Spike forms as recorded from a neuron in area V1 at different times and body positions.

bars were presented on a computer display that was masked by a circular aperture (diameter: 22 deg) and located at 57 cm from the animal. The background luminance was 25 cd/m², the luminance of the light and dark bars 65 cd/m² and 0.5 cd/m², respectively. We studied the responses of single neurons in striate (area V1) and prestriate cortex (areas V2 and V3/V3A). For details of our methods of animal preparation and recording see Peterhans and von der Heydt (1993). We controlled the stability of our experimental set-up by taking records of spike forms periodically for each neuron at different times and different body positions (Fig. 1B).

RESULTS

We have studied 106 orientation selective neurons both in striate and prestriate cortex. For 44 neurons we could compare the preferred orientation at 2-3 body positions. In striate cortex most neurons (19/23) were of a non-compensatory type showing a change in the preferred stimulus orientation according to the body tilt and the estimated counterrolling of the eye (Fig. 2). In our experimental condition with body tilts of 25-30 deg the estimated counterrolling of the eyes was about 4 deg (cf. Haslwanter et al., 1992). Fig. 2 shows two

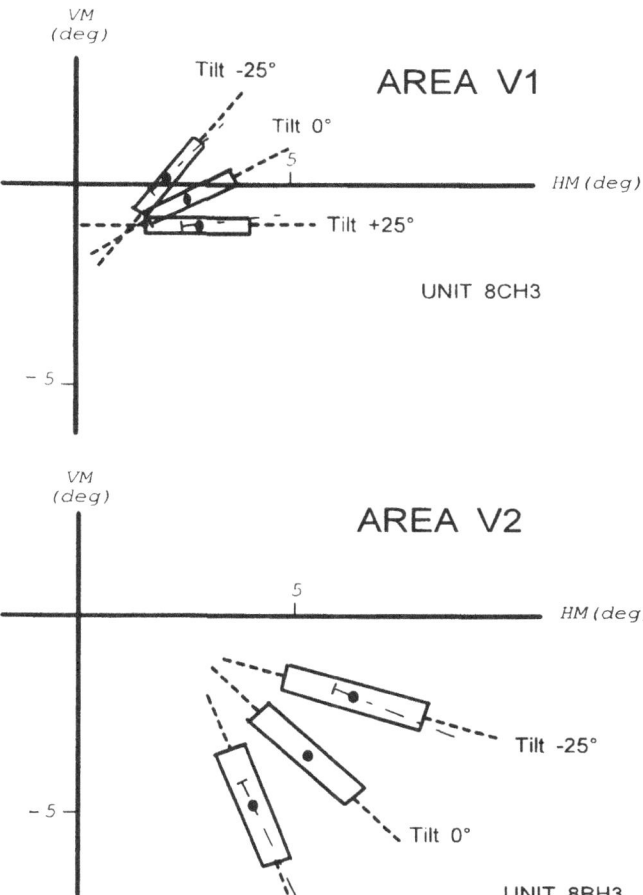

Figure 2. Response properties of 'non-compensatory' cells. The response fields of two neurons are shown, one was recorded in area V1 (8CH3), the other in area V2 (8BH3). The fields were plotted first in the upright position (tilt 0 deg), then in tilted positions (tilt ±25 deg). The observed preferred orientation (- - - - -) and the expected orientation (|— —) are shown for each neuron. VM and HM refer to the vertical and horizontal meridian, respectively.

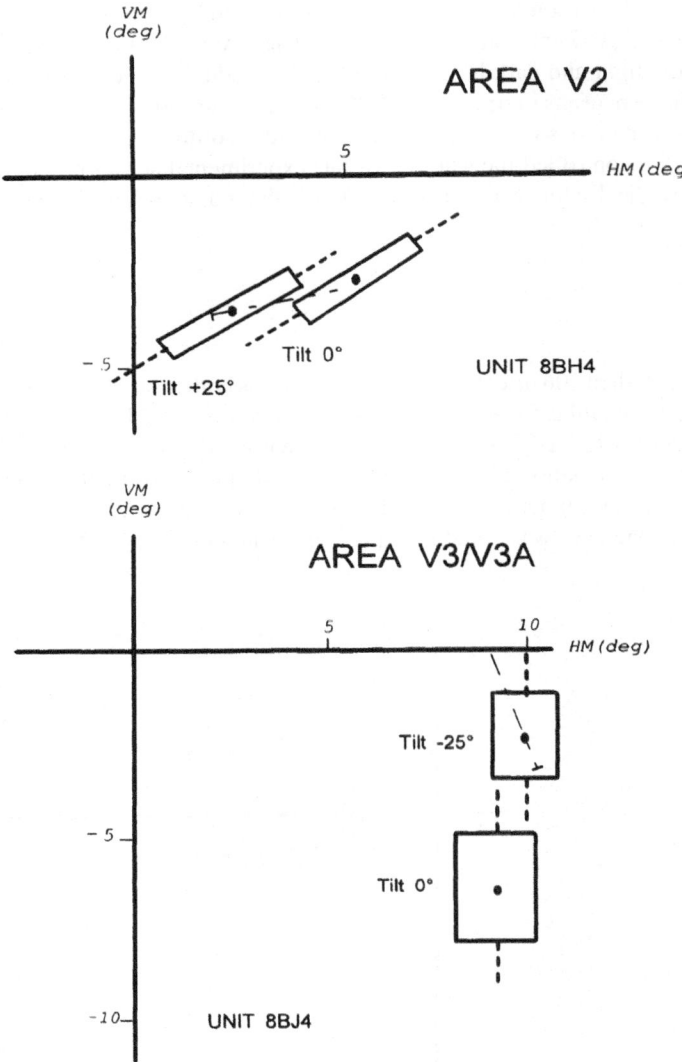

Figure 3. Response properties of 'compensatory' cells. The response fields of two neurons are shown, one was recorded in area V2 (8BH4), the other in the area V3/V3A complex (8BJ4). In contrast to the neurons of Fig. 2 these neurons preferred similar orientations in all body positions. Conventions as described for Fig. 2.

neurons of the non-compensatory type, one recorded in area V1 and another in area V2. These neurons showed orientation constancy in retinotopic coordinates. In each body position the preferred orientation corresponded to the orientation predicted from the body tilt and the counterrolling of the eye.

By contrast, about 40% of the neurons in prestriate cortex (8/21) were of a compensatory type (Fig. 3). These neurons, although showing a change in field position, preferred similar stimulus orientations at all body tilts. The predicted counterrolling of the eye (4 deg) was too small to account for this result. This indicates that the orientation selectivity of the compensatory neurons was invariant with respect to the direction of gravity. These neurons appear to encode contour orientation in space centered coordinates.

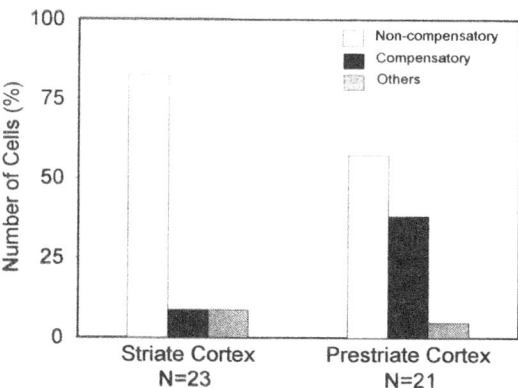

Figure 4. Frequencies of cell types studied. In striate cortex 19/23 neurons were of the non-compensatory type, in prestriate cortex 8/21 neurons were of the compensatory type. The difference between the frequencies of the two types of neurons in striate and prestriate cortex was significant (Kruskall-Wallis test, N=41, p<0.0001).

Fig. 4 shows the frequency distribution for all neurons recorded in striate and prestriate cortex for which we determined the preferred orientation in at least two body positions. One can see that the neurons which showed orientation preferences invariant with respect to the direction of gravity (filled bars) concentrated in prestriate cortex. These results suggest a step in contour processing between striate and prestriate cortex.

DISCUSSION

In prestriate cortex of the awake behaving monkey we found neurons which showed orientation preferences for bar stimuli that were invariant with respect to the direction of gravity. These neurons preferred similar orientations at all body tilts. Except for very few cells, such neurons were not found in striate cortex. The results suggest that in visual cortex information about the direction of gravity is first implemented at the level of prestriate cortex.

Evidence from behavioral studies suggests that information about the direction of gravity is used in judgments about body orientation and the orientation of objects in space (cf. Bischof, 1974). Little is known as for how information about the direction of gravity is integrated in mechanisms of form processing, and how this information may contribute to perceptual phenomena like orientation and size constancy. Recently, Leone et al. (1994) made an interesting observation concerning the perception of symmetry. They found that the preference of a vertical axis of symmetry over a horizontal one may be based on non-visual information since this phenomenon weakened in weightlessness. A similar weakening of the phenomenon was also observed when the subjects were tested in the horizontal supine posture, a position in which the gravity vector cannot serve as orientation reference for visual stimuli. Thus, Leone et al. (1994) concluded that otolith inputs may modify the orientation sensitivity of visual cortical neurons, possibly already at an early level of processing. Our results support this interpretation.

The results of the present paper also agree with earlier findings in cat visual cortex which showed that some neurons exhibited orientation constancy with respect to the direction of gravity (Denney et al., 1972; Horn et al., 1972; Tomko et al. 1981). These results suggest that in cat visual cortex information about the direction of gravity is implemented at the earliest level of cortical visual processing, in area 17. However, these results were in some respects difficult to interpret. The animals were in an anesthetized or paralyzed condition which makes it hard to determine the center of gaze. Also, the authors found it difficult to control for the eye drifts that occurred during the experiment. Furthermore, they noted spontaneous changes of receptive field properties over time, even when the animals were not tilted, but kept in a stationary, upright position. These parameters are much more easily controlled in the awake behaving monkey as used in the present study.

Information about changes of head or body position relative to the direction of gravity arises from the otolith system and proprioception. The otolith organs are stimulated when the head or the body is tilted about the naso-occipital axis in the roll plane, and their responses are more or less trigonometrically related to the angle of tilt (Fernandez et al., 1972). The route(s) by which signals from the otoliths reaches the visual cortex is (are) unclear. Vestibular projections to the parietal cortex have been found, but cortical vestibular neurons do not seem to respond to static stimulation of the otoliths (Grüsser at al., 1990, 1992). Parietal areas involved in central vestibular processing concern area 7 (Kawano et al., 1980; Ventre and Faugier-Grimaud, 1986), the parieto-insular vestibular cortex (PIVC), the "neck" region of area 3a, and area 2v (for a review see Grüsser et al., 1992). Projections to the prestriate cortex have, so far, only be found from area 7 (Cavada and Goldman-Rakic, 1993). Otolith signals may reach the occipital cortex also via the pulvinar where neuronal responses have been reported to be gated by eye position (Robinson et al., 1991). The proprioceptive information, on the other hand, may originate in the neck muscles, the joints, and in the gravity receptors of the trunk (Mittelstaedt, 1992).

In conclusion, our results suggest that in monkey visual cortex integration of visual information and information about the direction of gravity occurs at an early level of processing.

SUMMARY

We have investigated the effect of body tilt on mechanisms of contour processing in the visual cortex of the alert rhesus monkey. The responses of single neurons were studied while the monkey performed a visual fixation task. The animal worked either in the upright position or with its whole body tilted about the naso-occipital (roll) axis by 25-30 deg, clockwise or counterclockwise. In 44 neurons we plotted the response fields for light or dark bars in the upright position and in 1-2 positions in which the body was tilted. We compared the stimulus orientation preferred by cortical neurons in 2-3 body positions. In striate cortex, most neurons (19/23) were of a non-compensatory type showing a change in the preferred orientation according to the body tilt and the estimated counterrolling of the eye. This indicates orientation constancy in retina centered coordinates. By contrast, about 40% of the neurons in prestriate cortex (8/21) were of a compensatory type preferring similar orientations in all body positions. This indicates orientation constancy in space centered coordinates, invariant with respect to the direction of gravity. We conclude that information about the direction of gravity is implemented early in visual processing, in the monkey at the level of area V2.

ACKNOWLEDGMENTS

We thank R. Desimone for providing the software CORTEX and R. Baumann for preparing the microelectrodes. X.M.S. was supported by a grant ESPRIT Mucom-II 6615 to V. Henn. This research was further supported by the following grants to E. Peterhans: ESPRIT Insight-II 6019 and HFSP RG-31/93. These findings were presented by X.M. Sauvan in September 1994 at the symposium "Sensory Interaction in Posture and Movement Control". Correspondence should be addressed to E. Peterhans at the address given above.

REFERENCES

Andersen, R.A., and Mountcastle V.B., 1983, The influence of the angle of gaze upon the excitability of the light-sensitive neurons of the posterior parietal cortex, *J. Neurosci.* 3:532-548.

Bischof, N., 1974, Optic-vestibular orientation to the vertical, in: Handbook of Sensory Physiology, Vol. VI, Part 2 (H. H. Kornhuber, ed.), Springer-Verlag, Berlin, pp. 155-190.

Cavada, C., and Goldman-Rakic, 1993, Multiple visual areas in the posterior parietal cortex of primates, *Prog. Brain Res.* 95:123-137.

Denney, D., and Adorjani, C., 1972, Orientation specificity of visual cortical neurons after head tilt, *Exp. Brain Res.* 14:312-317.

Fernandez, C., Goldberg, J.M., and Abend, W.K., 1972, Response to static tilts of peripheral neurons innervating otolith organs of the squirrel monkey, *J. Neurophysiol.* 35:978-997.

Galletti, C., and Battaglini, P.P., 1989, Gaze-dependent visual neurons in area V3A of monkey prestriate cortex, *J. Neurosci.* 9:1112-1125.

Grüsser, O.-J., Pause M., and Schreiter, U., 1990, Localization and responses of neurones in the parieto-insular vestibular cortex of awake monkeys (Macaca fascicularis), *J. Physiol.* (London) 430:537-557.

Grüsser, O.-J., Guldin, W.O., Harris, L., Lefèbre, C., and Pause, M., 1992, Cortical representation of head-in-space movement and some psychophysical experiments on head movement, in: The Head-Neck Sensory Motor System (A. Berthoz, W. Graf, and P.P. Vidal, eds.), Oxford University Press, New York, pp. 497-509.

Haslwanter, Th., Straumann, D., Hess, B.J.M., and Henn, V., 1992, Static roll and pitch in the monkey: Shift and rotation of Listing's plane, *Vision Res.* 32:1341-1348.

Horn, G., Stechler, G., and Hill, R.M., 1972, Receptive fields of units in the visual cortex of the cat in the presence and absence of bodily tilt, *Exp. Brain Res.* 15:113-132.

Kawano, K., Sasaki, M., and Yamashita, M., 1980, Vestibular input to visual tracking neurons in the posterior parietal association cortex of the monkey, *Neurosci. Lett.* 17:55-60.

Leone G., Lipshits M., Gurfinkel V., and Duhamel, J.R., 1994, Influence of non visual cues on bilateral symmetry detection: The effect of prolonged weightlessness, *Soc. Neurosci. Abstr.* 20:1580.

Mittelstaedt, H., 1992, Somatic versus vestibular gravity reception in man, *Ann. N.Y. Acad. Sci.* 656:124-139.

Peterhans, E., and Baumann, R., 1994, Elements of form processing from motion in monkey prestriate cortex, *Soc. Neurosci. Abstr.* 20:1053.

Peterhans, E., and von der Heydt, R., 1991, Subjective contours-bridging the gap between psychophysics and physiology, *Trends Neurosci.* 14:112-119.

Peterhans, E., and von der Heydt, R., 1993, Functional organization of area V2 in the alert macaque, *Eur. J. Neurosci.* 5:509-524.

Robinson, D.L., McClurkin, J.W., Kertzman, C., and Petersen, S.E., 1991, Visual responses of pulvinar and collicular neurons during eye movements of awake, trained macaques, *J. Neurophysiol.* 66:485-496.

Sáry, G., Vogels, R., and Orban, G.A., 1993, Cue-invariant shape selectivity of macaque inferior temporal neurons, *Science* 260:995-997.

Schlag, J., Schlag-Rey, M., Peck, C.K., and Joseph, J.P., 1980, Visual responses of thalamic neurons depending on the direction of gaze and the position of the target in space, *Exp. Brain Res.* 40:170-184.

Schwartzkroin, P.A., 1972, The effect of body tilt on the directionality of units in cat visual cortex, *Exp. Neurol.* 36:498-506.

Tomko, D.L., Barbaro, N.M., and Ali, F.N., 1981, Effect of body tilt on receptive field orientation of simple visual cortical neurons in unanesthetized cats, *Exp. Brain Res.* 43:309-314.

Ventre, J., and Faugier-Grimaud, S., 1986, Effects of posterior parietal lesions (area 7) on VOR in monkeys, *Exp. Brain Res.* 62:654-658.

7

PRIMATE VESTIBULAR CORTICES AND SPATIAL ORIENTATION

O. -J. Grüsser and W. O. Guldin

Department of Physiology
Freie Universität
Arnimallee 22
14195 Berlin
Germany

INTRODUCTION

Orientation in the extrapersonal space is a multimodal achievement; in man and most primates the visual system predominates. The extrapersonal space is perceived by the "outer senses" and can be subdivided into four different compartments: The *grasping space*, the *near-distant action space*, the *far-distant action space*, and the *visual background* (Grüsser, 1983). The *perceived coordinates* of the extrapersonal space are closely related to the coordinates of the ego-space (spatiotopically organized egocentric coordinate system). For spatial orientation man uses an additional pictorial concept of the extrapersonal space, which is stored in memory and operates with *allocentric coordinates*: We are able to perceive and describe spatial relations relative to each other without referring to our actual egocentrically organized extrapersonal space.

Lesions of the inferior-posterior parietal areas disturb perception of the extrapersonal space and lead to the symptom of *contralateral spatial hemineglect*. Lesions of the occipito-parieto-temporal or mesial-occipito-temporal cortical regions produce *topographagnosia*, which is related predominantly to allocentric maps and appears in four subtypes: *perceptive, apperceptive, associative*, and *cognitive-emotional* (Grüsser, and Landis, 1991). The percepts related to the spatiotopically organized extrapersonal space are derived from multimodal sensory information including visual, tactile, auditory and olfactory cues. Hereby the contribution of the non-visual modalities is restricted mainly to objects appearing in the grasping or near-distant action space. Kinesthetic and proprioceptive signals together with efference copy signals, characterizing the movements of the body or body parts, are continuously integrated into the signals representing the extrapersonal space and contribute to the perceived stability of that space despite a continuous change in afferent sensory signals. Within this framework of spatial perception the *vestibular system* provides the necessary information about head position and head movement (rotatory or translatory) in space and also contributes to the "recalibration" of the egocentric coordinates of the extrapersonal space during and after motion. Furthermore cutaneous somatosensory and proprioceptive signals are continuously compared with vestibular signals and provide information about position of the body and extremities relative to the head and about the coordinates of the extrapersonal space (standing, sitting, lying etc.). Multimodal signal processing is a characteristic of vestibular responses recorded in single nerve cells of the brainstem vestibular nuclear complex (VNC), in vestibular structures of the cerebellum, the thalamus and the neocortex. We will demonstrate that this multimodal interaction and "updating" of vestibular signals occurs not only at the brainstem level but also in neurons of the different cortical areas forming the primate vestibular cortical system. The different vestibular cortices are interconnected with each other and form a cortical circuitry, which provides important information for the coordinates of the egocentric space perception and interacts simultaneously with the brainstem afferent vestibular input.

METHODS

Single nerve cell activity was recorded with tungsten or elgiloy microelectrodes in the brain of primates, one species of Old-World monkeys (*Macaca fascicularis*) and two species of New-World monkeys (*Saimiri sciureus, Callithrix jacchus*). The cortical recordings were performed in the parieto-insular vestibular cortex (PIVC, i.e. cytoarchitectonically mainly the medial area Ri), but nerve cell activity was also studied in area 3aV, the neck region of area 3a, in the medial part of a temporal area adjacent to PIVC in the occipital direction and designated area VTS (*Ventral Temporal Sylvic area*), and in the parietal area 7a. Vestibular stimulation was achieved by sitting the awake animal with its head fixed in a conventional monkey chair on an electronically controlled rotating platform.

After identification of cortical or subcortical structures by recording single nerve cell activity, the neuronal connections were studied with anterograde and retrograde tracer substances: *Horseradish peroxidase* HRP, *Fast Blue* FB, *Nuclear Yellow* NY, HRP-labelled *Wheat Germ Agglutinine* WGA-HRP, *Evans Blue*, *Rhodamine* or FIC, *Dextranes* of different molecular weights labelled with Rhodamine or FITC, Rhodamine- or FITC-labelled latex balls. Up to four different tracer substances were injected into different sites in one brain. After an adequate delay, histological processing was performed and the tracer substances were identified either by normal or fluorescence-microscopy. For neurophysiological or neuroanatomical technical details, the reader is referred to the following publications: Grüsser et al., 1982, 1990a,b, 1992, 1994; Akbarian et al., 1988, 1992, 1993, 1994; Guldin et al., 1992, 1993.

NEUROPHYSIOLOGY OF THE PARIETO-INSULAR VESTIBULAR CORTEX (PIVC)

The following summary is based on single unit recordings from 168 PIVC cells in the Squirrel monkey and 152 PIVC-cells in the Java monkey. In this region (Fig. 1) about 50 percent of nerve cells responded directionally selective to head-in-space-rotation, whereby one or more pairs of semicircular canals were involved. The majority of these "vestibular" units was characterized by a multimodal response pattern: optokinetic stimulation, somatosensory stimulation of the skin, proprioceptive stimulation of the neck region, the hip region, or the extremities evoked neuronal activation or inhibition. About 20 percent of the PIVC neurons were selectively somatosensory, 8 to 10 percent responded to visual stimulation only, predominantly to large moving fields. About 20 percent of the units could not be driven by any stimulus applied. No significant responses to steady tilt (otolith receptor

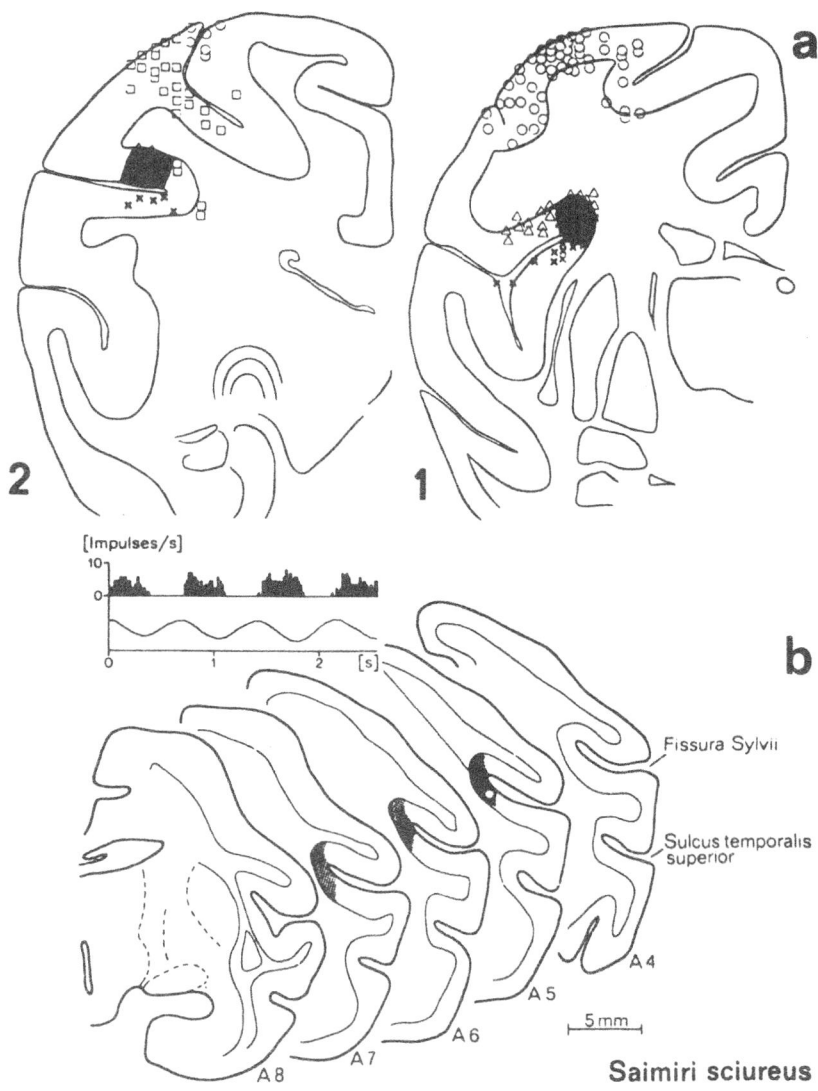

Figure 1. Location of the parieto-insular vestibular cortex (PIVC, dark) in Java monkeys (a) and squirrel monkeys (b). Coronal cortical sections. In Java monkeys the PIVC extends mainly along the *parietal* operculum, while in squirrel monkeys it is shifted chiefly to the fundus region of the retroinsular temporal operculum. Inset: response of a PIVC neuron to horizontal sinewave rotation. Neuron was located at the site of the white spot.

stimulation) were observed. The *vestibular* responses depended on the plane of rotation and were restricted to a certain response sector (Fig. 2). All response types (I, II and III according to Duensing and Schaefer, 1958) were found at the 3 main planes of rotation (yaw, pitch, roll) (Fig. 3a). The rotation planes leading to optimum responses were distributed throughout all possible planes in space. Most of the vestibular PIVC-neurons also responded to *optokinetic stimulation* (Fig. 3b). The optimum response planes for rotatory optokinetic and vestibular (semicircular canal) stimulation coincided approximately in space (Fig.2). In a

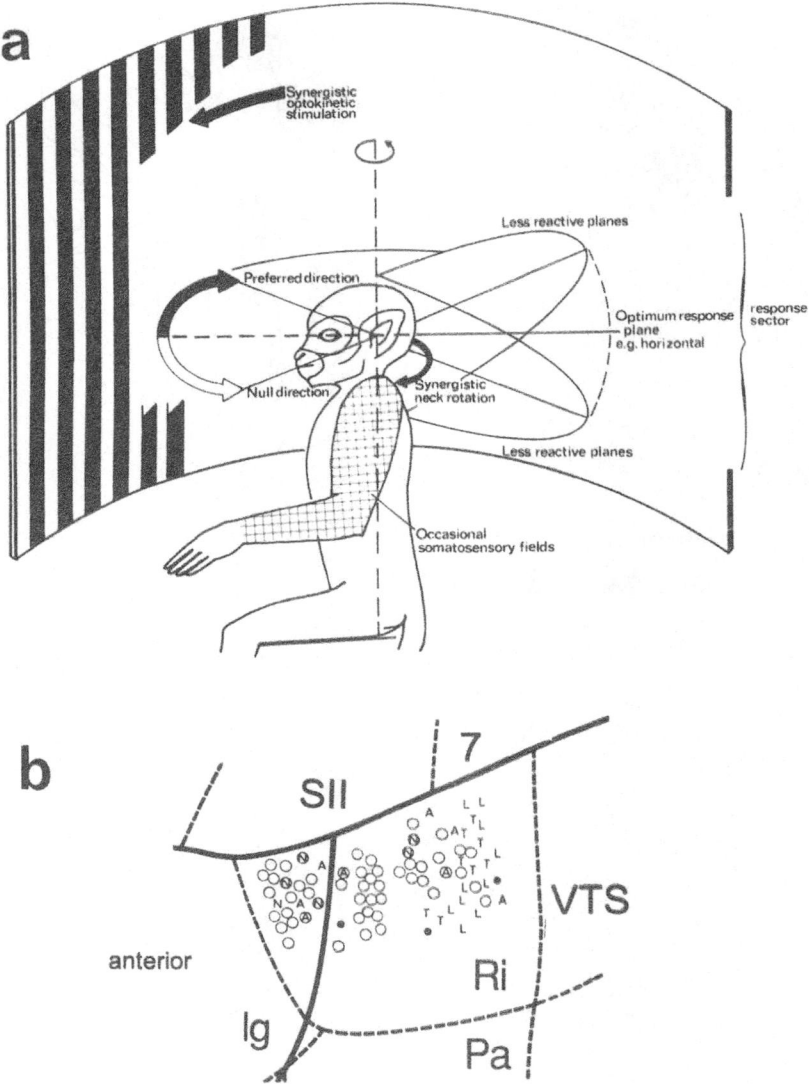

Figure 2. (a) Schematic drawing of a squirrel monkey sitting on a rotating platform. The vestibular dynamic responses of an arbitrarily selected PIVC neuron are restricted to a certain response sector (head-in-space rotation) and to optokinetic stimulation with a response sector predominantly similar to the vestibular response sector. "Synergistic" vestibular and optokinetic responses were activations appearing in opposite directions. In addition, proprioceptive inputs of the neck region (trunk rotation while the head was stationary in space), cutaneous somatosensory stimulation and in a few neurons also deep proprioceptive stimulation of muscles of the hip region and the extremities led to an activation of PIVC neurons. (b) Location of vestibular neurons (circles) within areas Ri and Ig of a squirrel monkey. Data were obtained by reconstruction of the electrode tip position.

few neurons the optimum optokinetic response plane and the optimum vestibular response plane were perpendicular to each other. Most of the vestibular neurons activated by sinusoidal rotation in yaw also responded to stimulation of the deep proprioceptive neck receptors, i.e. to sinusoidal trunk rotation while the head was fixed in space (Fig. 3c). Steady horizontal turning of the head on the trunk to the left or right led in some neurons to a modification of the responses to horizontal rotation. In several neurons hip-rotation also evoked a directionally selective activation. *Cutaneous somatosensory receptive fields* were usually very large and could occur on both sides of the body. In some of the neurons somatosensory receptive fields were restricted to the skin of a hand or foot. PIVC corresponded to the medial part of area Ri in squirrel monkeys and to area ript in Java monkeys (Pandya and Sanides, 1973).

Neuronal response patterns in old-world and new-world monkeys were very similar, suggesting that the principles of PIVC neuronal mechanisms had developed before the splitting during phylogenesis (about 60 million years ago). There is good evidence that the PIVC of the primate brain is homologous to the vestibular ASSS-region of the cat (Walzl and Mountcastle 1949; Grüsser et al., 1959; Pandya and Sanides, 1973; Mergner, 1979, Mergner et al., 1981a,b, 1985).

NEURONAL RESPONSE PATTERNS IN THE "VESTIBULAR NECK REGION" OF AREA 3a

In preparing tracer injections to area 3aV of squirrel monkeys (the vestibular region occupying the representation of the neck region in area 3a), single nerve cell recordings were performed and the same stimuli applied as in the study of PIVC-neurons. All the neurons in this region responded to somatosensory or proprioceptive stimulation of the neck region, in particular to trunk rotation, when the head was fixed in space. Many, however, also responded to semicircular canal stimulation, like PIVC-neurons (horizontal sinewave rotation in total darkness). A distinct response sector also characterized these vestibular responses of 3aV nerve cells. Responses to optokinetic stimulation were usually absent or very weak. Ödkvist et al. (1974) demonstrated with the evoked potential technique that electrical stimulation of the vestibular nerve led to responses in the arm field of area 3a in squirrel monkeys. Thus, one can assume that in addition to the neck region, other parts of area 3a may also receive vestibular input. From the neuroanatomical data, the most important candidate seems to be the *hand region* of area 3a (see below).

RESPONSES OF NERVE CELLS LOCATED IN AREA VTS AND AREA Ig

In addition to recordings from PIVC, responses were recorded in a few penetrations from nerve cells in the medial part of area T3, as described in the cytoarchitectonic studies of Burton and Jones (1976) and Jones and Burton (1976). A detailed cytoarchitectonic examination of this region revealed that a subdivision is plausible (Guldin, 1994). The medial part was named *Ventral Temporal Sylvian region* (VTS) in squirrel monkeys. Nerve cells in this region rarely responded to stimulation of the semicircular canals, but fairly regularly to *optokinetic stimulation*, whereby the units were characterized by preferred optokinetic directions and response planes. Somatosensory inputs were not studied for these nerve cells.

The most occipital part of the granular insula (Ig) adjacent to the PIVC of area Ri also contained nerve cells responding to vestibular stimulation. It is so far not clear whether

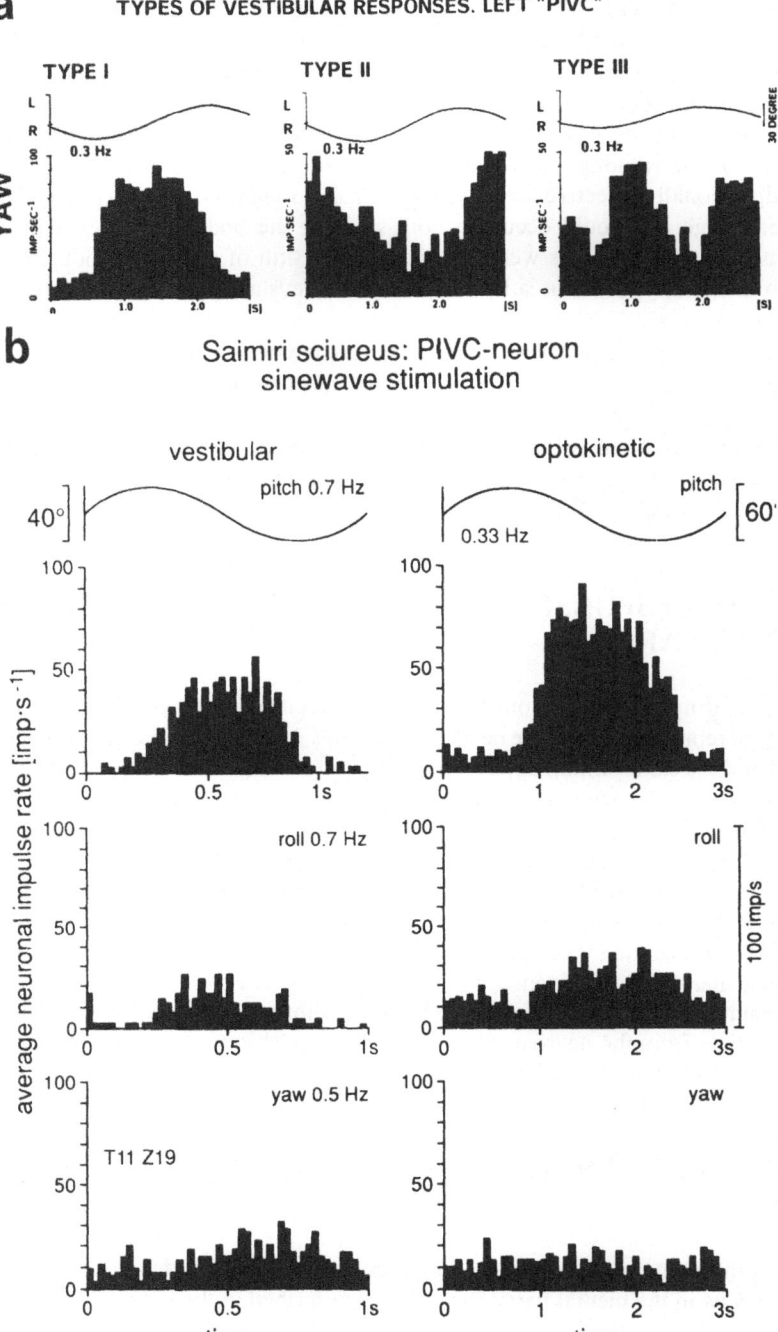

Figure 3. (a) Example of three response types of PIVC neurons: activation during ipsilateral sinusoidal rotation, type I, activation during rotation in contralateral direction, type II, and activation in both directions, type III (Duensing and Schaefer, 1958). (b) Examples of the responses of a PIVC neuron to vestibular and optokinetic sinewave stimulation evoked by a random dot pattern of light spots. (c) Responses of a PIVC neuron to horizontal sinewave rotation in the dark (V) and to horizontal trunk rotation while the head was fixed in space (N). All data presented are from squirrel monkeys, but similar responses were also obtained in Java and marmoset monkeys.

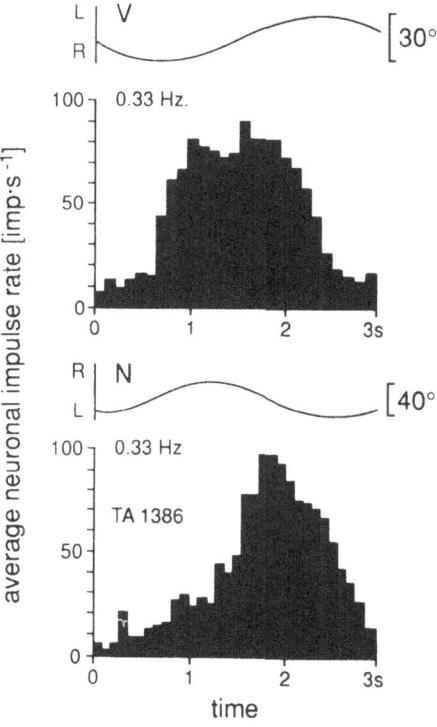

Figure 3. (Continued)

this region is an extension of the PIVC beyond the cytoarchitectonic borders of area Ri or a separate group of vestibular insular neurons.

THE CORTICAL CIRCUITRY OF VESTIBULAR REPRESENTATION

Fig. 4 summarizes schematically the three cortical regions from which vestibular responses have been recorded with microelectrodes so far: the *area 2v* (Büttner and Buettner, 1978) in macaques, which corresponds in new-world monkeys to a small anterior part of area 7, the *PIVC* and the *area 3aV*. These three cortical regions form the *inner vestibular cortical circuit* and are closely interconnected. Their nerve cells project *directly* to the brainstem vestibular nuclear complex (VNC). Closely associated and interconnected with these regions are four further cortical areas: *Area VTS* transmits mainly optokinetic signals to PIVC, most likely through afferent connections from the visual cortices V1 and V2 via areae MT and MST. In turn, area MST also seems to receive some vestibular input (Thier and Erickson, 1992). Part of *area 6a* projects not only to brainstem vestibular nuclei but also to area PIVC and 3aV. It probably transmits information about *gaze movements*. The

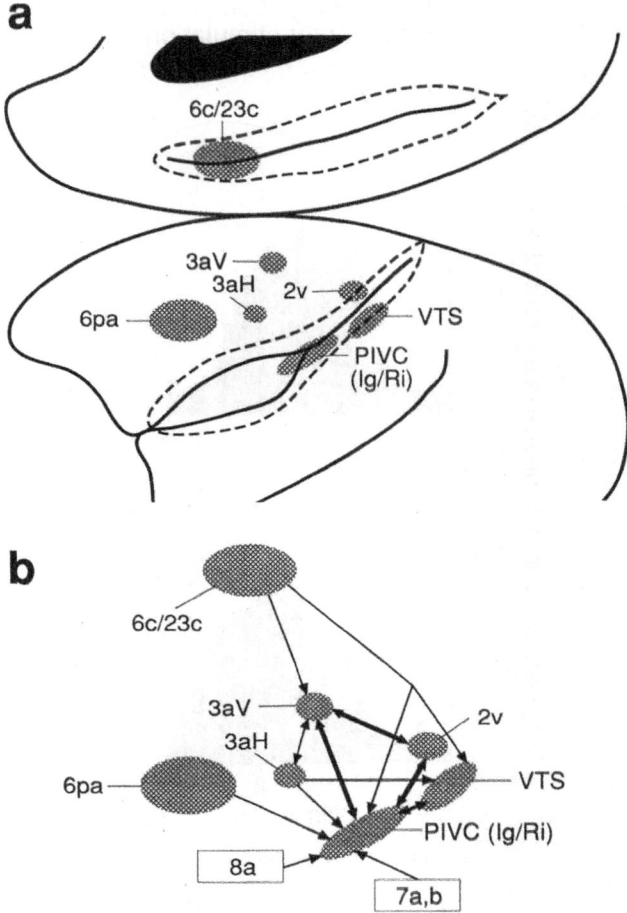

Figure 4. Schematic representation of the 7 cortical "vestibular" areas that project directly to the brainstem vestibular nuclear complex (a) and their interconnections (b). The areas 2v, PIVC (Ri) and 3aV form the *inner vestibular cortical circuit*. The neuroanatomical connections with the other cortical vestibular areas and with areas 7a,b and 8a that are known from tracer studies to date are shown in (b).

somatosensory hand area 3aH is also connected to PIVC and the VNC. Finally, a portion of area 23c of the cortex located along part of the cingular sulcus (CS) is also connected with PIVC and area 3aV and also projects directly to VNC. It is fairly probable that this area represents a cortical field in which abdominal mechano-receptor signals are integrated into the cortical vestibular system. These signals were demonstrated in psychophysical experiments in man (cf. H. Mittelstaedt, this volume, p. 147). This interpretation can only be corroborated or refuted, however, by future experimental evidence. Regarding the projections to the VNC from all these cortical areas mentioned, a detailed analysis revealed that those from the different "vestibular" cortical areas preferred certain substructures of the VNC (Akbarian et al., unpublished, 1993).

The *thalamic input* to the cortical vestibular fields was studied in squirrel monkeys only (Akbarian et al., 1992). Area 3aV and area PIVC receive signals from thalamic neurons located in the ventro-posterior complex, especially the pars posterior shell region of that complex. Some thalamic input to PIVC, VTS and area 7a originates in the *pulvinar*. Other thalamic regions projecting to the structures of the inner cortical vestibular circuit are nerve cells located in the ventro-lateral nucleus and the non-specific thalamic nuclei along the midline. The thalamic "vestibular" nerve cells (Büttner and Henn, 1976; Büttner et al., 1977) have similar response properties but a somewhat higher vestibular gain than PIVC neurons. The density of these vestibular thalamic nerve cells, however, does not seem to be very high and, correspondingly, the direct connections between VNC and the thalamic relay structures are not very dense (Lang et al., 1979). It seems likely that vestibular signals also reach thalamic relay neurons via the cerebellum and the cerebellar nuclei and perhaps also through neuronal structures of the brainstem related to gaze control.

The main task of the inner cortical vestibular circuit seems to be to provide a representation of head-in-space and body-in-space movement. It transmits signals about these movements to structures involved in representing the extracellular space, like area 7a, or the movement of the organism within that space, like area MST. In addition, the cortical vestibular system seems to control, at least in part, the afferent vestibular signal flow from the nerve cells of the VNC to the thalamus and cortex. It may, however, also modify the vestibular brainstem reflexes. The cortical vestibular system furnishes other cortical structures with exactly the signal necessary to update continuously the coordinates of the extrapersonal space or the signals about the relative position of the organism and the extrapersonal space. In the light of this idea, it is not surprising that vestibular and optokinetic stimulation modify spatial hemineglect in man (Rubens, 1985; Cappa et al., 1987; Pizzamiglio et al., 1990; Vallar et al., 1993).

CORTICAL OTOLITH SIGNAL REPRESENTATION?

In experiments to date, the nerve cells of the inner cortical vestibular circuit have only responded to dynamic vestibular stimulation and not at all to steady tilt of the head in space. These negative findings indicate that otolith signals have a separate cortical representation. So far it is only known that a considerable number of cat area 17 and 18 visual cortex neurons respond to labyrinth polarization, which activates not only the afferent synapses from semicircular canal receptors but also those from otolith receptors (Grüsser et al., 1959; Grüsser and Grüsser-Cornehls, 1960, 1972; Denny and Adorjani, 1972; Horn et al, 1972; Schwarzkroin, 1972; Tomko et al., 1981; Kleine and Grüsser, 1993, 1994). In a recent study, Sauvan et al. (1994) found that the receptive field orientation of visual area V2 neurons changed in a compensatory direction when the animal was tilted steadily sideways (cf. Sauvan et al., this volume, p. 43).

SUMMARY

In Java monkeys (*Macaca fascicularis*), squirrel monkeys (*Saimiri Sciureus*) and marmoset monkeys (*Callithrix jacchus*) neurophysiological and neuroanatomical investigations were performed to analyze function and connections of the cortical vestibular areas. Microelectrode recordings were made mainly from single nerve cells of the *parieto-insular vestibular cortex* (PIVC), *area 3 aV*, the neck region of area 3a, the *granular insula Ig*, and *area VTS* adjacent to PIVC in part of area T3, as defined by Burton and Jones (1976). Anterograde and retrograde tracer studies were conducted by injections into the brainstem vestibular nuclei and different cortical vestibular regions.

The vestibular core area of the primate cortex is PIVC, in which the majority of nerve cells responded to stimulation of semicircular canals, optokinetic flow fields, somatosensory stimulation and/or proprioceptive stimulation, especially of the neck and hip region. PIVC was found to be closely connected with *area 3aV*, the neck region of area 3a in which neurons also responded to semicircular canal stimulation but not to optokinetic stimulation. Close anatomical connections between area 3aV and PIVC and the small area 2v, located in squirrel monkeys in the anterior part of area 7 (*area 7ant*), revealed that PIVC, area 3aV and 2v form the closely interconnected *"inner cortical vestibular circuit"*. These structures receive cortical projections from the following areas: *area 6a*, which presumably transmits motor-related gaze signals, *area VST*, which carries optokinetic signals from area MST, *area 3aH*, which transports somatosensory signals from the hand region, and *area 23c* of the cingulate cortex. This latter area located around part of the cingular sulcus perhaps integrates abdominal signals into the cortical vestibular system.

All the structures mentioned project directly to vestibular nuclei and thus may modify the afferent vestibular signal flow and the vestibular brainstem reflexes. The afferent vestibular input to the cortical vestibular areas uses direct vestibulo-thalamic connections but presumably also a pathway that runs via the cerebellum and the cerebellar nuclei to the thalamic structures. Vestibular *thalamic relay neurons* are mainly located in a shell region of the *ventro-posterior complex*. Another important thalamic input to the inner cortical vestibular circuit comes from those parts of the pulvinar which most likely mediate visuo-motor information.

It is assumed that the cortical vestibular representation mediates signals on head-in-space position and movement as well as relative body position and movement in space. The structures of the inner cortical vestibular circuit update the representation of the extrapersonal space in area 7a and the signals about body movements within the extrapersonal space that are related to the activity of area MST neurons. The connections to cortical structures representing the extrapersonal space explain the clinical finding that *spatial hemineglect* is transiently modified during vestibular and optokinetic stimulation.

ACKNOWLEDGMENTS

The work was supported in part by a grant of the Deutsche Forschungsgemeinschaft (GR 161). We thank Mrs. J. Dames for her expert help in writing the manuscript, Mr. P. Holzner for photographic work, Mr. J. Lerch for his mechanical assistance and Dipl. Ing. L.-R. Weiß for writing computer programs.

REFERENCES

Akbarian, S., Berndl, K., Grüsser, O.-J., Guldin, W.O., Pause, M., and Schreiter, U., 1988, Responses of single neurons in the parietoinsular vestibular cortex of primates. *Ann. N.Y. Acad.Sci.* 545:187-202.

Akbarian, S., Grüsser, O.-J., and Guldin W.O., 1992, Thalamic connections of the vestibular cortical fields in the squirrel monkey (Saimiri sciureus). *J. Comp. Neurol.* 326:423-441.

Akbarian, S., Grüsser, O.-J., and Guldin, W.O., 1993, Corticofugal projections to the vestibular nuclei in the squirrel monkey -futher evidence of multiple cortical vestibular fields. *J. Comp. Neurol.* 332:89-104.

Akbarian, S., Grüsser, O.-J., and Guldin, W.O., 1994) Corticofugal connections between the cerebral cortex and brainstem vestibular nuclei in the macaque monkey. *J. Comp. Neurol.* 339:421-437.

Büttner, U., and Henn, V., 1976, Thalamic unit activity in the alert monkey during natural vestibular stimulation. *Brain Res.* 103:127-132.

Büttner, U., Henn, V., and Oswald, H.P., 1977, Vestibular related neuronal activity in the thalamus of the alert monkey during sinusoidal rotation in the dark. *Exp. Brain Res.* 30:435-444.

Büttner, U., and Buettner, U.W., 1978, Parietal cortex (2v) neuronal activity in the alert monkey during natural vestibular and optokinetic stimulation. *Brain Res.* 153:392-397.

Burton, H., and Jones, E.G., 1976, The posterior thalamic region and its projection in new world and old world monkey. *J. Comp. Neurol.* 168:249-301.

Cappa, S.F., Sterzi, R., Vallar, G., and Bisiach, E., 1987, Remission of hemineglect and anosognosia during vestibular stimulation. *Neuropsychologia*, 25:775-782.

Denny, D., and Adorjani, C., 1972, Orientation specificity of visual cortical neurons after head tilt. *Exp. Brain Res.* 14:312-317.

Duensing, F., and Schaefer, K.-P., 1958, Die Aktivität einzelner Neurone im Bereiche der Vestibulariskerne bei Horizontalbeschleunigungen unter besonderer Berücksichtigung des vestibulären Nystagmus. *Arch.Psychiat.Nervenkr.* 198:224-252.

Grüsser, O.-J., 1983, Multimodal structure of the extrapersonal space. In: A. Hein and M. Jeannerod (eds.), *Spatially Oriented Behavior*, Springer-Verlag New York Berlin Heidelberg Tokyo, p. 327-352.

Grüsser, O.-J., and Grüsser-Cornehls, U., 1960, Mikroelektrodenuntersuchungen zur Konvergenz vestibulärer und retinaler Afferenzen an einzelnen Neuronen des optischen Cortex der Katze. *Pflügers Arch.* 270:227-238.

Grüsser, O.-J., and Grüsser-Cornehls, U., 1972, Interaction of vestibular and visual inputs in the visual system. *Progress in Brain Research* 37:573-583.

Grüsser, O.-J., Grüsser-Cornehls, U., and G. Saur, G., 1959, Reaktionen einzelner Neurone im optischen Cortex der Katze nach elektrischer Polarisation des Labyrinths. *Pflügers Archiv* 269:593-612.

Grüsser, O.-J., and Landis, Th., 1991, Visual agnosias and other disturbances of visual perception and cognition. In: J. Cronly-Dillon (ed.) *Vision and Visual Dysfunction*, Vol. 12, Houndmills, Basingstoke, Hampshire and London: The Macmillan Press Ltd., 610 p.

Grüsser, O.-J., Pause, M., and Schreiter, U., 1982, Neuronal responses in the parieto-insular vestibular cortex of alert Java monkeys (Maccaca fascicularis). In A. Roucoux, and M. Crommelinck (eds): *Physiological and Pathological Aspects of Eye Movements*. The Hague, Boston, London: Dr. W. Junk.

Grüsser, O.-J., Pause, M., and Schreiter, U., 1990, Localization and responses of neurons in the parieto-insular vestibular cortex of awake monkeys (Macaca fascicularis). *J. Physiol.* 430:537-557.

Grüsser, O.-J., Pause, M., and Schreiter, U., 1990, Vestibular neurons in the parieto-insular cortex of monkeys (Macaca fascicularis): Visual and neck responses. *J. Physiol.* 430:559-583.

Grüsser, O.-J., Guldin, W., Harris, L., Lefebre, J.-C., and Pause, M., 1992, Cortical representation of head-in-space movement and some psychophysical experiments on head movement. In: A. Berthoz, W. Graf, P.P. Vidal (eds.), *The Head-Neck Sensory Motor System*, New York: Oxford University Press, pp. 497-509.

Grüsser, O.-J., Guldin, W.O., Mirring, S., and Salah-Eldin, A., 1994, Comparative physiological and anatomical studies of the primate vestibular cortex. In: (Hrsg. B. Albowitz, K.Albus, U.Kuhnt, H.-Ch. Nothdurft, P.Wahle), Structural and Functional Organization of the Neocortex, Proceedings of a Symposium in the Memory of Otto D. Creutzfeldt, May 1993, *Exp. Brain Res. Series,* 24: 358-371.

Guldin, W., 1994, Physiologie und Anatomie des cerebralen vestibulären Systems. Habilitationsschrift. Physiologie, FU-Berlin.

Guldin, W.O., Akbarian, S., and Grüsser, O.-J., 1992, Cortico-cortical connections and cytoarchitectonics of the primate vestibular cortex: A study in squirrel monkeys (Saimiri sciureus). *J. Comp. Neurol.* 326:375-401.

Guldin, W.O., Mirring, S., and Grüsser, O.-J., 1993, Connections from the neocortex to the vestibular brain stem nuclei in the common marmoset. *NeuroReport* 5:113-116.

Horn, G., Stechler, G., and Hill, R.M., 1972, Receptive fields of units in the visual cortex of the cat in the presence and absence of bodily tilt. *Exp. Brain Res.* 15:113-132.

Jones, E.G., and Burton, H., 1976, Areal differences in the laminar distribution of thalamic afferents in cortical fields of the insular, parietal and temporal regions of primates. *J. Comp. Neurol.* 168:197-248.

Kleine, J., and Grüsser, O.-J., 1993, Responses of rat vestibular nuclei neurons to natural stimulation and labyrinth polarization. *Soc. Neurosci. Abstr.*, Vol. 19, Part 2: p. 1491.

Kleine, J., and Grüsser, O.-J., 1994, Uniform responses of rat semicircular canal and otolith afferent neurons to trapezoidal and sinewave polarization of the labyrinth. *Soc. Neurosci. Abstr.*, Vol. 20, Part 2: p.970.

Lang, W., Buettner-Enever, J.A., and Büttner, U., 1979, Vestibular projections to the monkey thalamus: an autoradiographic study. *Brain Res.* 177:3-17.

Mergner, T., 1979, Vestibular influences on the cat's cerebral cortex. In R. Granit, and O. Pompeiano (eds): *Reflex control of posture and movement.* Amsterdam: Elsevier, pp. 567-579.

Mergner, T., Anastasopoulos, D., Becker, W., and Deecke, L., 1981a, Comparison of the modes of interaction of labyrinthine and neck afferents in the suprasylvian cortex and vestibular nuclei of the cat. In A.F. Fuchs, and W. Becker (eds): *Progress in oculomotor research.* Amsterdam: Elsevier, pp. 343-350.

Mergner, T., Deecke, L., and Wagner, H.-J., 1981b; Vestibulo-thalamic projection to the anterior suprasylvian cortex of the cat. *Exp. Brain Res.* 44:455-458.

Mergner, T., Becker, W., and Deecke, L., 1985, Canal-neck interaction in vestibular neurons of the cat's cerebral cortex. *Exp. Brain Res.* 61:94-108.

Mittelstaedt, H., 1995, this volume, p. 147-155.

Ödkvist, L.M., Schwarz, D.W.F., Fredrickson, J.M., and Hassler, R., 1974, Projection of the vestibular nerve to the area 3a arm field in the squirrel monkey. *Exp. Brain Res.* 21:97-105.

Pandya, D.N., and Sanides, F., 1973, Architectonic parcellation of the temporal operculum in rhesus monkey and its projection pattern. *Zschr. Anat. Entw.gesch.* 139:127-161.

Pizzamiglio, L., Frasca, R., Guariglia, C., Incoccia, C., and Antonucci, G., 1990, Effect of optokinetic stimulation in patients with visual neglect. *Cortex* 26: 535-540.

Rode, G., and Perenin, M.T., 1994, Temporary remission of representational hemineglect through vestibular stimulation. *NeuroReport* 5:869-872.

Rubens, A.B., 1985, Caloric stimulation and unilateral visual neglect. *Neurology* 35: 1019-1024.

Sauvan, X.M., Henn, V., and Peterhans, E., 1994, Mechanisms of contour processing in monkey prestriate cortex include information on direction of gravity. *Soc. Neurosci. Abstr.,* Vol. 20, Part 2: p.1740.

Sauvan, X.M. et al., 1995, this volume, p. 43-49.

Schwarztkroin, P.A., 1972, The effect of body tilt on the directionality of units in cat visual cortex. *Exp. Neurol.* 36:498-506.

Thier, P., and Erickson, R.G., 1992, Vestibular input to visual tracking neurons in area MST of awake rhesus monkeys. *Ann. NY Acad. Sci.* 656: 960-963.

Tomko, D.L., Barbarao, N.M., and Ali, F.N., 1981, Effect of body tilt on receptive orientation of simple visual cortical neurons in unanesthetized cats. *Exp. Brain Res.* 43:309-314.

Vallar, G., Antonucci, G., Guariglia, C., and Pizzamiglio, L., 1993, Deficits of position sense, unilateral neglect and optokinetic stimulation. *Neuropsychologia* 31: 1191-1200.

Walzl, E.M., and Mountcastle, V.B., 1949, Projection of the vestibular nerve to the cerebral cortex of the cat. *Am. J. Physiol.* 159: 595.

8

PROPRIOCEPTIVE AND CUTANEOUS FEEDBACK IN THE MODULATION OF CORTICAL OUTPUT IN MAN

P. H. Ellaway

Department of Physiology
Charing Cross and Westminster Medical School
London W6 8RF
United Kingdom

INTRODUCTION

The degree and sign of changes in motoneurone excitability and reflex responses evoked by peripheral stimulation of cutaneous and proprioceptive afferents varies with the muscle and location of the stimulus (Sherrington, 1906; Hagbarth, 1952). Our work has explored the influence of sensory input on the responses of skeletal muscles to stimulation of the motor cortex using transcranial magnetic stimulation (TMS) of the brain in man.

METHODS

Informed consent was obtained from the subjects and approval for this study was obtained from the local ethical committee. Electromyographic recordings were made from normal healthy subjects, with no history of neurological disorder, using surface electrodes. Transcranial magnetic stimulation (TMS) was achieved using a MagStim 200 1.5 Tesla machine with a 9 cm circular coil centered over the vertex (skin stimulation experiments) or the low threshold site for the muscle response (joint position experiments). Cutaneous stimulation was carried out using a stiff toothbrush and by brushing in a circular fashion using light pressure within 1 square cm at each point on a 1.5 cm grid drawn over (e.g.) the thenar eminence or over the finger pads or joints. TMS was carried out with the muscles of the hand relaxed. In experiments designed to test the effect of wrist joint position on the response of the extensor carpi ulnaris (ECU) muscle to TMS, the hand and forearm were placed in a manipulandum which could be adjusted to hold the hand at any angle between 60° flexion and extension. Subjects were asked to relax the arm so that the background EMG recorded from ECU was absent or kept to a low level.

Multisensory Control of Posture, Edited by T. Mergner and F. Hlavačka
Plenum Press, New York, 1995

RESULTS

Cutaneous Stimulation

We have examined the changes in excitability to TMS of a number of muscles in the hand (abductor digiti minimi - ADM, thenar and first dorsal interosseus) (Davey et al., 1991; Ellaway et al., 1991). The motor cortex was stimulated at a strength just supra-threshold for a consistent motor evoked potential in the muscle at rest. Contralateral motor evoked responses (cMEPs) could be either augmented or decreased by mechanical stimulation of the skin of the hand. The locations that produced facilitation and inhibition were repeatable within the same subject but differed between subjects. However, a common finding was strong facilitation from skin overlying the muscle, and over the joints relevant to the action of that muscle. Fig. 1A shows averaged responses in thenar muscles to eight magnetic stimuli either with (below) or without (above) stimulation of the skin over the metacarpophalangeal (MCP) joint of the thumb. The actual protocol involved four control recordings followed by four recordings during skin stimulation. There was then a 20 sec interval to allow any persisting skin sensation to die away before the sequence was repeated. In this subject, skin stimulation over the MCP joint facilitated the response to TMS fourfold. Fig. 1B presents results from nine subjects and shows that skin stimulation over the MCP joint never reduced the response to TMS. The responses during skin stimulation showed increases that were significantly different (Students t-test, $P < 0.01$) from control recordings in six subjects. Inhibition of the response of the thenar muscles to TMS occurred during skin stimulation but less frequently than facilitation. Nevertheless, inhibition was elicited at one or more sites in 13 out of 14 subjects. The sites producing inhibition had a different distribution to those producing facilitation. Inhibition was not seen on stimulating skin over the MCP joint but was more evident at sites towards the base of the thumb. These results suggest that there is

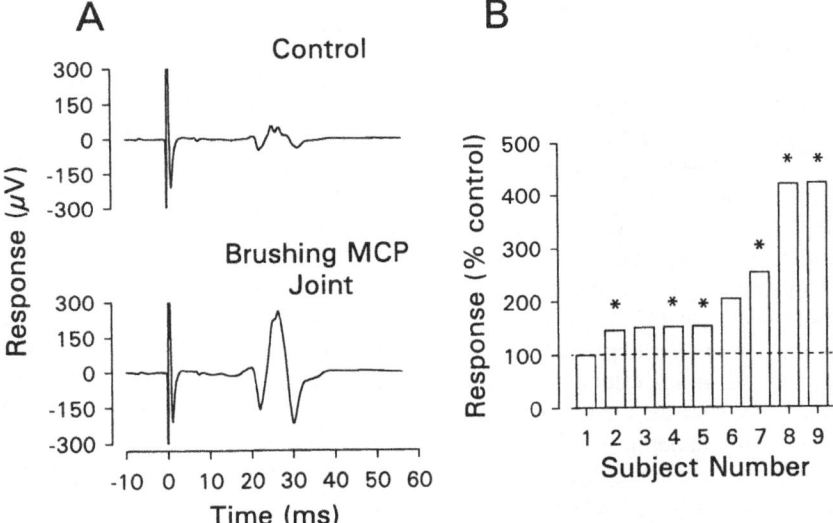

Figure 1. (A) Averages of eight EMG responses of the thenar muscles to TMS recorded from one subject (#8) without intentional skin stimulation (above) and during brushing of the skin over the metacarpophalangeal (MCP) joint (below). (B) Changes in thenar muscle responses to TMS during skin stimulation over the MCP joint of the thumb in 9 subjects. Asterisks indicate significant ($P < 0.05$) increases.

a complex somatotopic organization of peripheral sensory inputs, probably including both cutaneous and proprioceptive feedback, that modulate cortico-spinal excitability.

Electrical stimulation of digital nerves of the little finger, up to 5 times perceptual threshold, tended to inhibit cMEPs elicited in ADM by TMS. Similar results were obtained by Day et al. (1988) using electrical stimulation of the digital nerves of the index finger and recording responses to TMS of the first dorsal interosseus muscle. However, we found that brushing the skin on the pad of the little finger facilitated rather than inhibited responses of ADM to TMS. Thus, the pattern of afferent activity elicited by natural stimulation has a central effect on motor responses that may not be mimicked by electrical stimulation of peripheral nerves.

Joint Manipulation

We have investigated the influence of passive joint position on the response of muscles acting at that joint, to TMS of the motor cortex in twelve normal subjects (Pope et al., 1993). Three different angles of the wrist joint (60° flexion, neutral 0° and 60° extension) were selected in random order and the cMEP responses in extensor carpi ulnaris (ECU) to fifteen TMS were recorded, full wave rectified and averaged. Fig. 2 shows results from one subject. Passive positioning of the wrist joint in 60° extension facilitated cMEPs in ECU in this subject (Fig. 2A) whereas any alteration in the response at 60° flexion was equivocal. The mean of the three averaged responses at 60° extension was statistically (Students t-test) significantly greater (P = 0.025) than the mean value in the neutral wrist position (0°). The size of the cMEP response to TMS was not related to the presence or amount of background EMG in the muscle (Fig. 2B). Thus, when the response to a constant TMS was facilitated during wrist extension the change was not simply a consequence of increased depolarization of motoneurones resulting from voluntary, or involuntary, contraction (Mazzocchio et al., 1994). The average cMEP response for twelve subjects (ages 20-25 years), as a percentage

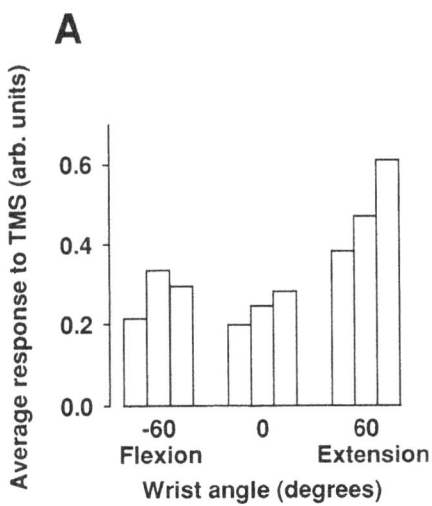 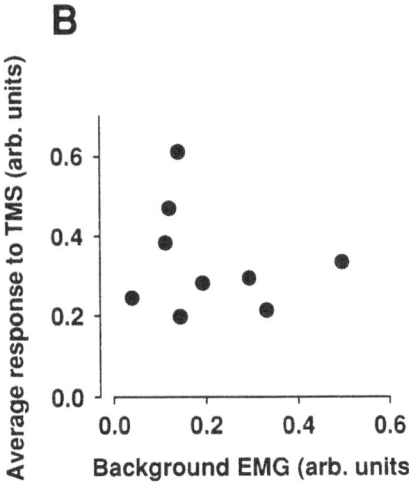

Figure 2. (A) Size of the EMG response recorded from the extensor carpi ulnaris (ECU) muscle in response to TMS. Each bar represents the average of the responses to 15 stimuli. Three sets of 15 stimuli were obtained for each position of the wrist: flexion, -60°; control, neutral position, 0°; extension, 60°. (B) The average response in ECU to TMS, plotted against the amount of background EMG. Note the lack of correlation between these two parameters.

of control responses, was $163 \pm 14\%$ (mean $\pm$ SE) with the wrist in 60° extension. Passive flexion failed to influence cMEPs in the ECU muscles with any consistency. The average response in flexion was $93 \pm 8\%$. Only in extension was the mean response significantly different (Students paired t-test, $P < 0.001$) from the control, neutral position with wrist held straight (0°).

To avoid the possibility that changes in cMEP responses might have been due simply to alterations in the amount of EMG signal, as a result of movement of the muscle relative to the over-lying EMG electrodes, the experiment was repeated with the cMEP responses normalized to the direct MEP responses of the muscle to a fixed electrical stimulus applied to the muscle nerve. The mean ($\pm$ SE) normalized response of the ECU at 60° extension of the wrist ($420 \pm 81\%$) was significantly greater ($P < 0.01$) than the response in a neutral position (0°) in a further four subjects (ages 33-49 years). Passive wrist positioning at 60° flexion produced no significant difference in mean, normalized cMEP amplitude compared with the neutral position. Bearing in mind that wrist flexion (stretch of ECU) rather than extension would be expected to facilitate ECU motoneurones via stretch reflex afferents, we suggest that extension of the wrist joint facilitates corticospinal activation of the ECU muscle at a supraspinal level which is likely to be in the motor cortex.

CONCLUSION

The execution of accurate movements relies on feedback from cutaneous and proprioceptive sense organs in all but the fastest of ballistic movements. The experiments reported here show that this feedback influences the motor command to skeletal muscle at the level of the corticospinal pathway in that sensory input modifies the response of a muscle to transcranial magnetic stimulation of the motor cortex. Indirect evidence provides a pointer that the action of skin afferents may be at the level of the cortex rather than the motoneurones in the spinal cord. Continuous natural stimulation of the skin predominantly facilitated the response of a muscle to TMS whereas the short latency, segmental action of afferents excited by electrical stimulation of sensory nerves innervating that skin tended to inhibit. When the wrist joint was extended and the wrist extensor muscle held at a short length, the response of that muscle was facilitated. This did not happen on lengthening the muscle even though such a maneuver would have been expected to facilitate the wrist extensor motoneurones via the spinal segmental action of spindle primary afferents.

The actions of cutaneous and proprioceptive afferents on the corticospinal responses of muscles clearly has a complex organization. Some features were found in the majority of subjects. For example, the tendency for skin stimulation to be most effective in eliciting facilitation of thenar muscles when applied over the metacarpophalangeal joint. However, it was also clear that there was a precise topographical organization of facilitatory and inhibitory afferent inputs that was peculiar to the individual. It would be interesting to know whether this organization develops with the acquisition of motor skills or is inherited.

SUMMARY

Non-noxious mechanical stimulation of the skin can augment or decrease the motor evoked responses (cMEPs) of hand muscles to transcranial magnetic stimulation (TMS) of the contralateral motor cortex in man. Individual subjects show a topographic organization of the sites from which increases and decreases in cMEP amplitude may be elicited. The most consistent finding between subjects was that brushing the skin over the joints of digits engaged by the muscle responding to TMS produced the most powerful facilitation. In other experiments, static position of the wrist joint was found to influence the cMEP response of the wrist extensor muscle, extensor carpi ulnaris (ECU) to TMS. Passive extension of up to 60° increased the size of the response in ECU whereas passive flexion produced no significant change. These results suggest that the particular cutaneous and/or proprioceptors that influence the degree of activation of a muscle via corticospinal pathways arise in structures closely related to the insertion or origin of that muscle, and that their action may be related to specific joint positions or movements.

ACKNOWLEDGMENTS

This work was carried out in collaboration with N. J. Davey, M. McD. Lewis, D. W. Maskill, A. J. Pope, A. G. O. Redman and P. Romaiguère.

REFERENCES

Davey, N.J., Maskill, D.W., and Ellaway, P.H., 1991, Facilitation by mechanical cutaneous stimulation of muscle responses to transcranial magnetic stimulation in man, *J. Physiol.* 438:7P.

Davey, N.J., Maskill, D.W., Romaiguère, P., and Ellaway, P.H., 1993, Facilitatory and inhibitory effects of discrete cutaneous bushing on the responses of thenar muscles to transcranial magnetic stimulation in man, Abstracts XXXII Congress for the IUPS, 248.20P.

Day, BL., Dressler, D., Maertens de Noordhout, A., Marsden, C.D., Nakashima, K., Rothwell, J.C., and Thompson, P.D., 1988, Differential effect of cutaneous stimuli on responses to transcranial magnetic stimulation, *J. Physiol.* 399:68P.

Ellaway, P.H., Davey, N.J., Maskill, D.W., and Romaiguère, P., 1991, Organisation of cutaneous facilitation of hand muscles responses to transcranial magnetic stimulation, Abstracts Soc. for Neuroscience 17:553.14.

Hagbarth, H.-E., 1952, Excitatory and inhibitory skin areas for flexor and extensor motoneurones, *Acta Physiol. Scand.* 26, Suppl. 94:1-58.

Mazzocchio, R., Rothwell, J.C., Day, B.L., and Thompson, P.D., 1994, Effect of tonic voluntary activity on the excitability of human motor cortex *J. Physiol.* 474:261-267.

Pope, A.J., Redman, A.G.O., Lewis, M.McD., Davey, N.J., and Ellaway, P.H., 1993, The effect of wrist angle on the response of a wrist extensor muscle to transcranial magnetic stimulation in man, *J. Physiol. 459:470P.*

Sherrington, C.S., 1906, The integrative action of the nervous system, Scribner's, New York.

9

THE ROLE OF PROPRIOCEPTIVE AND VESTIBULAR INPUTS IN TRIGGERING HUMAN BALANCE CORRECTIONS

J. H. J. Allum and F. Honegger

University HNO-Klinik
Petersgraben 4
CH-4031 Basel
Switzerland

INTRODUCTION

The relative contribution of proprioceptive and vestibulo-spinal signals to automatic balance corrections can be examined in terms of triggered patterns of muscle activity which are then amplitude modulated by sensory inputs. For both the triggering and modulation processes the question arises whether sensory weighting employed is similar across all joints or different for the different groups of coordinated agonist-antagonist muscle activity acting across each joint.

It has been known for some time that the relative influences of vestibulo-spinal and proprioceptive inputs are different on the flexor and extensor muscles of the legs (Wilson and Melvill Jones, 1979) causing an animal to adopt a crouch position with heightened proprioceptive reflex gains the moment it is perturbed (Hulliger, 1989). In fact, a number of spinal neural circuits could be postulated as the basis for varying proprioceptive and vestibulo-spinal inputs to α-motoneurons of specific groups of muscles. The most powerful may well be Renshaw inhibition shaping muscle action at the final common output path of α-motoneurones (Windhorst, 1989). Inhibitory Renshaw cells receive a triad of inputs; descending vestibulo-spinal influences via the locus coeruleus nucleus (Pompeiano et al., 1989), ascending muscle spindle afferents signals (specifically via Ia pathways, see Hultborn et al., 1979), and recurrent feedback from the same motoneurons (Renshaw, 1941). The very dynamic nature of Renshaw action (Christakos et al., 1987) would, as other authors have speculated (Iles, 1992), make Renshaw circuits an ideal medium to regulate motoneural output in response to sudden body perturbations when a brisk muscle contraction is required to stabilize stance.

Variations in the weighting and gating provided by spinal interneuronal circuits on motor output could well explain some of the reported controversies (Forssberg and Hirschfeld, 1994) on the role of vestibular and proprioceptive inputs in triggering and modulating human balance corrections. For example, differences in the influence of vestibulo-spinal inputs on ankle flexor and extensor muscles could underlie weak vestibular modulation in triceps surae muscles (Horak et al., 1990,1994; Dietz et al., 1988). In contrast, ankle flexor muscles receive a strong influence from the vestibular system (Keshner et al., 1987; Fries et al., 1993). However, because the latency of these latter responses was not affected by vestibular loss, the possibility could not be excluded that vestibular influences on leg muscles requires the presence of a proprioceptive triggering signal to set up the appropriate muscle response timing synergy which is then modulated by a combination of visual, vestibular and proprioceptive inputs (Keshner et al., 1987; Allum et al., 1994).

The question of which proprioceptive signals are responsible for triggering automatic balance responses is very difficult to answer because of the methodological problems involved with eliminating proprioceptive input at one joint (e.g., the ankle) without affecting proprioceptive input at the next most proximal joint (e.g., at the knee) and the relative rarity of subjects with bilateral proprioceptive deficits limited to one set of joints, e.g. both ankle joints. Previous attempts to determine the relative weighting of proprioceptive and vestibular inputs in triggering balance corrections have led to similar findings, namely that vestibulo-spinal signals increase muscle activity some 60 ms after the onset of head acceleration (Bussel et al., 1980; Dietz et al., 1988; Fries et al., 1993; Greenwood and Hopkins, 1976), but the interpretations of the significance of vestibular versus proprioceptive cues in the rapid regulation of balance perturbations to stance and gait differed considerably (Allum et al., 1994; Allum et al., 1995a; Horak et al., 1994).

Alternative explanations for the seeming incongruity of findings when head (vestibular) and ankle (proprioceptive) movements are tested as trigger signals (Allum et al., 1994; Horak et al., 1994) have evolved into two different roles for vestibulo-spinal signals. One role, based on a substitution effect, assumes that vestibular signals are only used when proprioceptive signals from the ankle joint are not reliable (Nashner et al., 1982; Horak et al., 1994). An alternative is to assume that only proprioceptive signals such as pelvis rotation, consistently present during all balance perturbations, underlie the triggering of a centrally stored pattern of muscle-response timing (Forssberg and Hirschfeld, 1994; Allum et al., 1995a). A corollary to this later triggering mode is the participation of other proprioceptive inputs, such as knee inputs, as well as vestibular signals, in triggering variations to the centrally stored pattern of response activation. Discrete timing synergies with subsequent continuous amplitude modulation by sensory signals is a known characteristic of balance responses (Allum et al., 1993). Furthermore, timing patterns appear to be similar for muscle responses across many segments in contrast to quite varied weighting of different sensory signals contributing to amplitude modulation of different muscles. For example, only two types of timing synergies were observed for five different types of balance perturbation with the same amplitude of ankle dorsiflexion (Allum et al., 1993). Such observations indicate that investigations using controlled balance perturbations which restrict the amount of ankle and knee joint rotation, will provide insights into triggering mechanisms underlying balance corrections. The aim of the current work was then to explore these mechanisms in normal subjects, those with functionally absent vestibular inputs, and those with bilateral proprioceptive loss in the lower legs.

METHODS

Three combinations of support-surface rotation and rearward translation were used to induce balance corrections reestablishing perturbed upright stance. The balance perturbations were presented randomly and consisted either of a 4 deg dorsiflexion rotation, a simultaneous 4 cm rearward translation and 4 deg dorsiflexion rotation to yield a total of 6 deg of ankle dorsiflexion, or a rearward translation combined with 4 deg of plantar flexion to yield negligible dorsiflexion of the ankle joint. Stimulus durations were always 150 ms. A goniometer system was used to register the angle of the lower leg. Because the angle of the support surface was also measured the profile of ankle rotation could be calculated. This later signal was controlled by a separate servo controller-system driving the support-surface rotation to ensure that the profile of ankle dorsiflexion of all test subjects was regulated to match an average normal profile over the first 240 ms after stimulus onset.

As in previous studies (Allum et al., 1993,1994) subjects were required to respond to the randomly presented stimuli as quickly as possible once perturbed from normal upright stance. Prior to each stimulus, subjects could monitor their deviation from their individual "upright" with visual and acoustic feedback. Each stimulus type was presented 10 times for a total of 30 stimuli in series. One series was presented under eyes-open conditions, the second under eyes-closed conditions.

Three populations of subjects were tested. A group of 14 normal subjects, a group of 5 otherwise healthy subjects who had a bilateral peripheral vestibular deficit according to clinical criteria (described in Allum et al., 1994), and 2 subjects who had bilaterally absent Achilles and patella tendon reflexes (due to a peripheral neuropathy of diabetic origin) without significant leg muscle weakness. In order to confirm the bilaterality of the vestibular and proprioceptive loss electromyographic responses were recorded with surface electrodes from the left and right tibialis anterior and soleus muscles. Otherwise recordings were taken from the right medial gastrocnemius, quadriceps, paraspinals and upper trapezius muscles. These recordings were sampled at 1 KHz after filtering, rectification and smoothing. Simultaneously, biomechanical recordings were sampled at 500 Hz. These recordings included trunk and upper-leg angular velocity (bandwidth 50 Hz), and ankle torque responses from the right and left foot calculated from the outputs of strain gauges imbedded in the support system.

Biomechanical and EMG responses were first averaged across all similar stimuli in a series except that the first 3 responses were ignored to avoid adaptation effects (Keshner et al., 1987) influencing the data significantly. All averages used the first inflection in the velocity of ankle dorsiflexion as stimulus onset, thereby compensating for differences in the onset of support-surface movement for rotation and translation stimuli. Once responses of an individual subject to all identical stimuli were combined, response data were averaged across subjects in the same population.

RESULTS

Dorsiflexion of the support surface forces the body to move as if it were basically a two-link structure (Allum and Honegger, 1992); the legs as one link and the trunk as a second. Little movement occurs at the knees except for a slight hyperflexion (see knee angular velocity traces in Figures 1 and 2). As a result the legs are pitched backwards, aided by a stretch reflex response at 50 ms in Soleus and Gastrocnemius, and the trunk is pitched forwards. To regain an upright stance, normal subjects activate the tibialis anterior and quadriceps muscles at approximately 120 ms, thereby rotating the legs forwards about the ankle joint, and simultaneously activate the paraspinal muscles, rotating the trunk backwards about the pelvis (see thin traces in figures 1-4).

In subjects with vestibular loss, activation amplitudes in the tibialis anterior (TA) and quadriceps (QUAD) muscles were reduced in response to support-surface dorsiflexion (Fig. 1), whereas the same muscle responses were delayed but not reduced in amplitude in subjects with proprioceptive loss (Fig. 2). The 30-50 ms shift in the latencies of the TA and QUAD responses in Figure 2 was consistent with the shift in response latency observed for the soleus (SOL) and gastrocnemius (GASTROC) muscles from 50 to approximately 90 ms. Alterations

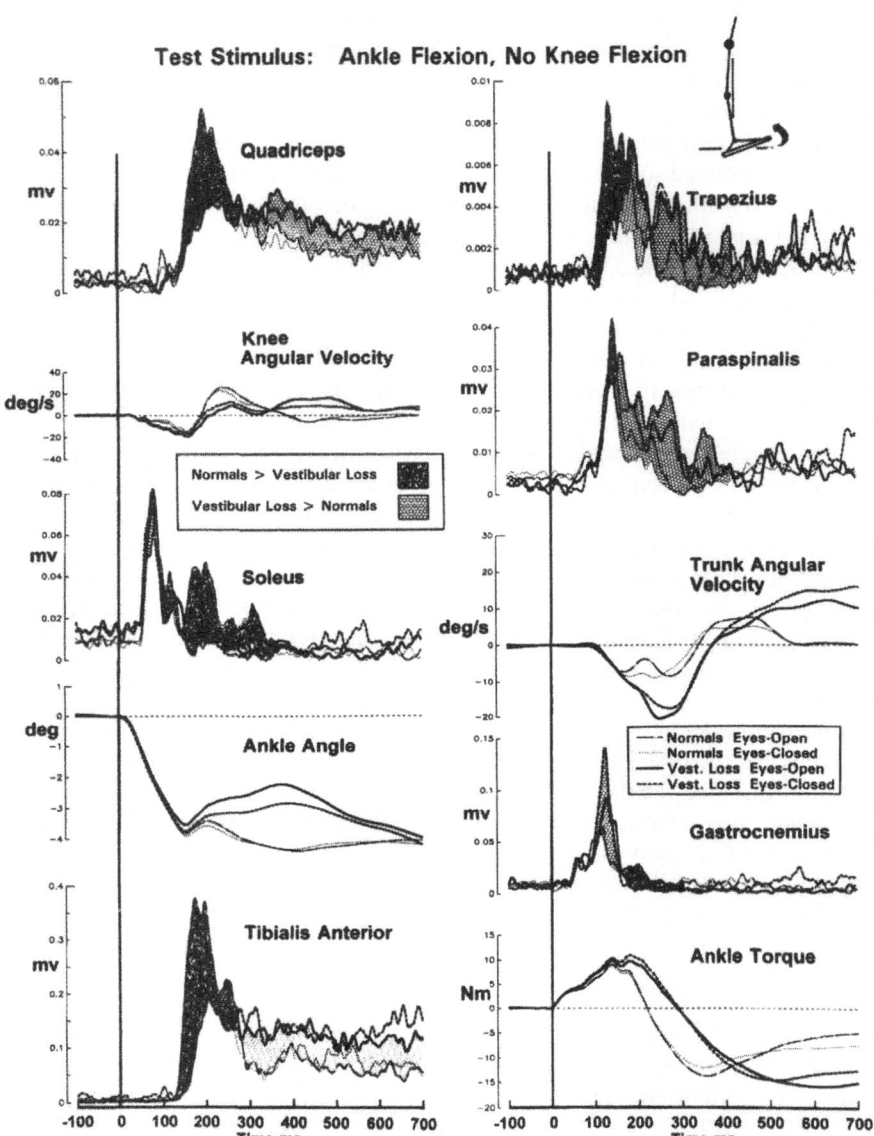

Figure 1. Differences between normal and vestibular-loss responses to a dorsiflexion of the support surface. Muscle activation patterns of subjects with bilateral peripheral vestibular loss (thick lines) are compared to the average responses of normal subjects (thin lines). Eyes-open responses are shown by full lines, eyes-closed responses by interrupted lines. All traces are aligned with the first stimulus-induced deflection of ankle angular-velocity (thick vertical line at 0 ms). Rearward rotation of the trunk (with respect to earth fixed coordinates), ankle plantar-flexion and knee flexion are plotted upwards, as is increased plantar-flexion torque imposed by the foot on the support surface. The insert label indicates significant differences between the responses. Note the changes in response modulation after vestibular loss (from Allum and Honegger, 1995).

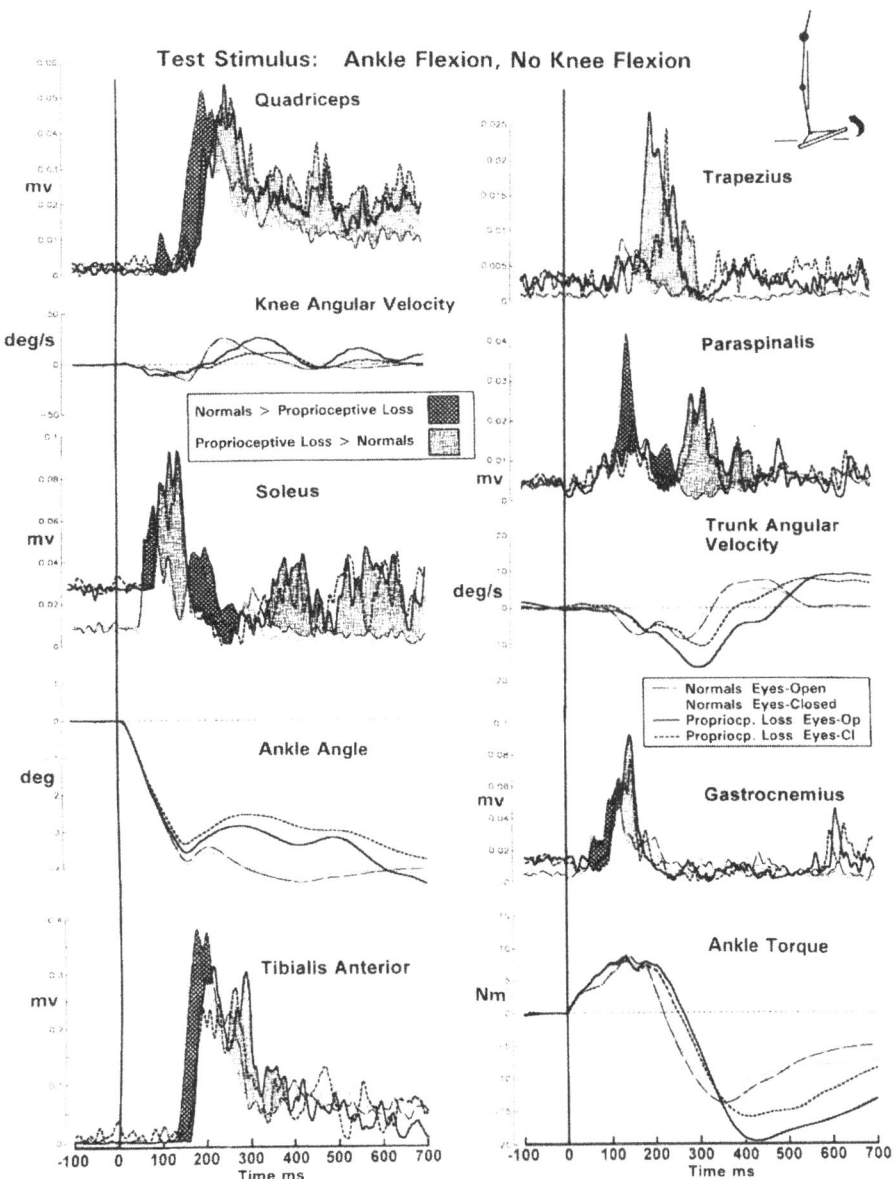

Figure 2. Differences between normal and proprioceptive-loss responses to a dorsiflexion of the support surface. The average responses of subjects with bilateral absence of Achilles and patella tendon reflexes are shown by the thick lines. For other details refer to the legend of Fig. 1. Note the changes in tibialis anterior and quadriceps response onsets, consistent with the response onset shift in soleus.

to the profiles of ankle torque and trunk angular velocity traces followed trends expected from the alterations to muscle responses in the two peripheral deficit populations. Ankle torque changed to a dorsiflexion direction at about the same time after the onset of support rotation, 170 ms, in vestibular-loss subjects as observed for normals. However, the slope of the torque trace was significantly less steep (see also Allum and Pfaltz, 1985; Keshner et al., 1987; Allum et al., 1994). Ankle torque profiles of proprioceptive-loss subjects post 170 ms were characterized by a delay in the onset of dorsiflexing torque. After this delay, the slope of the torque change was equally strong as that of normals. The effect on ankle angle was similar in both deficit populations. A slight increase in ankle dorsiflexion, consistent with a rotation of the leg forward about the ankle joint, was noted for normal subjects. Instead, a reduction in the stimulus imposed dorsiflextion occurred in the deficit populations. In other words, the legs started rotating backwards, placing the subjects in danger of falling. This instability was enhanced by the influence of vestibular-loss and proprioceptive-loss on trunk angular-velocity profiles.

Trunk angular velocity profiles confirmed the observations of larger forward pitching rotations than normal angular velocity profiles in both deficit populations (see traces in figures 1 and 2). Differences in the profiles of angular velocity with respect to normal and associated changes in paraspinal (PARAS) EMG activity indicated that the altered profiles of EMG responses were probably related to changed muscle stretch profiles in back muscles. Trunk forward pitching velocity started later and reached a maximum later in the proprioceptive deficit population, whereas the trunk of vestibular-loss subjects pitched forward at the same time as normals, reaching a maximum later than normals, but earlier than the proprioceptive loss patients. The changes in PARAS EMG activity with respect to normal responses in Figures 1 and 2 reflect these differences. In proprioceptive-deficit subjects, first a decrease in activity was observed followed by a later increase (Fig. 2). Vestibular-loss subjects showed an increased modulation in PARAS activity to the normal activity pattern (figure 1). In all three populations, the initial increase of PARAS activity between 80 and 120 ms was not altered.

Two initial conclusions could be drawn from Figures 1 and 2 concerning the triggering of balance responses in the absence of significant knee proprioceptive inputs. First, the absence of vestibular inputs did not significantly influence response timing. Second, proprioceptive inputs at the ankle joint are only influencing the initial part of the response in leg muscles. Despite the absence of vestibular and lower-leg proprioceptive inputs, balance responses remained relatively intact. Data in Figures 3 and 4 reinforce these initial conclusions.

Rearward translation forces the body to move as a three-link structure (Allum and Honegger, 1992). Even when perturbation velocities are as low as 15 cm/s (Allum et al., 1995, unpublished observations), there is considerable flexion about the knee joint, though this is seldom acknowledged (Horak and Nashner, 1986; Kuo 1995). In addition to the ankle dorsiflexion, the trunk first rotates backwards and then tips forwards. The simultaneous rotation of the support surface imposed in the current experiments changed the amplitudes of trunk and ankle rotations without significant influencing knee rotation profiles. In Figures 3 and 4 the effect of a simultaneous rearward translation and plantarflexion of the support surface is shown. The ankle-angle movements were limited by the precise control on ankle flexion to be within ± 1 deg of ankle angle values prior to support-surface movements.

Nulling out ankle movements only eliminated the first part of the SOL and GAS-TROC response. At 100 ms a strong response was seen in all three populations tested. The profile of knee angular velocity shown in Figures 3 and 4 provided information on the origin of these responses. Rearward translation first forced the knee into flexion and elicited a stretch-reflex response at 50 ms in QUAD. This stretch reflex response was absent in the proprioceptive-loss subjects and considerably reduced in vestibular-loss subjects. Sub-

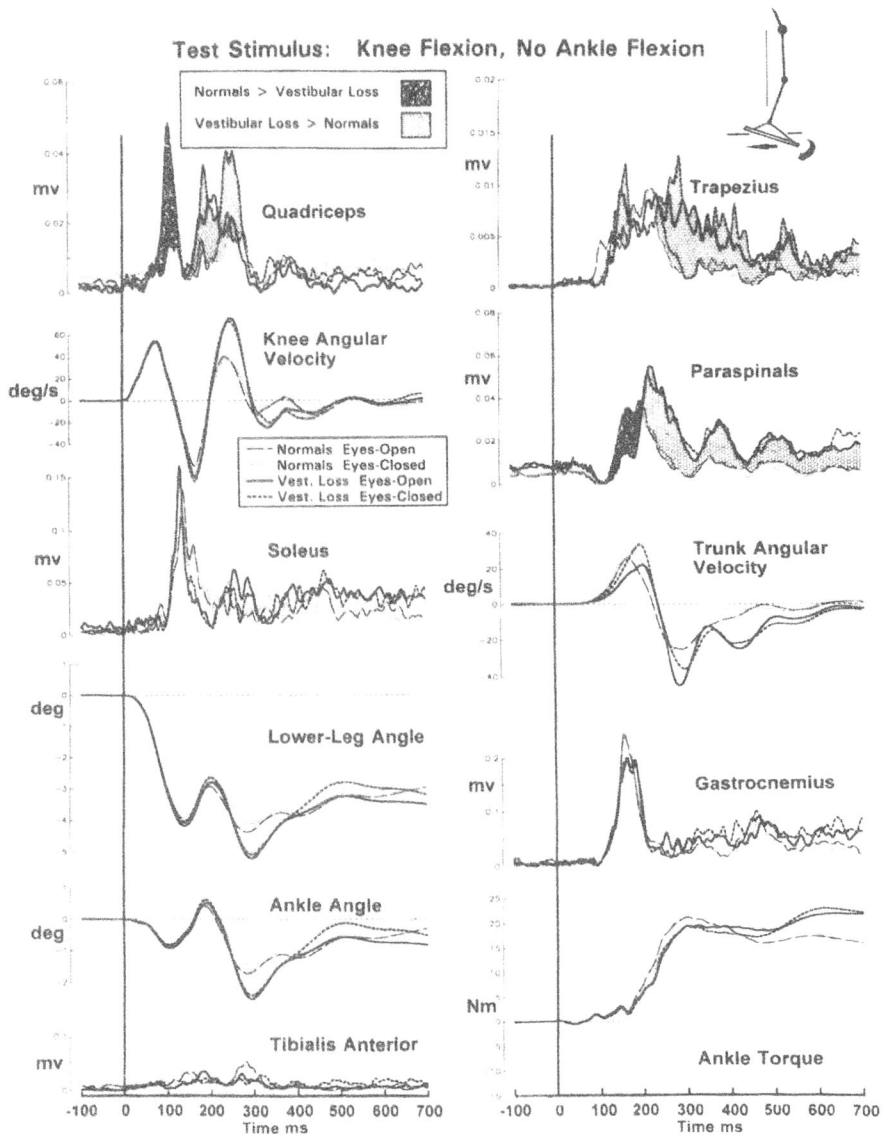

Figure 3. Differences between normal and vestibular-loss responses to a simultaneous rearward translation and plantar-flexion rotation of the support surface. The profile of support-surface rotation was designed to minimize ankle angle rotations over the first 240 ms from stimulus onset. Details of the traces are described in the legend to Fig. 1. Note the absence of a tibialis anterior response in both populations and the presence of responses at 100 ms in soleus and gastrocnemius muscles, which are not altered by vestibular loss (from Allum and Honegger, 1995).

Test Stimulus: Knee Flexion, No Ankle Flexion

Test Stimulus: Knee Flexion, No Ankle Flexion

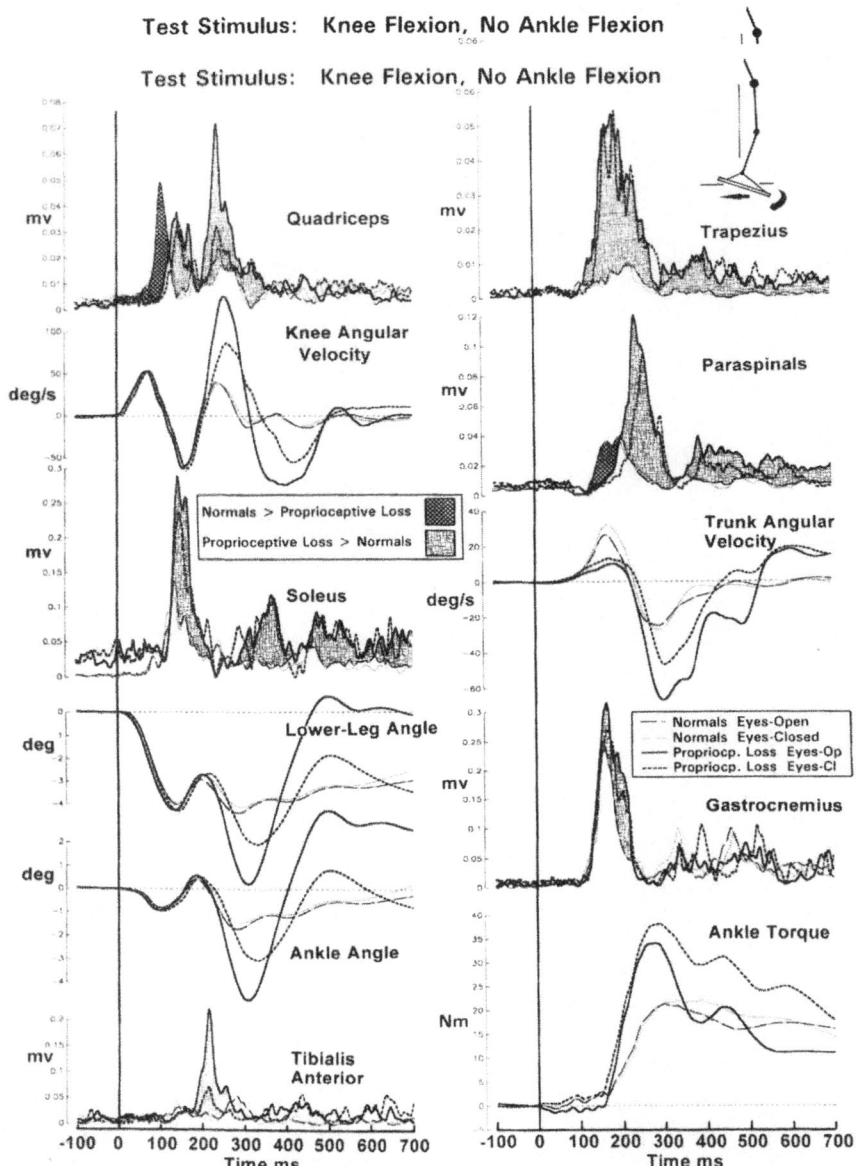

Figure 4. Differences between normal and proprioceptive-loss response to a rearward translation of the support surface combined with a rotation which nulls ankle-angle movements. Note the changed profile of ankle torque, consistent with increased activity in the soleus and gastrocnemius muscles.

sequently, the pattern of EMG activity in QUAD was considerably different from normal in both deficit populations. The divergence from the normal QUAD EMG response pattern was even greater for rearward translation combined with dorsiflexion for vestibular-loss subjects (not shown) even though knee angular movements were very similar. After the peak knee flexion of 50 deg/s at ca 70 ms, the knee was rapidly extended reaching a velocity of 60 deg/s at approximately 150 ms. Presumably, this rapid extension caused a stretch response in the triceps surae muscles SOL and GASTROC at 100 ms.

As figures 3 and 4 illustrate, responses at 100 ms in SOL and GASTROC were not altered by vestibular loss, but were enhanced by proprioceptive loss. Proprioceptive loss caused an approximately 100 % increase in the SOL response and a lesser increase (ca. 25 %) in GASTROC response between 100 and 200 ms. Similarly, there was no alteration to the basicallly absent response in TA, however, a burst of activity commencing at ca 180 ms was observed in the TA responses of proprioceptive-loss subjects. In fact, TA was the only muscle response which was practically eliminated in the absence of an ankle joint rotation in the normal and vestibular-loss subjects. As a result of the unchanged ankle muscle responses in vestibular-loss subjects with respect to normals, the ankle-torque profiles post 170 ms were identical to those of normals except for a minor decrease, probably related to changes in upper body EMG responses described below. Prior to 170 ms changes in torque profiles are not expected because the servo-controlled nulling of ankle angle effectively reduced any torque change with respect to normal. Ankle torque profiles of proprioceptive-loss subjects showed a considerable increase in plantar flexion ankle torque post 170 ms with respect to normals, presumably following increased activity in the triceps surae muscles with respect to that of normal subjects.

In the absence of ankle inputs (confirmed by the lack of a stretch reflex at 50 ms in SOL and GASTROC) and in the absence of a normal functioning stretch reflex in QUAD in the two deficit populations, an early response to trunk rearwards rotation was still observed prior to responses at 100 ms in SOL and GASTROC. It consisted of a reduction of PARAS EMG activity at ca 80 ms (see Figs. 3 and 4) and corresponded to the increase in PARAS at the same latency when the trunk pitched forwards first (see Figs. 1 and 2). For the two deficit populations, the duration of the inhibition of PARAS backward activity was extended with respect to normal. The increase in backwards-pitching trunk velocity was slower and the changeover to forward-pitching trunk-acceleration was also delayed in the deficit popula-tions. For both types of rearward translation, nulled ankle flexion obtained with simultaneous support-surface plantarflexion (Fig. 3) and enhanced ankle flexion using dorsiflexion of the support surface (not shown), the pattern of PARAS activity had a similar timing in the normal and vestibular loss subjects, albeit slightly delayed in the vestibular loss subjects with respect to normal response (see Fig. 3). The extent of the delayed timing in PARAS responses was similar in proprioceptive-loss subjects (see Fig. 4). Such timing delays raise the question whether PARAS activity patterns are changed because of changes in the stretch profiles modulating proprioceptive inputs to trunk muscles or because of differences in timing patterns following a sensory deficit.

Translation of the support surface induces in normal subjects faster trunk pitching velocities than observed for rotation (Allum et al., 1993). In the deficit populations the trunk velocities are even faster than normal, thereby inducing greater stretch velocities on the back muscles. In an attempt to keep the head stable in space, increased trapezius (TRAP) EMG activity with respect to normal was recorded in the deficit populations (see Figs. 3 and 4). Interestingly the increased TRAP activity preceded the increased PARAS activity in proprio-ceptive-loss subjects whereas in vestibular-loss subjects the increased TRAP activity fol-lowed that of PARAS. Such results support previous results indicating a delay in the triggering of neck muscle activity following vestibular loss (Allum et al., 1994; Horak et al., 1994).

DISCUSSION

Previous studies have concentrated on the role of lower-leg proprioceptive inputs on triggering balance corrections (Nashner 1976; Nashner et al., 1982; Horak et al., 1994; Inglis et al., 1995). Only recently has the possible participation of trunk muscle and knee joint proprioception in eliciting balance corrections been considered as an alternative triggering mechanism (Allum et al., 1993; Forssberg and Hirschfeld, 1994). Our observations on the initiation of balance corrections support the hypothesis that trunk rotation about the pelvis is the primary signal for triggering automatic activation of muscle responses along many body segments. Because we have already established that at least two discrete muscle timing synergies occur, depending on whether the support surface is rotated or translated (Allum et al., 1993), triggering signals other than trunk rotation are presumably used to aid selection of the appropriate timing synergy. Our recordings of knee angular velocities in the absence of ankle proprioceptive inputs (both mechanically, using nulling servo-techniques, and neuro-sensorially, with proprioceptive-loss patients) indicate that knee flexion and extension is used as an auxiliary trigger signal supporting trunk movements in the timing selection process.

In a multi-linked structure, such as the human body in which a coupling of movement across joints transmitted via the elastic structure of muscles can easily occur, a perturbation at one link will consistently produce a movement at adjoining links. It has been documented that an acceleration of the support surface is mechanically transferred up the body and causes an acceleration to the head some 10 ms later (Allum, 1983). Furthermore, such accelerations were estimated to be supra-threshold for vertical semi-circular canal and otolith afferents (Allum and Pfaltz, 1985). Thus it appears very likely that a displacement at either the foot, shoulders or head will elicit a chain reaction of joint movements throughout the body. Because of the known sensitivities of vestibular and proprioceptive afferent systems (see Wilson and Melvill Jones; 1979; Matthews, 1972) movements at any of the adjacent links to a perturbed link could, at times dependent on the length of the afferent transmission arc, provide an appropriate sensory trigger signal for eliciting automatic balance corrections. Under these circumstances, it is very difficult to reach conclusions about appropriate sensory triggering signals merely from the order in which muscles are activated.

The current set of experiments have demonstrated that the absence of a stretch reflex in soleus muscles at 50 ms, which is the first observed response following support-surface movements, does not alter the pattern of activity in upper-leg and trunk muscles. In fact, the absence of ankle inputs affected only the timing of TA activity. An exception to this finding occurred when knee inputs were also reduced. Then, QUAD timing was delayed but not eliminated. Thus the hypothesis that balance corrections are triggered by ankle inputs and then ascended the body in a distal to proximal order (Nashner, 1976; Horak and Nashner, 1986; Inglis et al., 1995) runs counter-intuitive to the data presented in Figures 1-4. In fact, the several elements of this hypothesis appear to be invalid. As described above, other proprioceptive and vestibular signals which can excite muscles with much shorter reflex arcs than that of the triceps surae muscle could act as trigger signals, for example, knee flexion on translation stretches the QUAD muscle, and trunk movements cause early PARAS activity (see Figures 3 and 4). Second, the observation of a distal to proximal activation order is only valid for muscles on the ventral surface of the legs and trunk in response translation (Horak and Nashner, 1986). If other muscle responses to translation are examined, such as quadriceps and the rectus abdominis, it appears that balance correcting responses in these muscles are activated as early as those in triceps surae (Keshner et al., 1988; Allum et al., 1993), that is the activation order is more simultaneous across many segments. In other words, the triggered sequential activation in triceps surae, hamstrings, and paraspinals may be part of an overall centrally triggered timing pattern ensuring stability of the legs simultaneous with the activation of dorsal leg and trunk muscles.

Apart from the changes induced in the timing patterns in quadriceps and abdominal muscles, the contribution of vestibulo-spinal mechanisms to balance corrections appear to be limited to modulating the amplitude of triggered patterns of muscle activity. Previous reports comparing responses of normal subjects with those of vestibular-loss subjects have emphasized the modulating rather than triggering role of vestibular inputs (Keshner et al., 1987; Allum et al., 1994). It is possible to postulate a triggering role for vestibulo-spinal inputs in the case of the quadriceps and rectus abdominal muscles. Interestingly, the triggering role for quadriceps was most evident for the early stretch reflex in quadriceps. The absence of an early vestibular input appeared to gate stretch reflex excitation in this muscle (see Fig. 3). The signals responsible for vestibulo-spinal modulation of muscle activity could be either semicircular canal afferents or otolith receptors sensitive to pitch angular and linear accelerations, respectively. Both angular and linear accelerations of the head are known to commence some 20 ms after the onset of support-surface movements and therefore both could be involved in the vestibular modulation of muscle activity (Allum et al., 1993). Attempts to independently modulate the amplitudes of linear and angular head accelerations using different combinations of support-surface translation and rotation (see Allum et al., 1994 and this report) have not produced, in any experimental protocol used to date a significant elimination of the normal vestibular modulation of balance corrections as judged from differences between responses of normal and vestibular-loss subjects. Thus, it appears that a combination of semicircular canal and otolith inputs underlies vestibulo-spinal modulation of human balance corrections and not just one type of vestibular signal.

Different combinations of support-surface movement always caused either an early excitation or early inhibition of paraspinal muscles at approximately 80 ms (Figs. 1-4, and Allum et al., 1993). The relationship of trunk muscle activity to trunk motion appears to be independent of motion at other more distal joints. The early excitation was present in Figures 1 and 3 in the absence of knee inputs, vestibular inputs, and proprioceptive inputs from the lower legs. Similarly, the early inhibition was not influenced by vestibular or proprioceptive loss nor by absence of ankle inputs. Thus, the rotation of trunk, stretching or releasing paraspinal muscles, would appear to be an excellent signal for triggering appropriate timing patterns for balance corrections. Such a sensory signal would have several advantages for initiating stabilizing balance movements. First, trunk proprioception is presumably directly related to motion of the heaviest and the most vulnerable parts of the body, the trunk and head, respectively. Secondly, trunk inputs are not directly influenced by a vestibular or proprioceptive loss (almost all deficits in the proprioceptive system occur in the extremities). Moreover, as described in this report, indirect influences of peripheral vestibular and leg proprioceptive loss leading to increased trunk instability are registered by increased stretch to trunk muscles. Finally, the major advantage of trunk inputs as a trigger signal for balance corrections is its applicability to balance control during sitting (Forssberg and Hirschfeld, 1994). Essentially, the same signal can be used during standing and sitting to generate an initial timing template for balance corrections. In this respect, a trunk signal would have the advantage of being utilized from a very early age. Additional trigger signals would then emerge as the infant develops control of stance. Other signals are presumably required by the CNS in order to predict the overall motion of the body, because, as Figures 1-4 illustrate, the initial motion of the trunk following a support-surface movement is opposite to the direction in which the body will eventually fall. Our results indicate that knee proprioceptive inputs provide a second input which would then modify the timing pattern dependent on CNS recognition that body motion was primarily pitching forwards because the knees were significantly flexed requiring control of the legs as two links, as opposed to falling backwards with the knees extended, when the legs could be controlled as one link.

Once the temporal pattern of balance corrections has been triggered, we postulate on the basis of our results that the ensuring modulation of muscle activity generating an adequate set of joint torques to correct for the perturbation is primarily dependent on local-proprioceptive and vestibular inputs whose weightings are pre-determined centrally. The question may be raised whether a default level of muscle activity acts as a reference response onto which sensory modulation is added (see Diener et al., 1988; Beckley et al., 1992). Previous results (Allum et al., 1993) have, however, documented that the level of this default response must be quite low, because most of the response variation as link velocities are changed can be described by concomitant changes in sensory input amplitudes. In addition, the default response, if present, cannot be preprogrammed in amplitude by the CNS (Allum et al., 1995b). One overriding aspect of the sensory modulation documented by our results is the strong modulation of dorsal leg-muscle activity by vestibulo-spinal inputs. This difference in the vestibulo-spinal and proprioceptive modulation of dorsal and ventral leg muscle activity appears to be related to the different stability requirements of falling backwards compared to falling forwards.

SUMMARY

Falling over backwards is potentially more dangerous for standing humans than falling forwards. For backward stability there is a limited posterior base of support and restricted knee extension. Consistent with these dissimilar task-dependent stability requirements, the CNS may well trigger balance corrections based on an unambiguous set of sensory signals, and subsequently shape postural adjustments preventing a fall according to the directional effect of muscle action on body motion. To elucidate these concepts, balance corrections to falling backwards or to falling forwards with different sets of possible somatosensory trigger signals were examined in normal subjects, and those with either bilateral peripheral vestibular loss, or bilateral somatosensory loss (in the lower legs). Backwards falling was induced using a dorsiflexion of the support surface which produces little knee rotation. Forwards falling was elicited by simultaneous rearward translation and dorsiflexion of the support surface causing large ankle and hip joint rotations. An intermediate perturbation was also employed, simultaneous rearward translation and plantar flexion rotation which practically eliminated ankle joint rotation without significantly influencing the profile of knee joint rotation present during rearward translation. The results, expressed by quantifying link velocities and EMG responses induced by these three types of balance perturbations, each presented under eyes-open and eyes-closed conditions, indicated that although every leg muscle may have its distinct set of sensory modulation influences, there are only a discrete number of common triggering modes. The primary trigger signal for most muscle responses appeared to be trunk rotation about the pelvis. This signal is supplemented by knee extension in the development of the appropriate muscle-timing synergy. Vestibulo-spinal modulating influences were strongest on ventral leg muscles, whereas proprioceptive influences, specifically knee inputs, dominated dorsal leg muscle activity. Most of the increased instability of the trunk following vestibular or somatosensory loss was attributable to the absence of appropriate sensory modulation on leg muscles rather than a change in sensory contributions to trunk muscles. Visual and ankle proprioceptive inputs had a minor influence on muscle responses to sudden balance perturbations. We speculate that the major influence on centrally generated patterns of stabilizing muscle activity is the predicted profile of trunk angular rotation and whether or not a knee extension can be employed to recreate upright stance. The centrally generated pattern is then amplitude modulated by vestibular-spinal and proprioceptive contributions.

ACKNOWLEDGMENTS

Support for this work was provided by Swiss National Research Foundation grant 31.30863.91 to J.H.J. Allum. We thank Mrs. E. Clarke for preparing the computer script.

REFERENCES

Allum J.H.J., 1983, Organization of stabilizing reflex responses in tibialis anterior muscles following ankle flexion perturbations of standing man. Brain Res. 264:297-301.

Allum J.H.J., Pfaltz C.R., 1985, Visual and vestibular contributions to pitch sway stabilization in the ankle muscles of normals and patients with bilateral peripheral vestibular deficits. Exp. Brain Res. 58:82-94.

Allum J.H.J., Honegger F., 1992, A postural model of balance correcting movement strategies. J. Vest. Res. 2:323-347.

Allum J.H.J., Honegger F., Schicks H., 1993, Vestibular and proprioceptive modulation of postural synergies in normal subjects. J. Vest. Res. 3:59-85.

Allum J.H.J., Honegger F., Schicks H., 1994, The influence of a bilateral peripheral vestibular deficit on postural synergies. J. Vest. Res. 4:49-70.

Allum J.H.J., Honegger F., Acuña H., 1995a, Differential control of leg and trunk muscle activity by vestibulo-spinal and proprioceptive signals during human balance corrections. Acta Otolaryngol. (Stockh.) 115:124-129.

Allum J.H.J., Honegger F., Huwiler M., 1995b, Prior intention to mimic a balance disorder: Does central set influence normal balance correcting responses. Gait and Posture (in press).

Allum J.H.J., Honegger F., 1995, Interactions between vestibular and proprioceptive signals in triggering and modulating human balance-correcting responses. (in preparation).

Beckley D.J., Bloem B.R., Remler M.P., Ross R.A.C., van Dijk J.G., 1991, Long latency postural responses are funtionally modified by cognitive set. Electroencephal. clin. Neurophysiol. 81:353-358.

Bussel B., Katz R., Pierrot-Deseilligny E., Bergego C., Hayat A., 1980, Vestibular and proprioceptive influences on the postural reactions to a sudden body displacement in man. In spinal and supraspinal mechanisms of voluntary motor control and locomation. Prog. Clin. Neurophys. 8:310-322.

Christakos C.N., Windhorst V., Rissing R., Meyer-Lohmann J., 1987, Frequency response of spinal Renshaw cells activated by stochastic motor axon stimulation. Neuroscience 23:613-623.

Diener H.C., Horak F,B., Nashner L.M., 1988, Influence of stimulus parameters on human postural responses. J. Neurophysiol 59:1888-1905.

Dietz V., Horstmann G.A., Berger W., 1988, Fast head tilt has only a minor effect on quick compensatory reactions during the regulation of stance and gait. Exp. Brain Res. 73:470-476.

Forssberg H., Hirschfeld H., 1994, Postural adjustments in sitting humans following external perturbations: muscle activity and kinematics. Exp. Brain Res. 97:515-527.

Fries W., Dietrich M., Brandt Th., 1993, Otolithic contributions to postural control in man: short latency motor responses in a case of otolithic Tullio phenomenon. Gait and Posture 1:145-153:

Greenwood R., Hopkins A., 1976, Muscle responses during sudden falls in man. J. Physiol. (Lond) 254:507-518.

Horak F.B., Nashner L.M., 1986, Central programming of postural movements: adaptation to altered support-surface configurations. J. Neurophysiol. 55:1369-1381.

Horak F.B., Nashner L.M., Diener H.C., 1990, Postural strategies associated with somatosensory and vestibular loss. Exp. Brain Res. 82:167-177.

Horak F.B., Shupert C.L., Dietz V., Horstmann G., 1994, Vestibular and somatosensory contributions to head and body displacements in stance. Exp. Brain Res. 100:93-106.

Hulliger M., Dürmüller N., Prochazka A., Trend P., 1989, Flexible fusimotor control of muscle spindle feedback during a variety of natural movements. PBR 80:87-101.

Hultborn H., Lindström S., Wigstrőm H., 1979, On the function of recurrent inhibition in the spinal cord of the cat. Exp. Brain Res. 37:399-403.

Iles J.F., Pisini J.V., 1992, Vestibular evoked postural reactions in man and modulation of transmission in spinal reflex pathways. J. Physiol. (London) 455:407-424.

Inglis J.T., Horak F.B., Shupert C.L., Jones-Rycewicz C., 1995, The importance of somatosensory information in triggering and scaling automatic postural responses in humans. Exp. Brain Res. (submitted).

Keshner E.A., Allum J.H.J., Pfaltz C.R., 1987, Postural coactivation and adaptation in the sway stabilizing responses of normals and patients with bilateral peripheral vestibular deficit. Exp. Brain Res. 69:77-92

Keshner E.A., Woollacott M.H., Bebu B., 1988, Neck and trunk muscle responses during postural perturbations in humans. Exp. Brain Res. 71:455-466.

Kuo A.D., 1995, An optimal control model for analyzing human postural balance. IEEE Trans. BME 42:87-101.

Mathews P.B.C., 1972, Mammalian muscle receptors and their central actions. London Edward Arnold.

Nashner L.M., 1976, Adapting reflexes controlling the human posture. Exp. Brain Res. 26:59-72.

Nashner L.M., Black F.O.and Wall C. III, 1982, Adaptation to altered support surface and visual conditions during stance: patients with vestibular deficits. J. Neurosci. 2:536-544.

Pompeiano O., 1989, Relationship of noradrenergic locus coeruleus neurons to vestibulospinal reflexes reflexes. PBR 80:329-343.

Renshaw B., 1941, Influence of discharge of motoneurons upon excitation of neighboring motoneurons. J. Neurophysiol. 4:167-183.

Wilson V.J., Melvill Jones G., 1979, Mammalian vestibular physiology. New York: Plenum Press.

Windhorst V., 1989, Do Renshaw cells tell spinal neurones how to interpret muscle spindle signals? PBR 80:283-294.

10

CHARACTERISTICS OF HEAD AND NECK STABILIZATION IN TWO PLANES OF MOTION

E. A. Keshner,[1] G. Peng,[2] T. Hain,[3] and B. W. Peterson[2]

[1] Sensory Motor Performance Program
Rehabilitation Institute of Chicago
345 East Superior Street
Chicago Illinois 60611
[2] Department of Physiology
Northwestern University Medical School
303 East Chicago Avenue
Chicago Illinois 60611
[3] Department of Neurology and Otolaryngology
Northwestern University
645 N. Michigan Avenue
Chicago Illinois 60611

INTRODUCTION

The head serves as a sensory platform for detection of motion by the visual and vestibular systems and is held in a stereotyped position relative to the environment during natural movements like locomotion (Gibson, 1966; Pozzo et al., 1990; Winter, 1991). This stability of head position helps to maintain the orientation of the head's special sensory receptors relative to the environment, and regulates the attitude of the head on the trunk as part of overall postural control (Nashner, 1971). Vestibular and cervical reflex responses participate in stabilizing the head and neck by exciting neck muscles to compensate for movements of the head and body (Outerbridge and Melvill Jones, 1971; Schor et al., 1988). The vestibulocollic response (VCR) stabilizes the head in relation to the environment. The cervicocollic response (CCR) is responsible for aligning the head with respect to the body. It is initiated by activation of cervical proprioceptors, and its actions complement those of the VCR during movement (Peterson et al., 1985). In cats, if the body is stationary and the head moves, the VCR and CCR sum their actions to stabilize the head. During body rotation, however, the CCR opposes the compensatory movements generated by the VCR (Goldberg and Peterson, 1986). Data of Goldberg and Peterson (1986) suggest that the VCR and CCR, or related neurally mediated responses, dominate head stabilization in the alert cat up to frequencies of 3-4 Hz where biomechanics become more important.

Multisensory Control of Posture, Edited by T. Mergner and F. Hlavačka
Plenum Press, New York, 1995

Even in the study of Goldberg and Peterson (1986), where voluntary movements arguably may have assisted the VCR and CCR in stabilizing the head, compensatory, stabilizing movements amounted to only about half the applied body motions. In fact, the bulk of evidence suggests that neck reflexes offer an insignificant contribution to stabilization of the head. Most of the evidence suggests a primary role of voluntary mechanisms in stabilization of the head with respect to the trunk. Stabilization of the head at high frequencies has been seen to occur through anticipatory presetting of the static and dynamic sensitivity of the postural control system (Jeannerod, 1984; Viviani and Berthoz, 1975). Anticipatory responses to predictable head perturbations and postural perturbations have also been observed (Bouisset and Zattara, 1990). Guitton et al. (1986) employed latency measurements to reveal that triggered reactions or voluntary movements were the primary mechanisms involved in stabilizing the head against random sinusoidal rotations of the trunk at frequencies up to 1 Hz. In studies of passive head angular accelerations in humans, Barnes and Rance (1974, 1975) were unable to identify any reflex compensation in the horizontal and vertical planes during predictable sinusoidal rotations up to 20 Hz with subjects seated in the dark.

We have now performed a series of studies in an attempt to reveal the relative influence of the multiple mechanisms capable of producing stabilization at the head and neck. Our approach has been to control the predictability and frequency range of the stimulus, and the attention of the subject. We have found that several control mechanisms operate simultaneously to produce the final motor output. In the horizontal plane, subjects rely on reflex mechanisms to smooth the transition between control by voluntary mechanisms to that by system mechanics at frequencies greater than 1.5 Hz, avoiding resonant instabilities that would otherwise occur (Keshner and Peterson, 1995). In the vertical plane, reflex mechanisms emerge even at frequencies below 1 Hz where voluntary mechanisms are optimally operating (Keshner et al., 1995). Apparently there are several pathways available for control of the head and neck, and their relative participation is indicative of the multiple roles this motor system plays in natural movement.

STABILIZATION OF THE HEAD IN THE HORIZONTAL PLANE

We first assessed the influence of mental set, visual, and vestibular cues on head stabilization in subjects seated in the dark and rotated about the vertical axis of the head in the horizontal (yaw) plane (Keshner and Peterson, 1988, 1992, 1995). Subjects were seated in a rigid, molded chair that provided support to the whole body. The head was free to rotate in any plane, but measures of angular velocity in the pitch and roll planes proved to be insignificant. A triaxial angular rate sensor and a laser pointer, affixed to a helmet worn by the subjects, were positioned at the rotational axis of the head. Angular velocity of the head with respect to the trunk (neck) and myoelectric activity of two neck muscles (splenius capitis (SPL) and sternocleidomastoid (SCM)) were recorded during rotations with a random sum-of-sines stimulus at frequencies ranging from 0.185 to 4.18 Hz. Gain and phase of head velocity and EMG responses were calculated using a fast fourier transform and analyzed with respect to chair velocity.

Four experimental paradigms were used. Voluntary stabilization (VS) required that the subject keep the head-referenced light spot coincident with a stationary target spot while the chair was rotated. The no vision condition (NV) was performed in the dark while the subject was given the task of stabilizing the head by imagining both the stationary target spot and the head-referenced light signal. During mental arithmetic (MA), a mental calculation task was provided so that the subject's attention was removed from the task of stabilization while rotation in the dark was ongoing. System dynamics during voluntary movement as opposed to head stabilization was also assessed via a visual tracking task (VT).

In this case the target spot was moved, the body remained stationary, and subjects were instructed to follow the moving target with the head-referenced light spot.

Responses to a Perturbation

A gain of one and phase of -180° for the neck with respect to trunk response represents perfect compensation of the head for the motion of the trunk in this paradigm. As seen in Fig. 1, our subjects produced good compensatory gains and phases in the voluntary stabilization conditions at frequencies up to 1 Hz (NV and VS were almost identical except for slightly lower gains in NV). Above 1 Hz, the gains began to drop and a phase lead appeared. At 3 Hz and greater, gains increased beyond unity gain and phases dropped off so that the head moved more than the trunk as would occur in a mechanically resonant response.

A different pattern was observed in the MA condition. Here, gains were very low until about 0.5 Hz, and then a steep increase was observed. Just as the gains began to increase, phases leveled off. Around 1 Hz, the slope of the gains and phases flattened (note arrow in Fig. 1) suggesting a damping of head movement prior to the resonant response at 3 Hz. If the responses in NV were overlaid with those in MA, a crossing over of gains and phases would occur around 1 - 1.5 Hz. We believe that this is the frequency range at which the voluntary mechanisms begin to fail to compensate for the perturbation, and the reflexes

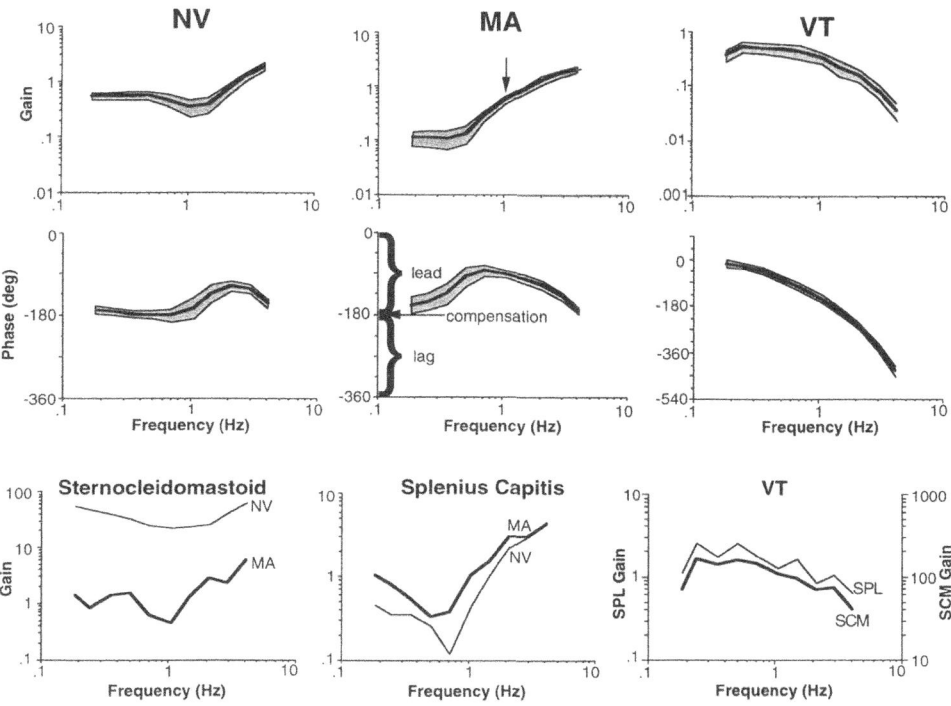

Figure 1. Top 2 Rows: Bode plots of vectorially averaged gains and phases ± 1 SE (shaded area) of the neck with respect to trunk responses of seven subjects during horizontal rotations in the NV, MA, and VT conditions. Downward arrow on MA gain plot is explained in the text. y-axis in MA phase plot is labeled to indicate whether the head is leading, lagging, or fully compensating for movement of the trunk. Bottom Row: Gain plots of muscle EMG responses (in arbitrary units) for one subject during horizontal rotations in the NV, MA, and VT conditions.

emerge to smooth the movement of the head. Beyond this point, even the reflexes become unable to stabilize the head and a mechanical response of the system emerges.

Further evidence for neural control of head compensation at this frequency range is found in the neck muscle EMG responses. Gains of the two neck muscles recorded here were observed to diminish from the initial level of activation up until 1 Hz. Then, the muscles slowly began to increase their activity so that a U-shaped response pattern was consistently observed (Fig. 1). Muscle EMG response phases were more complex and suggested that more than one neural mechanism contributed to the output at the neck. SCM exhibited a steady 0° to 90° response phase (between peak rightward velocity and position) in all three conditions that would be expected of a muscle responding preferentially to vestibular input (via voluntary mechanisms in VS and NV as indicated by much higher gains there). SPL, however, exhibited complex frequency related shifts that we attribute to the interaction of VCR and CCR signals (Keshner et al., 1995).

Voluntary Tracking Responses

When the head tracks a moving visual stimulus, perfect head tracking can be described by a gain of one and a phase of 0°. Our subjects were most successful in matching the head to the visual target at frequencies below 1 Hz (see Fig. 1). Frequencies greater than 1 Hz produced diminished gains and progressive phase lags. Head movement phase lags were echoed in the neck muscle responses. The right-sided muscles stayed in phase with leftward head velocity at low frequencies, but then exhibited steadily increasing lags as frequencies increased.

A DYNAMIC MODEL OF HEAD AND NECK CONTROL

Our experimental results suggest that the reflexes do participate in the head stabilization response, primarily as a damping function to smooth the transition between control by voluntary mechanisms and mechanics. We have now started to develop a dynamic model to ascertain, in a more quantitative way, the neck reflex contributions to stabilizing the head in space. One issue to be considered when determining the functional contribution of reflexes is the comparative anatomy between humans and species with smaller heads. Given that the goal of the VCR is to stabilize the head in space, animals with larger heads may have less need for a VCR for stabilization owing to the greater inertia of the head. One might hypothesize then that conclusions drawn from studies of small-headed animals such as monkeys and cats might be of less relevance to humans. Another related concern is that most known experimental data about head stability are drawn from studies of lower animals. Reflex transfer functions from these animals might not necessarily be appropriate for humans. For example, the VCR transfer function must at some level incorporate the mechanical time constant of the semicircular canals, which is likely to be larger in humans than in cats.

From decerebrate (Berthoz and Anderson, 1971; Ezure and Sasaki, 1978; Bilotto et al., 1982; Dutia and Hunter, 1985) and alert (Goldberg and Peterson, 1986) cat studies, we know that contribution of the VCR and CCR to the maintenance of head stability emerges in the form of EMG activation or neck torque responding to vestibular or neck proprioceptive input. These signals can travel along any of several neuronal pathways that lie between the vestibular nucleus and neck motoneurons (Wilson and Peterson, 1988). Other than the basic existence of the reflexes and pathways, however, very little is known about the details of how the these reflexes function during human head movement. A systems model provides a method of making trial modifications in systems parameters for human subjects, and then

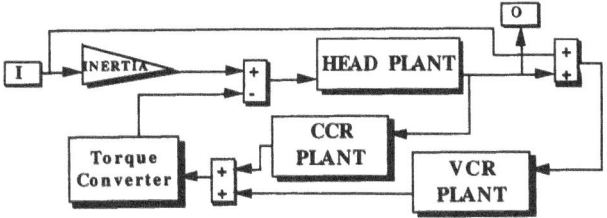

Figure 2. Bode plots predicted by the model displaying gains and phases of neck movements with respect to an input perturbation to the trunk. Data from the horizontal plane paradigm presented in Fig. 1 are superimposed. Vestibular patient data are taken from Guitton et al., 1986.

predicting experimental results. We have chosen to approach this problem by simulating the head mechanics, the VCR, and the CCR, and then determining what reflex contributions would be needed to reproduce experimental data.

The block diagram in Fig. 2 presents the MA condition in the horizontal plane as a lumped-parameter model created in the Matlab/Simulink environment. An input perturbation of trunk movement in space (I) is converted to torque input with respect to the trunk and applied to the head plant which, in turn, supplies input to the neural reflexes. The VCR and CCR controllers provide negative feedback torque to the head plant. Negative feedback is typically used in control system design to minimize steady-state error and effects of noise disturbances, so it is appropriate to model the reflexes as negative feedback controllers or actuators. The output under consideration is head movement with respect to the trunk (O). The input-output relationship of this closed loop feedback system is determined by perturbing the system around the zero state operating point in which the torques and angular velocities are zero. The frequency response is generated and presented in terms of gain and phase in relation to the trunk movement in space. Perfect reflex compensation or stabilization produces a gain of 0 dB and a 180° phase lag.

The passive yaw plane head plant is modeled as a simple 2nd order mechanical system. The passive inertia, viscosity, and stiffness parameters, taken from biomechanical measurements, produce a frequency response of a mildly underdamped system with a 9 dB resonance around 3 Hz. Above that frequency, the passive head is relatively stable due to the inertial and viscoelastic properties of the head-neck system. The reflex plants are modeled with transfer functions determined from animal studies (Ezure and Sasaki 1978; Goldberg and Fernandez, 1971), with adjustments made for human data where available. In these studies, EMG responses to head or neck movements closely approximated open loop VCR and CCR frequency responses. The frequency response patterns, produced from adding the reflex feedback controllers, show VCR time constants influencing low to mid-frequency (0.01-3 Hz) dynamics and CCR time constants affecting high frequencies (1-10 Hz). At low frequencies (0.01-1 Hz) the VCR pushes the gain and phase towards compensation but never reaches complete head stabilization. At mid-frequencies (1-3 Hz) VCR feedback acts to stabilize the head in space as an adjunct to the head mechanics. Specifically, it is involved in decreasing the amplitude of the resonant peak, bringing the head closer to being critically damped, which reduces the amount of head wobble. At the same time, the CCR feedback effectively damps the high frequency response characteristics, by bringing the gain and phase towards an overdamped system. The gain is pushed down and the phase is pulled up from 180° phase lag, counteracting the VCR. In this sense, the CCR is counterproductive (cf.: Goldberg and Peterson, 1986).

These patterns suggest that the CCR is mainly useful during high-energy motion when damping is needed to prevent injury, and the VCR is mainly involved in fine tuning the natural effects of the passive head mechanics at high frequencies. We believe that voluntary control adjusts for the difference in compensation at low frequencies.

Model predictions can be made to fit closely with experimental behavior for the MA condition in the horizontal plane, as shown by the solid lines in Fig. 3. The fit was made by adjusting the gain factors for the VCR and CCR transfer functions. Loop gain analysis indicates that head movements in this condition require a VCR that is approximately 14 times stronger than the CCR. The dotted lines in the Bode plots display model behavior with VCR feedback taken out (CCR feedback still intact). Comparing to gain data from a single vestibular deficient subject (Guitton et al., 1986), the model does fit the data down to approximately 0.2 Hz. This exercise demonstrates the versatility of this model and the possibilities for future studies.

Limitations and Future Development of the Model

The head and neck is a complex system and this model is but a first step towards representing its behavior. An obvious flaw is that the head plant is modeled as a simple one joint system, even though we know that more than one joint moves during horizontal rotations (Thomson et al., 1994). The importance of multiple joint movements in the horizontal plane is yet to be assessed, and multiple joint kinematics may eventually be needed to strengthen this model. Our model also does not incorporate known nonlinearities (e.g.,

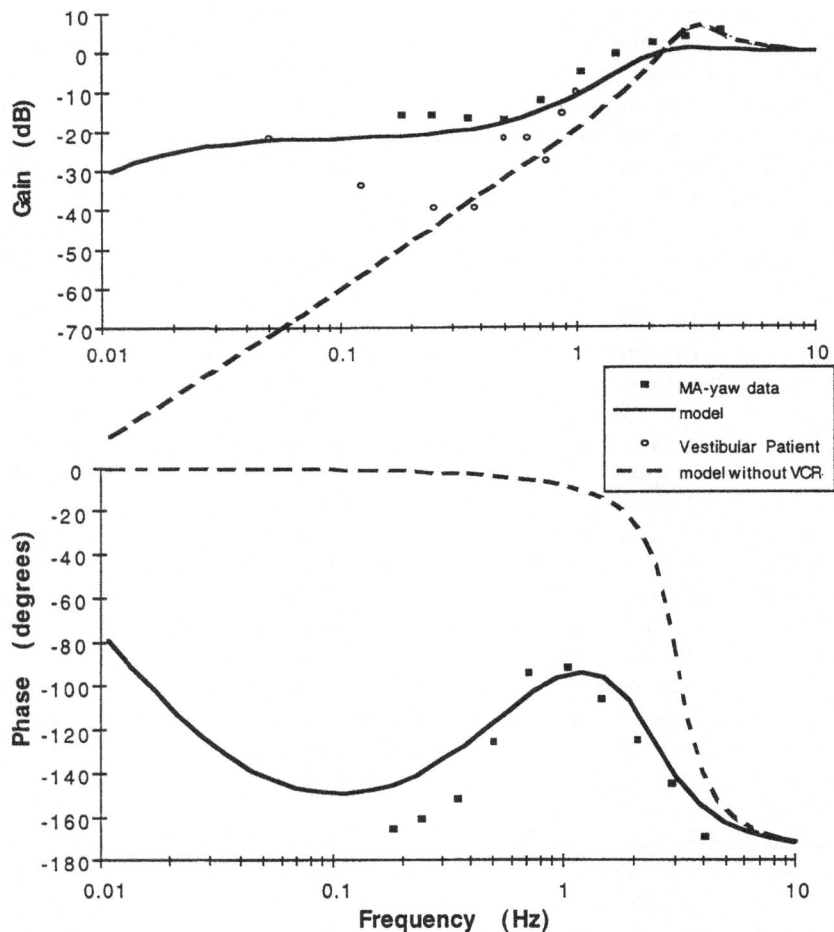

Figure 3. Block diagram of a lumped-parameter model of the head and neck system in the horizontal plane (modified from Schor et al., 1988 and Goldberg and Peterson, 1986).

stiffness for the passive head plant). We do plan to add these nonlinearities, to the extent that good human data are available.

In this first generation of the model, muscle torque is grossly approximated from EMG activation with a first order low-pass system representing muscle contraction dynamics. Clearly, only relative contributions of reflex activity can be modeled at this point. We have chosen to "lump" the reflex transfer functions although they represent any number of vestibular-to-neck pathways, either directly from the vestibular nucleus or through more complex paths which include the cerebellum. In the future, the lumped elements will be broken down as more information becomes available. Our model also does not include any voluntary inputs. We know from our experimental data that voluntary input contributes significantly to normal head stabilization. Both the MA data and model predictions indicate that at low frequencies, where voluntary movements exist, the involuntary behavior generates movements with gains of only 0.1. Voluntary control of neck stiffness via cocontraction (as, for example, when using a telescope) is also likely to be an important input to the system. We do plan to explore these issues within the model, and to compare model predictions with future experimental data.

A BIOMECHANICAL MODEL OF THE HEAD AND NECK

Our colleagues, Dr. Scott Delp and Dr. Siping Li, have begun to develop a detailed biomechanical model of the human head and neck musculoskeletal system to investigate the contributions of the neck muscles to head stabilization. This model is implemented in SIMM (Software for Interactive Musculoskeletal Modeling), which allows one to create anatomically-based models of many different musculoskeletal structures (Delp et al., 1990, Delp and Loan, 1995). The model is based on the bony anatomy including the skull, cervical, thoracic, and lumbar spine, the ribcage, clavicle, and scapula. The geometry of the bones was obtained by digitizing the bones of a skeleton and then scaling the bones to represent the fiftieth percentile of the male population. Eleven pairs of neck muscles representing the posterior, anterior-transverse, posterior-transverse, and lateral muscles have been built into the model. Attachment sites of the muscles were defined based on data in the literature (Seireg and Arvikar, 1989) and the landmarks of the computer representations of the bones. The current model consists of two joints, each of which allows for pitch, yaw, and roll motions. The ranges of motions of three axial motions representing flexion-extension, axial rotation, and lateral bending were defined according to data reported from in vivo and in vitro experiments (White and Panjabi, 1990).

This biomechanical model computes the lengths and moment arms of each of the modeled neck muscles over a range of head positions. Given the force-generating parameters of the muscles, the model calculates the joint torque contributed by each muscle as a function of head-neck position. The preliminary model simulation indicates that the major dorsal muscles provide large moment arms in both pitch and yaw movements, but that all of the modeled muscles have greater moment arms for the pitch axis than for the yaw axis.

STABILIZATION OF THE HEAD IN THE VERTICAL PLANE

Our dynamic model predicts that the VCR is an active component of horizontal head stabilization, particularly at frequencies above 1 Hz, where experimental data suggest that voluntary mechanisms begin to drop out. We have also examined the actions of the VCR and CCR in the vertical plane, anticipating that responses of stabilizing mechanisms would be more robust in a plane where head mechanics and muscle receptors must compensate for the effects of gravity as well as for the imposed angular velocities. Vertical plane movements have been described as more functionally important because of the gravitational demands in this plane of motion (Allum and Pfaltz, 1985; Keshner et al., 1987; Nashner, 1977; Pozzo et al., 1990; Winter, 1991).

Employing the same experimental paradigms (VS, NV, MA, and VT) used in the horizontal plane, we examined the mechanisms responsible for stabilization of the head in the vertical plane. In this experiment, subjects sat in the rigid, molded chair and were rotated about the interaural axis of the head. The triaxial angular rate sensor and laser pointer were fixed to the helmet at the level of the temporal lobe. Angular velocity of the head with respect to the trunk (neck) and myoelectric activity of the semispinalis capitis (SEMI) and sterno-cleidomastoid (SCM) muscles were recorded during rotations with a random sum-of-sines stimulus at frequencies ranging from 0.35 - 3.05 Hz. Gain and phase of head velocity and EMG responses were calculated using a fast fourier transform and analyzed with respect to chair (trunk) velocity.

Responses to Perturbation

The primary difference in the response characteristics of the vertical plane is that at no time in any of the three instructional sets were patterns characteristic of resonance (as indicated by response gains greater than one and steeply descending phases) observed in the average group bode plots. In fact, comparison of these plots to the bode plots in the horizontal paradigm suggests that the dynamic characteristics have been shifted toward a higher frequency. Gain drops and phase shifts observed between 1-2 Hz in the horizontal plane do not begin until 2-3 Hz in the vertical plane, and the resonant gain peak seen with horizontal rotations does not fall within the frequency range tested here.

Compensation in the voluntary stabilizing paradigms (VS and NV) was best at lower frequencies as demonstrated by phases of -180° and gains around 0.7 (see Fig. 4). Although the vector averages of VS and NV were very similar, increased intersubject scatter in NV as compared to VS implied that some subjects were more dependent on visual inputs to produce voluntary head stabilization. Below 2.15 Hz, gains dropped gradually as a function of frequency. Above 2.15 Hz gains dropped more steeply and the compensatory phases observed below 1.45 Hz were lost. Thus subjects were able to maintain reasonable head stabilization up until 1.45-2.15 Hz. Anticompensatory phases or low gains seen above 2.15 Hz imply poor stabilization of the head at the high frequency end.

Gains in MA fell below 0.5 and exhibited more intersubject variability at frequencies below 1 Hz than in VS and NV. Gains and phases varied widely at 0.35 Hz, but the presence of phases around -180° for all but two of the subjects indicated the presence of compensatory activity even in the low frequency range. Just as in the horizontal rotation paradigm, response gains during MA increased and phases clustered more closely about -180° as frequency increased up to 2 Hz, implying improved head stabilization. Above 2 Hz, gains were maintained but phases became scattered. Phases in most subjects remained close to a compensatory -180°, however, in contrast to the lags that appeared at 2.15 Hz when subjects attempted to voluntarily stabilize the head in the VS and NV conditions.

Muscle EMG response gains continued to exhibit the U-shaped pattern of response observed in the horizontal plane, from which we infer increasing neural participation at frequencies above 1 Hz (Fig. 4). SEMI EMG activity was slightly lower in MA than in the other conditions at frequencies below 1 Hz, but above 2 Hz no differences were observed. Both muscles' EMG responses began with phases close to trunk angular position. By 1.5 Hz both muscles advanced to a phase related to velocity. The phase advance with increasing frequency resembles that seen in the VCR of decerebrate (Baker et al., 1985; Bilotto et al., 1982) and alert (Goldberg and Peterson, 1986; Keshner et al., 1992) cats. The larger EMG

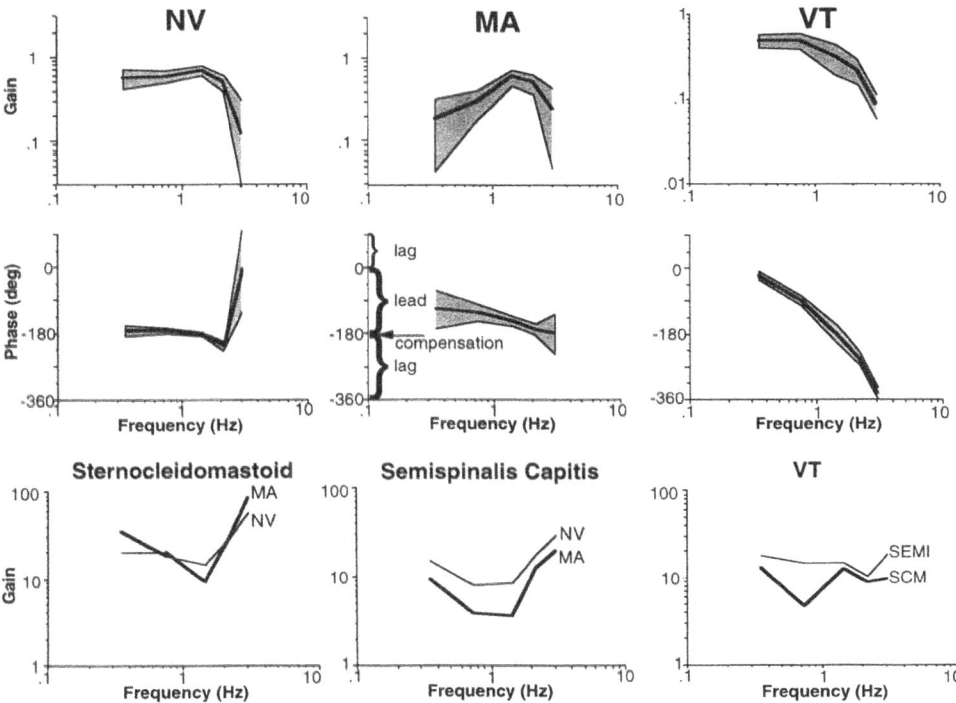

Figure 4. Top 2 Rows: Bode plots of vectorially averaged gains and phases ± 1 SE (shaded area) of the neck with respect to trunk responses of eight subjects during vertical rotations in the NV, MA, and VT conditions. y-axis in MA phase plot is labeled to indicate where the head is relative to the trunk. Bottom Row: Gain plots of muscle EMG responses (in arbitrary units) for one subject during vertical rotations in the NV, MA, and VT conditions.

response gains and advancing phases at higher frequencies suggest continued neuromuscular participation in the compensatory action of the head and neck.

Voluntary Tracking Responses

Gain and phase responses for VT indicate that, just as in the horizontal plane, voluntary mechanisms failed to track the vertically oriented stimulus as frequency increased (Fig. 4). At the lowest frequency, the head was moving in phase with the target as indicated by phases near 0°. Gains around 0.5 and below indicate, however, that subjects did not fully match the target velocity and thus tracked less accurately in the vertical than in the horizontal plane. As frequency increased, response gains continued to decrease and intersubject variability to increase. Phase lags between the projected spot and head velocity increased with frequency so that the response was 180° out of phase at 1 Hz. Despite the decrements in gain and increasing phase lags above 1 Hz, continued efforts to track the stimulus can be inferred at the highest frequencies because the EMG response gains exhibited continued modulation of the muscles (Fig. 4). EMG response phases presented a muscle activation pattern unique to the VT condition. At 1.5 Hz, SCM produced an almost 180° phase advance. SEMI, however, failed to phase advance at high frequencies, producing activation of the extensor muscle in phase with the flexor muscle. Some subjects demonstrated this tactic of phase shifting only one muscle, while others kept both muscles in phase with each other across the frequency range. Concurrent activation of the flexor and extensor muscles is indicative of a change in strategy for the tracking task from one of reciprocal activation to one of coactivation. Coactivation would change the mechanical properties of the head-neck system by increasing the stiffness and resonant frequency, thereby explaining the greatly diminished head velocities observed through the decreased gains. A change in control strategy could represent a centrally generated phase lag in SEMI output that would act to partially offset the lagging behavior of the head at frequencies above 1 Hz.

CONCLUSION

In response to rotational perturbations of the trunk, differences between the two planes of motion emerged at the lowest frequency (0.35 Hz) and at frequencies greater than 2 Hz. At the lowest frequency, differences were due to the higher gains, and thus improved stabilization, observed during vertical rotations. Our dynamic model supports the data in suggesting that the activity in MA reflects activation of the VCR and CCR. The rising gains and advancing phases of EMG output, which almost certainly arise from reflex dynamics, would allow neural mechanisms to maintain control of the head. Enhanced reflex contributions in the pitch plane significantly smoothed the performance of the head stabilizing system. It is probable that the inertial demands made by gravity and the additional inputs from the otoliths contributed to increasing the functional bandwidth of the neck reflexes in this plane. Significant differences above 2.0 Hz reflect the fact that the rising gains and decreasing phases associated with the resonant response appeared only during horizontal rotations. Differences in the head-neck biomechanical plant for pitch and yaw are likely to play a major role in the differences in high frequency responses. The modeling effort described above is intended to lead to quantitative dynamic models that will explain these differences. Our data suggest that more than one mechanism contributed to the control of head stabilization, and that dominance by either the neural or biomechanical parameters of the system depended upon the frequency content and orientation of a random external perturbation.

ACKNOWLEDGMENTS

This work was supported by grants NS22490 and DC01125.

REFERENCES

Allum, J.H.J., and Pfaltz, C.R., 1985, Visual and vestibular contributions to pitch sway stabilization in the ankle muscles of normals and patients with bilateral peripheral vestibular deficits, *Exp. Brain Res.* 58:82-94.

Baker, J., Goldberg, J., and Peterson, B., 1985, Spatial and temporal response properties of the vestibulocollic reflex in decerebrate cats, *J. Neurophysiol.* 54:735-756.

Barnes, G.R., and Rance, B.H., 1974, Transmission of angular acceleration to the head in the seated human subject, *Aerospace Med.* 45:4121-4126.

Barnes, G.R., and Rance, B.H., 1975, Head movement induced by angular oscillation of the body in the pitch and roll axes, *Aviation Space Environ. Med.*, 46:987-993.

Berthoz, A., and Anderson, J.H., 1971, Frequency analysis of vestibular influence on extensor motoneurons. II. Relationship between neck and forelimb extensors, *Brain Res.* 34:376-380.

Bilotto, G., Goldberg, J., Peterson, B.W., and Wilson, V.J., 1982, Dynamic properties of vestibular reflexes in the decerebrate cat, *Exp. Brain Res.* 47:343-352.

Bouisset, S., and Zattara, M., 1990, Segmental movement as a perturbation to balance? Facts and concepts. In: Winters, J. M., and Woo, S. L-Y. (eds.), Multiple muscle systems: Biomechanics and movement organization. New York, NY: Springer-Verlag, pp 498-506.

Delp, S.L., Loan, J.P., Hoy, M.G., Zajac, F.E., Topp, E.L., and Rosen, J.M., 1990, An interactive graphics-based model of the lower extremity to study orthopaedic surgical procedures, *IEEE Trans. on Biomed. Eng.* 37:757-767.

Delp, S.L., and Loan, JP., 1995, A software system to develop and analyze models of musculoskeletal structures, *Computers in Biol. and Med.* (in press).

Dutia, M.B., and Hunter, M.J., 1985, The sagittal vestibulocollic reflex and its interaction with neck proprioceptive afferents in the decerebrate cat, *J. Physiol.* 359:17-29.

Ezure, K., and Sasaki, S., 1978, Frequency-response analysis of vestibular-induced neck reflex in cat. I. Characteristics of neural transmission from horizontal semicircular canal to neck motoneurons, *J. Neurophysiol.* 41:445-458.

Gibson, J.J., 1966, The senses considered as perceptual systems. Boston, MA: Houghton Mifflin.

Goldberg, J., and Peterson, B.W., 1986, Reflex and mechanical contributions to head stabilization in alert cats, *J. Neurophysiol.* 56:857-875.

Goldberg, J.M., and Fernandez, C., 1971, Physiology of peripheral neurons innervating semicircular canals of the Squirrel Monkey. III. Variations among units in their discharge properties, *J. Neurophysiol.* 34:676-684.

Guitton, D., Kearney, R.E., Wereley, N., and Peterson, B.W., 1986, Visual, vestibular and voluntary contributions to human head stabilization, *Exp. Brain Res.* 64:59-69.

Jeannerod, M., 1984, The contribution of open-loop and closed-loop control modes in prehension movements. In: Kornblum, S., and Requin, J., (eds.), Preparatory states and processes. New Jersey: Erlbaum Assoc., pp 323-338.

Keshner, E.A., and Peterson, B.W., 1988, Motor control strategies underlying head stabilization and voluntary head movements in humans and cats. In: Pompeiano, O., and Allum, J.H.J., (eds.), Vestibulospinal control of posture and movement. Progress in Brain Research. Amsterdam: Elsevier, pp 329-339.

Keshner, E.A., and Peterson, B.W., 1992, Neural and mechanical contributions to voluntary control of the head and neck. In: Berthoz, A., Graf, W., and Vidal, P.P., (eds.), The Head-Neck Sensory-Motor System. Oxford University Press, New York, pp 381-386.

Keshner, E.A., and Peterson, B.W., 1995, Mechanisms controlling human head stability: I. Head-neck dynamics during random rotations in the vertical plane, *J. Neurophysiol.* (in press).

Keshner, E.A., Allum, J.H.J., and Pfaltz, C.R., 1987, Postural coactivation and adaptation in the sway stabilizing responses of normals and patients with bilateral peripheral vestibular deficit, *Exp. Brain Res.*, 69:66-72.

Keshner, E.A., Baker, J.F., Banovetz, J. and Peterson, B.W., 1992, Patterns of neck muscle activation in cats during reflex and voluntary head movements, *Exp. Brain Res.*, 88: 361-374.

Keshner, E.A., Cromwell, R., and Peterson, B.W., 1995. Mechanisms controlling human head stability: II. Head-neck dynamics during random rotations in the horizontal plane, *J. Neurophysiol.* (in press).

Nashner, L.M., 1971, A model describing vestibular detection of body sway motion, *Acta Otolarygol. (Stockh.).* 72:429-436.

Nashner, L.M., 1977, Fixed patterns of rapid postural responses among leg muscles during stance, *Exp. Brain Res.* 30:13-24.

Outerbridge, J.S., and Melvill Jones, G., 1971, Reflex vestibular control of head movements in man, *Aerospace Med.* 42:935-940.

Peterson, B.W., Goldberg, J., Bilotto, G., and Fuller, J.H., 1985, The cervicocollic reflex: Its dynamic properties and interaction with vestibular reflexes, *J. Neurophysiol.* 54:90-109.

Pozzo, T., Berthoz, A., and Lefort, L., 1990, Head stabilization during various locomotor tasks in humans. I. Normal subjects, *Exp. Brain Res.* 82:97-106.

Schor, R.H., Kearney, R.E., and Dieringer, N., 1988, Reflex stabilization of the head. In: Peterson, B. W., and Richmond, F. J. (eds.), Control of Head Movement. New York: Oxford University Press, pp. 141-166.

Seireg, A., and Arvikar, R., 1989, Biomechanical Analysis of the Musculoskeletal Structure for Medicine and Sports. New York: Hemisphere Publishing Corp.

Thomson, D.B., Loeb, G.E., and Richmond, F.J.R., 1994, Effect of neck posture on the activation of feline neck muscles during voluntary head turns, *J. Neurophysiol.* 72:2004-2014.

Viviani, P., and Berthoz, A., 1985, Dynamics of the head-neck system in response to small perturbations: Analysis and modelling in the frequency domain, *Biol. Cybern.* 19:19-37.

White, A.A., and Panjabi, M.M., 1990, Clinical Biomechanics of the Spine. Philadelphia: J.B. Lippincott Co.

Wilson, V.J., and Peterson, B.W., 1988, Vestibular and reticular projections to the neck. In: B.W. Peterson, B.W., and Richmond, F.J. (eds.), Control of Head Movement. New York: Oxford University Press, pp. 129-140.

Winter, D.A., 1991, The Biomechanics and Motor Control of Human Gait: Normal, Elderly and Pathological (2nd Ed). Waterloo: University of Waterloo Press.

11

PRINCIPAL COMPONENT ANALYSIS OF AXIAL SYNERGIES DURING UPPER TRUNK FORWARD BENDING IN HUMAN

A. Alexandrov,[1] A. Frolov,[1] and J. Massion[2]

[1] Institute of Higher Nervous Activity and Neurophysiology
Russian Academy of Science
5A, Butlerov
Moscow, Russia, 117865
[2] Laboratory of Neurobiology and Movement, CNRS
31, chemin Joseph Aiguier
13402 Marseille Cedex 20, France

INTRODUCTION

Upper trunk bending in humans is always accompanied by lower segment movements in the opposite direction, as first described qualitatively by Babinski (1899). He called this phenomenon "axial synergy" and suggested that it may serve to maintain equilibrium.

Experiments carried out during the cosmic flight have shown, however, that even under microgravity conditions, when it is not necessary to maintain equilibrium, the axial synergy during upper trunk bending is preserved (Massion et al., 1993). This fact suggests that the observed kinematic pattern is not solely related to the permanent control of the centre of gravity (CG) sagittal displacement. This pattern may be learned under normal gravity conditions and then automatically reproduced under microgravity conditions.

If we take the human body to be a multijoint chain of rigid segments, then during upper trunk bending in the sagittal plane, a covariation of the 4 joint angles - those of the ankle, knee, hip and neck - can be observed. It is reasonable to assume that, during forward bending, the subject simultaneously controls two behaviourally significant kinematic parameters: (a) the angle of the trunk with respect to the vertical axis in order to perform the behavioural task, and (b) the sagittal shift of the CG (stabilizing it within the support area) in order to maintain equilibrium. This control therefore imposes only 2 kinematic constraints on the 4 otherwise independent joint angles. The bending task is therefore kinematically redundant. If the observed angle changes turn out to be reduced to the temporal variation of only one independent variable without loss of information, this will suggest that the movement regulation is based on the central voluntary control of this variable and on some intrinsic kinematic constraints which can be automatically reproduced during the movement. Moreover, if the kinematic constraints turn out to be stable and reproducible in various trials

Multisensory Control of Posture, Edited by T. Mergner and F. Hlavačka
Plenum Press, New York, 1995

and/or in various subjects and to be preserved even when the duration or amplitude of the movement is voluntarily changed by the subject, this will confirm the hypothesis that the movement control is based on some well-learned stereotyped kinematic synergies.

In order to examine whether the trunk bending movement reveals the additional kinematic constraints reducing the number of independent variables, and if so, in order to quantify the intertrial and intersubject stability of the observed synergies under various experimental conditions (movement duration and amplitude), we analyzed the joint angle kinematics using principal components (PC) method.

MATERIALS AND METHODS

Experiments were performed on 5 healthy subjects (3 males and 2 females).

Experimental Paradigm and Setup

The subjects were instructed to stand quietly with their hands clasped behind their back and their eyes open. Immediately after a sound signal, they were asked to make a forward trunk inclination in the sagittal plane and then to hold this final position. The movement was performed either as fast as possible, or as slowly as possible, for about 2 s. The amplitude of the trunk bending had to be about 45°. One subject was asked to repeat the fast and slow series with a trunk inclination of about half the previous amplitude. The ELITE system was used to make kinematic recordings.

Principal Component Analysis of the Data

We carried out PC analysis to examine the linear covariation in time between 4 joint angles (ankle, knee, hip and neck). Usually this method is applied to the statistical analysis of sampled data (Van Egeren, 1973; Lacquaniti and Maioli, 1994) rather than temporal trajectories. In our approach, which was similar to the PC analysis of gait [Mah et al., 1994], one temporal frame in the data acquisition was taken as a sampling point.

The PC analysis gives the observed temporal variation of the joint angles as the sum of minimum number of statistically independent linear compounds (a sum of PCs), each accounting for a progressively smaller portion of the total trial variance of all the joint angles. The n-th PC time course in the trial is:

$$PC_n(t) = \sum_{m=1,...,4} v_{nm}(\phi_{m\,av}), \qquad n = 1,...,4 \tag{1}$$

where v_{nm} is the weight of the variance of the m-th joint angle $f_m(t)$ around its average value $f_{m\,av}$ in the n-th PC. The set of signed ratios v_{nm} (m=1,...,4) are the elements of eigenvector v_n corresponding to the n-th eigenvalue of the covariance matrix C (the matrix elements c_{mn} are the time averaged products of variations of the m-th and n-th joint angles around their mean trial values and the total trial variance is the trace of C). Each set of ratios is normalized:

$$\sum_{m=1,...,4} v^2_{nm} = 1, \qquad n = 1,...,4 \tag{2}$$

If all the PCs are calculated according to (1), then the whole reconstruction of the raw angle trajectories can be performed by the equation:

$$\phi_n(t) = \phi_{n\,av} + \sum_{m=1,\ldots,4} v_{mn}PC_m(t), \qquad n = 1,\ldots,4$$

(3)

The number of PCs equals the number of degrees of freedom (4 joint angles in our case). But if some intrinsic kinematic constraints are involved in the movement, then the number of PCs necessary for perfect kinematic reconstruction by (3) will be reduced.

4 normalized sets of signed ratios v_{nm} and 4 corresponding time courses $PC_n(t)$ $(n=1,\ldots,4)$ were calculated for every trial according to (1). The average value and the standard deviation (SD) of each ratio were then calculated in each trial series. The SDs were used to estimate the intertrial stability of the sets of ratios in each of the 4 PCs.

If the set of ratios in any PC varies greatly from trial to trial, then the averaging procedure distorts the condition of normalization (2) for the set of mean ratios. The value of this distortion is the additional estimation of PC stability.

RESULTS

Tab. 1 shows the ratios calculated of PC1 and PC2 and the percentage of the total variance explained by each PC.

In every trial, in all 5 subjects, the raw data reveals strong linear dependence between the angle changes, so that the variation of 4 angles was coded as the variation of only one time course of the first principal component PC1(t), which accounted for more than 99% of the total changes in all the joints. One can see that the ratios in PC1 have strikingly smaller

Table 1. PC1 and PC2: angle ratios and % of explained total variance for fast and slow bending

Subj no.			No. trials	Fast Ankle	Knee	Hip	Neck	% of total variance	No. trials	Slow Ankle	Knee	Hip	Neck	% of total variance
				Ratios x 100						Ratios x 100				
1	PC1	mean	n=9	-20.7	-7.3	97.2	-7.8	99.68	n=8	-18.7	-9.3	96.7	-14.5	99.88
		SD		0.9	1.3	0.2	1.7	0.08		1.6	0.7	0.4	2.3	0.06
	PC2	mean		42.9	72.9	17.6	40.6	0.23		33.9	26.3	12.2	27.4	0.09
		SD		12.4	12.9	2.6	23.8	0.07		48.0	20.0	7.5	67.1	0.07
2	PC1	mean	n=10	-20.6	-5.5	95.6	-18.6	99.40	n=10	-19.9	-10.3	95.8	-17.3	99.65
		SD		1.5	1.8	1.0	6.9	0.22		1.7	1.4	1.1	3.9	0.17
	PC2	mean		23.3	79.8	18.1	45.3	0.49		18.5	38.1	22.1	76.7	0.31
		SD		8.1	12.1	6.2	21.4	0.18		15.0	25.8	4.0	30.6	0.16
3	PC1	mean	n=9	-13.2	-11.5	89.4	-40.9	99.53	n=10	-11.9	-14.4	93.9	-27.9	99.83
		SD		1.6	2.4	1.3	3.6	0.11		1.0	1.8	2.4	7.5	0.08
	PC2	mean		34.9	65.9	36.3	49.1	0.39		-12.2	-0.8	25.4	90.9	0.14
		SD		7.0	17.9	5.0	17.6	0.09		20.1	19.5	9.0	8.2	0.08
4	PC1	mean	n=7	-19.7	-13.3	88.1	-40.8	99.54	n=6	-19.0	-14.3	92.6	-28.6	99.61
		SD		1.9	1.3	1.4	2.4	0.18		1.7	3.0	1.5	4.8	0.24
	PC2	mean		52.4	68.2	31.5	20.4	0.32		36.2	30.2	34.2	73.5	0.33
		SD		14.6	14.6	9.2	26.0	0.10		16.4	23.3	4.8	19.0	0.24
5	PC1	mean	n=5	-22.1	-11.9	95.7	-14.1	99.73	n=6	-19.2	-14.2	97.0	1.4	99.78
		SD		0.7	0.9	0.5	2.7	0.11		1.3	1.3	0.4	3.0	0.09
	PC2	mean		54.4	19.0	24.5	64.8	0.18		22.1	0.6	6.8	-4.3	0.16
		SD		20.4	24.1	2.8	29.6	0.11		21.5	22.2	4.5	92.1	0.11

Table 2. Bending with various duration and amplitude

		No. trials	ratios of PC1 x 100				% variance expl. by PC1
			ankle	knee	hip	neck	
fast	high amplitude	9	-24.2	-8.5	92.8	-27.1	99.79
	low amplitude	10	-25.6	-8.2	91.4	-30.5	99.84
slow	high amplitude	10	-23.5	-8.0	94.3	-22.0	99.97
	low amplitude	8	-26.8	-9.8	93.1	-22.8	99.92

SDs than those of PC2 (the results for PC3 and PC4 are not shown because they played a negligible role in kinematic reconstruction according to (3)).

The positive sign of the ratio in the PC means flexion of the corresponding joint. As can be seen from Tab. 1, the pattern of the movement described by PC1 was the same in all the subjects: simultaneous ankle, knee and neck extension and hip flexion.

On the whole, the ratios in PC1 reveal relative intertrial stability (small SD). In the ankle, knee and hip each subject consistently produced approximately the same ratios in both the fast and slow movements. In all the series, the largest SDs were observed for the ratio in PC1 corresponding to the neck variance. In subjects 1,3,4, and 5 the neck ratio in PC1 also differed greatly between the fast and slow series.

Subject 1 was asked to perform the forward bending with high and low amplitudes as fast as possible and then slowly. The trials in each of 4 series (fast with high and low amplitudes and slow with high and low amplitudes) were then pooled together and the ratios and 4 time courses $PC_n(t)$ were calculated separately for each of 4 averaged trials. Tab. 2 gives the ratios of the PC1 and the percentage of the total angle variance explained by this component.

It can be seen that PC1 accounts for more than 99% of the total angle variance under all the experimental conditions. As regards the ankle, knee and hip angle variations, the ratios in PC1 did not differ greatly between the series. Relatively large changes occurred between the neck ratios for the neck under fast versus slow conditions. Fig.1 gives the superimposed time courses PC1(t) in each of 4 averaged trials.

It can be seen from Fig. 1 that the voluntary regulation of the desired movement duration and amplitude was performed by respective regulating of the time course of the

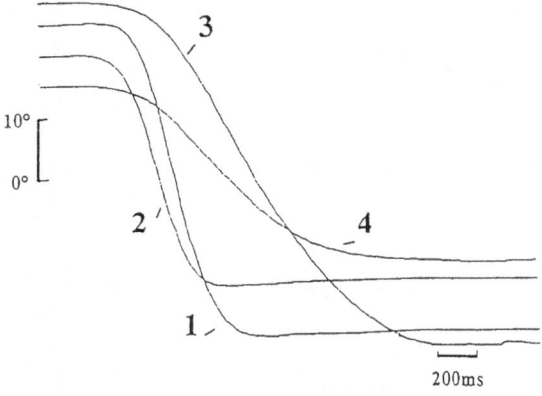

10°

0°

2

3

4

1

200ms

Figure 1. Superimposed time courses of PC1 in the averaged trials of subject 1. The pooling was performed separately under 4 experimental movement performance conditions: 1 - fast with high amplitude, 2 - fast with low amplitude, 3 -slow with high amplitude, 4 - slow with low amplitude. The respective PC1 ratios are shown in Tab. 2.

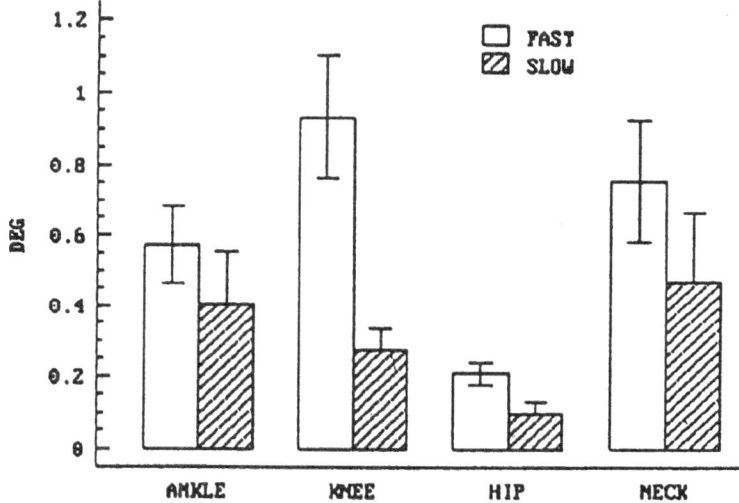

Figure 2. The mean SDs between original data and data obtained by kinematic reconstruction based only on PC1 in the fast and slow series shown in Tab. 1 in the case of subject 1.

PC1 scale factor (it had a higher excursion in the high-amplitude series and a higher velocity in the fast series), while the set of ratios in PC1 showed no significant changes (Tab. 2).

The intertrial and intersubject stability of the neck ratio were lower probably because the head position has relatively smaller influence on the displacement of the body CG.

The present results suggest that the desired voluntary movement was well represented by PC1. The following data show that the high-order PCs (those of the order higher than 1) exhibit the properties of random noise and that they seem to be mainly related to dynamic perturbations of the desired movement:

    a. As mentioned above, the ratios in the high-order PCs showed large intertrial SDs.

    b. The random properties of high-order PCs are also demonstrated by the fact that the averaging procedure (pooling all the trials in the series) always increased the angle variance portion explained by PC1 in comparison with the individual trials (the results are not shown).

    c. It can be seen from Tab. 1 that the percentage of the variance explained by PC1 was always higher in the case of slow movements than in that of fast movements. The contribution of PC1 to the angle variance was therefore relatively higher in the case of slow movements where fewer dynamic disturbances were caused by the passive interaction between inertial body segments than in the case of fast movements. As can also be seen in Fig.2 (which gives the results for subject 1), the error in the kinematic reconstruction based only on PC1 for all the joints decreased in the slow series (t-test, $p<0.05$). A similar result was observed in all the other subjects.

DISCUSSION

In the PC analysis method any PC represents a multijoint movement with fixed ratios between the joint angle changes. It is reasonable to consider a kinematic synergy as a movement described by any PC.[5]

By PC analysis it was found that (1) in the bending movement in humans, there exists a strong linear dependence between the joint angle changes examined, (2) the kinematic synergy described by PC1 accounted for more than 99% of all the joint angle changes in every trial, (3) each ratio in PC1 remained approximately the same for each subject, regardless of the required movement duration and amplitude, while the instruction mainly influenced the time course of the PC1 scaling, (4) the pattern of the main kinematic synergy described by PC1, involving ankle, knee and neck extension and hip flexion, was the same in all the subjects, and (5) the 3 remaining PCs (which accounted for less than 1% of the joint angle changes) had a large intertrial and intersubject variability and seemed to reflect mainly the random dynamic disturbances of the main synergy.

Relatively small intertrial and intersubject variability of the ratios between the angle changes represented in PC1 and their lack of dependence on the movement amplitude and velocity suggest that these are well-learned constraints of neural and biomechanical origin which are preserved under a single central control. The voluntary regulation of the amplitude and velocity of the movement is performed mainly by suitably adopting the time course of the PC1 scaling factor. In particular, once it has been defined for the whole trial, this main set of well-learned fixed ratios between the angle changes makes it possible to simultaneously perform two kinematic tasks (trunk bending and stabilizing the CG sagittal shift) by regulating only one parameter (scaling factor).

The present results can be said to provide quantitative evidence supporting the Bernsteinian concept that motor control learning in the biomechanical system with a redundant number of degrees of freedom is a process in which the kinematic redundancy is overcome by elaborating well-learned internal movement constraints (Bernstein, 1967).

SUMMARY

Healthy subjects were instructed to bend the upper trunk forward with various duration and amplitude. These movements were accompanied by backward movements of the hip and knee (axial synergy providing equilibrium by stabilizing the center of gravity). The covariation between ankle, knee, hip and neck angle joints was examined by principal component (PC) analysis in order to quantify the synergies. A strong correlation between these angle changes was found. The 1-st PC (PC1) accounted for more than 99% of the angle variance. The synergy described by PC1 revealed relative intertrial and intersubject stability. These results suggest that axial synergy is under a central control which regulates the movement duration and amplitude of the different segments mainly by scaling the set of fixed ratios between the joint angles.

ACKNOWLEDGMENTS

The authors thank Jean Claude Fabre for his help in the experimental procedure and Roselyne Aurenty for her technical assistance. The research was partly supported by the CNES and the International Science Foundation (project No. M91000). Alexandrov A. received a fellowship from the French Ministry of Research and Technology.

REFERENCES

Babinski, J., 1899, De l'asynergie cérébelleuse, *Rev. Neurol.* 7:806-816.
Bernstein, N.A., 1967, The co-ordination and regulation of movements, Oxford, Pergamon Press.

Lacquaniti, F. and Maioli, C., 1994, Coordinate transformations in the control of posture. *J. Neurophysiol.* 72:1496-1515.

Mah, C.D., Hulliger, M., Lee, R.G., and O'Callaghan, I.S., 1994, Quantitative analysis of human movement synergies: constructive pattern analysis for gait, *J. Motor Behav.* 26:83-102.

Massion, J., Gurfinkel, V., Lipshits, M., Obadia, A., and Popov, K., 1993, Axial synergies under microgravity conditions, *J. Vestibular Res.* 3:275-287.

Van Egeren, L. F., 1973, Multivariate statistical analysis. *Psychophysiol.* 10:517-532.

12

DO EQUILIBRIUM CONSTRAINTS DETERMINE THE CENTER OF MASS POSITION DURING MOVEMENT?

J. Massion,[1] L. Mouchnino,[2] and S. Vernazza[1]

[1] Laboratory of Neurobiology and Movements
C.N.R.S.
31, chemin Joseph Aiguier
13402 Marseille Cedex 20
France
[2] Faculté des Sciences du Sport
Université d'Aix-Marseille II
163, avenue de Luminy
13288 Marseille Cedex 9
France

INTRODUCTION

Maintaining equilibrium is one of the main constraints involved in the performance of movements. Moving the arms or the upper trunk in the sagittal plane while standing tends to shift the center of gravity (CG) in the same direction as the movement, thus resulting in imbalance. As first observed by Babinski (1899), simultaneous shifts of hip and knee in opposite directions stabilize the CG during an upper trunk movement, and prevent the subject from falling. Maintaining equilibrium during movement is not only achieved by stabilizing the CG. When movements involving a leg are performed, the CG is transferred toward the supporting leg before the movement onset, so that equilibrium is preserved during the movement. This occurs, for example, during gait initiation (see Brenière and Do, 1991) or leg raising (Rogers and Pai, 1990; Mouchnino et al., 1992). Two modes of CG control during movement can thus be identified, the first consisting of stabilizing the CG in the case of arm or trunk movements, and the second, of shifting it prior to a leg movement.

Some important questions about equilibrium control have not yet been elucidated. For example, is the CG the value which is regulated during movements, or is it the body geometry, as proposed by Lacquaniti et al. (1984) in the cat, the CG stabilization being a "by-product" of the body geometry control? In the case of the CG stabilization, do the kinematic synergies which stabilize the CG during arm or trunk movements depend on sensory cues specifying the distribution of the body masses, or are these synergies the result of learned automatic controls regulating the CG whatever the equilibrium constraints?

Similar questions can be asked about the CG shift which occurs prior to a leg movement. In the present survey, an attempt will be made to answer these questions.

IS THE CG THE VALUE WHICH IS REGULATED DURING ARM AND UPPER TRUNK MOVEMENTS?

Stance in humans is very different from stance in quadrupeds, due to the narrow supporting surface and the height of the CG above the ground. There exists some evidence that the CG is actually regulated. When an upper trunk movement is performed, movements of lower segments in the opposite direction can be observed, which stabilize the CG. A similar adjustment occurs in the case of arm movements: When the arms are raised forward, a backward shift of the trunk was noted (Martin, 1967). With both upper arm and trunk movements, the body geometry therefore changes in order to stabilize the CG. In a previous study, we investigated upper trunk movements performed with a 10-kg load placed on the subjects shoulders (Vernazza, Alexandrov and Massion, in preparation), in order to determine whether, during upper trunk movements, the potential effects of the additional mass on the CG shift occurring during the movement were taken into account. The results indicate that during forward movements (3 subjects) and during backward movements (3 subjects), the CG shift remained the same whether or not the load was applied. This indicates that the effects on the CG shift of an additional mass were taken into account and that a change in the kinematic synergy occurred. In two other subjects, in each movement direction, the CG shift with the mass on the shoulder was increased by 3-4 cm. In this case, the effects of the additional mass were therefore not taken into account.

All in all, these results indicate that the effects of an additional mass can be taken into account and that the adaptation of the CG control to the additional load involves changes of the kinematic synergies used.

HOW IS THE CG STABILIZED DURING UPPER TRUNK MOVEMENTS?

The stabilization of the CG during the upper trunk movements involves the use of the kinematic synergies described by Babinski (1899).

The following findings support the hypothesis that these synergies result from automatic controls which are learned in childhood, and are maintained regardless of the gravity constraints.

During upper trunk movements, a set of muscles is activated simultaneously before the movement onset (Crenna et al., 1987). This set of muscles is located in the back in the case of backward movements and in the front in that of forward movements. The muscles involved act on the hip, knee and ankle joint and probably contribute to the organization of the kinematic synergy.

The peak velocity of all the markers placed on various segments such as the trunk, the hip or the knee occurs synchronously (Crenna et al., 1987). This finding contrasts with the velocity curves predicted in the case where the kinematic synergy might result from passive interactions between the trunk movement and the lower segments (Ramos and Stark, 1990).

The kinematic synergy, which is characterized by simultaneous hip, knee and ankle joint changes, was analyzed using the principal component analysis method. With this mathematical method, the 3 degrees of freedom corresponding to the hip, knee and ankle joint angles are replaced by 3 other degrees of freedom, each consisting of a fixed ratio between the angles. As stated by Alexandrov et al. (this volume), the PC analysis shows that the first principal component accounts for 99 % of the kinematic synergy involved in forward movement. This suggests that the central control of the synergy defines the time course and amplitude of the kinematic changes and sets a fixed ratio between the angles.

The effects of microgravity on the axial synergies was tested in order to determine whether or not these synergies are gravity dependent and hence, whether they depend on the equilibrium constraints. This test was carried out on one subject during a long term space flight (Massion et al., 1993) and on two subjects during parabolic flights.

The results indicate that the center of mass (CM) is stabilized in a similar way under both 1G and 0G conditions. Secondly, the kinematic synergies used under 0G and 1G are the same. The kinematic synergies simultaneously controlling the movement and the CM are therefore an invariant aspect of this motor act. The EMG pattern on the contrary changes as a function of the environmental constraints. As shown in Fig. 1, under 1G conditions, an inhibition of the erector spinae (ES) and an activation of the quadriceps (Q) and tibialis anterior (TA) occur prior to the movement onset. An activation of the antagonists then occurs during the braking phase of the movement. Under 0G, a lack of activity can be observed in the trunk muscles, and tonic activity in the Q and TA which increases before the movement onset, while the antagonists are not activated. Unlike the kinematic synergy, the EMG pattern is therefore deeply modified under weightlessness.

To conclude, the CG regulation involving kinematic synergies is not dependent on equilibrium constraints and probably constitutes a highly stable automatic motor control mechanism, which is learned in childhood. Some adaptation of the kinematic synergies is possible however under normogravity, when the upper trunk movement is performed with an additional load on the shoulder.

THE CG SHIFT

The leg raising movement is preceded by an external rotation of the leg around the ankle joint, which shifts the CG toward the supporting leg. By comparing the force platform data and the kinematic analysis, two phases can be identified during the motor act. During the first phase, the CG is transferred toward the supporting leg. The CP first moves toward the moving leg (horizontal thrust) and then toward the supporting leg. The new CG position is then stabilized during leg raising (Mouchnino et al., 1992).

This movement was also tested in three subjects during parabolic flight (Mouchnino et al., submitted). The feet were fixed to the floor of the cabin by an electromagnet and the magnet on the side of the moving leg was released concomitantly with a tone signal triggering the movement. The main result here was that prior to leg raising, there was no CM shift toward the supporting leg. Random shifts occurred toward either the moving leg or the supporting leg, but the mean value was around 0. The CG shift which occurred prior to the leg movement can therefore be said to be mainly gravity dependent (Fig. 2).

Along with the gravity dependent CG displacement, a stabilization of the CG position was also found to take place during leg raising. In most subjects, this involved an external rotation of the trunk in the opposite direction to that of the leg raising. Interestingly, this external trunk inclination was found to still occur under microgravity, which confirms the previous finding made with upper trunk movements that the stabilization of the CG depends on automatic control processes which continue to operate in the absence of equilibrium constraints. During leg raising, two opposite effects of microgravity were therefore observed: the abolition of the CG shift prior to leg raising, and the stabilization of the new CG position during leg raising.

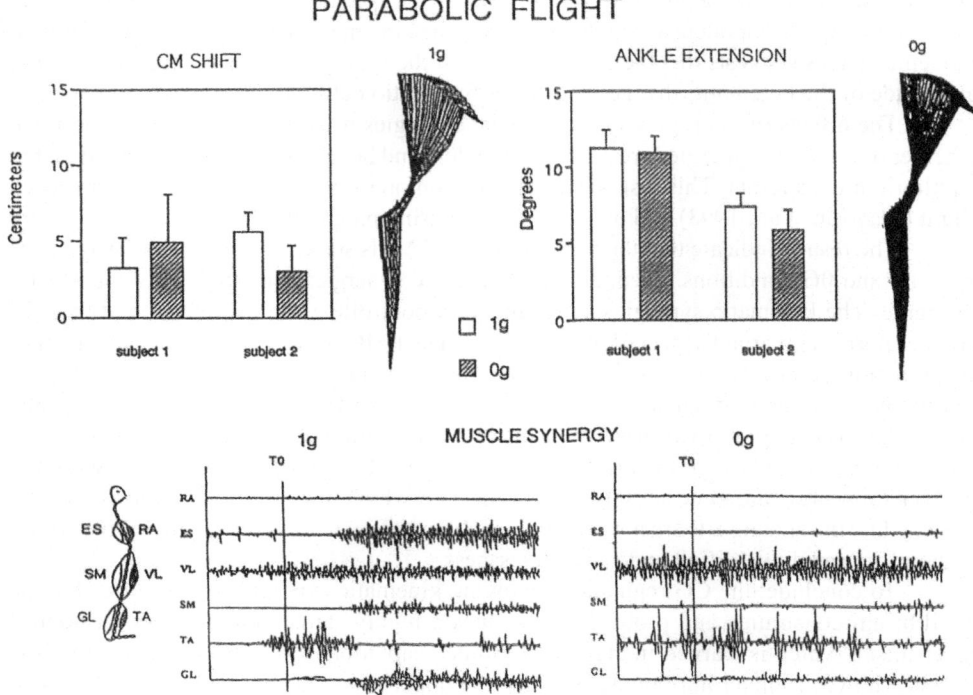

Figure 1. Forward upper trunk movement. Subject standing with the feet fixed to the floor. Comparison between normogravity and microgravity during parabolic flight. Upper left, center of mass (CM) transfer at the end of the upper trunk movement, reconstructed from the body kinematics and the anthrometric model by Winter (1990). Note the similarity between the CM displacements observed under normo- and microgravity. Upper right, ankle extension during upper trunk movement, which can be compared with the stick diagram showing the transfer of the body segments under both conditions. Note the similarities between conditions 1G and 0G. Lower trace, muscle synergy during upper trunk forward bending. T0: onset of movement measured on the acromion marker velocity trace. On the left, 1G. Note that the movement onset was preceded by a reduced activity in the erector spinae (ES) and a burst in the vastus lateralis (VL) and tibialis anterior (TA). After the movement onset, a braking activity occurred in the antagonistic muscles (VL), semimembranosus (SM), and gastrocnemius lateralis (GL). On the right, 0G, VL and TA were activated prior to movement onset. No clearcut braking phase was observed. Superimposed recordings from 5 trials.

CONCLUSIONS

1. The CG position is a reference value which is controlled during movement.
2. During movement two modes of CG control were found to exist. One involving CG stabilization (arm, trunk movements) and the other one a CG shift prior to leg movements.
3. The CG stabilization which occurs during arm or trunk movements depends on "kinematic synergies" which are stable automatic control mechanisms. These mechanisms persist under microgravity, and therefore do not depend on the equilibrium constraints.
4. The CG shift observed prior to leg movements is gravity dependent, and therefore depends on the equilibrium constraints.

NORMOGRAVITY   MICROGRAVITY

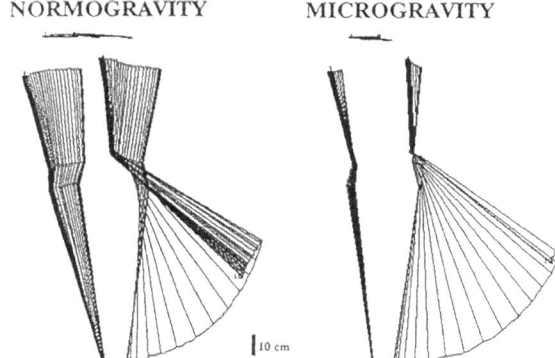

Figure 2. Stick diagram of the body during lateral leg raising under 1G and 0G. Note that under 1G an external rotation of the body axis occurred around the ankle joint that was lacking under microgravity. Bar = 10 cm.

Concluding, it is suggested, that these two modes of CG control, which involve CG stabilization and CG displacement, may be organized by different brain structures.

ACKNOWLEDGMENTS

The authors want to thank the Centre National d'Etudes Spatiales for their support to the present work.

REFERENCES

Alexandrov, A., Frolov, A., and Massion, J., 1994, Principal component analysis of axial synergies during upper trunk forward bending in human. This volume.

Babinski, J., 1899, De L'asynergie cérébelleuse, *Rev. Neurol.* 7:806-816.

Brenière, Y., Do, M.C., and Bouisset, S., 1987, Are dynamic phenomena prior to stepping essential to walking?, *J. Motor Behav.* 19:67-76.

Crenna, P., Frigo, C., Massion, J., and Pedotti, A., 1987, Forward and backward axial synergies in man, *Exp. Brain Res.* 65:538-548.

Lacquaniti, F., Maioli, C., and Fava, E., 1984, Cat posture on a tilted platform, *Exp. Brain Res.* 57:82-88.

Martin, J.P., 1967, The basal ganglia and Posture. Pitman, London.

Massion, J., Gurfinkel, V., Lipshits, M., Obadia, A., and Popov, K., 1993, Axial synergies under microgravity conditions, *J. Vestibular Res.* 3:275-287.

Mouchnino, L., Aurenty, R., Massion, J., and Pedotti A., 1992, Coordination between equilibrium and head-trunk orientation during leg movement: a new strategy built up by training, *J. Neurophysiol.* 67:1587-1598.

Ramos, C.F., and Stark, L.W., 1990, Simulation experiments can shed light on the functional aspects of postural adjustments related to voluntary movements. In: Multiple muscle systems: biomechanics and movement organization. J.M. Winters and L.-Y. Woo. Springer, New York, pp. 507-517.

Rogers, M.W., and Pai, Y.C., 1990, Dynamic transitions in stance support accompanying leg flexion movements in man, *Exp. Brain Res.* 81:398-402.

Winter, D.A., 1990, Biomechanics and motor control of human movement. John Wiley and Sons, New York, pp. 80-84.

13

SELECTION OF POSTURAL ADJUSTMENTS IN SITTING INFANTS
Effect of Maturation and Training

Mijna Hadders-Algra[1,2*] and Hans Forssberg[1]

[1] Department of Woman and Child Health and
Department of Neuroscience Karolinska Institute
Stockholm, Sweden
[2] Department of Medical Physiology
Bloemsingel 10
9712 KZ Groningen, The Netherlands

INTRODUCTION

Motor development results from an interaction between genetic programs and environmental signals (Waddington, 1962; Brauth, 1991; Jacobson, 1991). However, the relative importance of these two sources of information during the development of automatic movement patterns in human infancy is a matter of controversy. Thelen and coworkers (Thelen and Cooke, 1987; Thelen, 1988; Ulrich, 1989), who discussed development of infant locomotion from a dynamical systems point of view, claimed a large effect of self-organization induced by learning-by-doing. Others (e.g. Gesell, 1940; McGraw, 1943; Forssberg, 1985; Hirschfeld and Forssberg, 1994) suggested that motor development is primarily guided by endogenous maturation of predetermined neuronal connections. The latter notion could be in line with Edelman's 'neural group selection' theory (Edelman, 1989; Sporns and Edelman, 1993). According to this theory, development starts with the formation of epigenetically determined primary ensembles of variant neuronal groups. Experiential selection, mediated by changes in synaptic strength of intra- and intergroup connections, leads to secondary, more adapted repertoires of neuronal groups.

Development of postural control plays a pivotal role in infant motor development. Postural control involves adjustments to self-induced shifts accompanying voluntary movements and responses to external perturbations. Recently it has been suggested that postural adjustments to external perturbations are generated by central pattern generators organized in two levels (Forssberg and Hirschfeld, 1994). At the first level afferent input triggers a specific muscle activation pattern. For example: a backward (BW) sway of the body during

---

[*] Address correspondence to Dr. Mijna Hadders-Algra, Dept. Medical Physiology, Bloemsingel 10, 9712 KZ Groningen, The Netherlands. Tel: +31 50 614247; Fax: +31 50 633000.

standing or sitting elicits activity in the muscles at the 'ventral' side of the body, while a forward (FW) sway induces activity in the 'dorsal' muscles (e.g., Nashner, 1977; Horak and Nashner, 1986). At the second level the basic pattern gets fine-tuned in interaction with multisensorial inputs (e.g., Diener et al., 1983, 1988; Horak and Nashner, 1986; Kehsner et al., 1988; Macpherson, 1988; 1994; Hirschfeld and Forssberg, 1991; Allum et al., 1993).

To improve insight into developmental mechanisms involved in postural control, postural adjustments during sitting on a moveable platform were assessed longitudinally in nine healthy infants, who were subjected to daily balance training, and 11 healthy untrained infants.

METHODS

Postural adjustments during sitting on a moveable platform were assessed two times with an interval of two months in 20 healthy infants at the ages of 5-6 and 7-8 months (for a description of the apparatus see Forssberg and Hirschfeld, 1994). During each session surface EMGs of the sternocleidomastoideus (neck flexor, NF), rectus abdominis (RA), rectus femoris (RF), neck-, thoracal- and lumbar extensor muscles (NE, TE, LE) and hamstrings (HAM) were recorded while the infants were exposed to a random series of horizontal FW and BW displacements of the platform, which evoked a sway of the body in the opposite direction. A testing series started with 16 slow perturbations, followed by 16 fast ones. A perturbation related EMG response was considered to be present when a burst occurred exceeding by two standard deviations the base line activity, averaged during 500 ms prior to each perturbation.

At the first session none of the infants was able to sit independently. At this occasion the parents of 10 infants were asked to train their child's sitting balance three times a day during five minutes for the whole study period. Training consisted of toy presentation in the border zone of reaching-without-falling. Special attention was put on sidewards- and semi-backward reaches. The parents of one girl erroneously trained their daughter only till she was able to sit independently, which happened already one week after the first assessment. This infant was regarded as untrained. In the other nine study-infants training was carried out appropriately.

RESULTS

From the youngest testing age onwards FW-translations resulted in a predominant activation of the 'ventral' muscles (the flexors NF, RA and RF), while the 'dorsal' muscles (the extensors NE, TE, LE and HAM) showed varying amounts of inhibition (E-INH). During BW-translations the extensor muscles were mainly activated.

At 5-6 months of age both FW- and BW-translations elicited a large variation in response patterns consisting of any combination of the direction-specific muscles (Fig 1). The muscle activation patterns were independent of platform velocity.

At 7-8 months of age the variation in muscle activation patterns had decreased, resulting in a selection of the most complete patterns (FW-translations: E-INH+NF±RA+RF, Fig. 2; BW-translations: NE+TE+LE+HAM). Selection of these complete response patterns was promoted by the higher platform velocity. Interestingly, training facilitated the selection of the most complete patterns.

A salient training effect was found in the developmental trajectories of the preference pattern, i.e., the pattern the infant used most frequently in a specific condition. At the youngest age both groups of infants showed a large variation in preference pattern. A

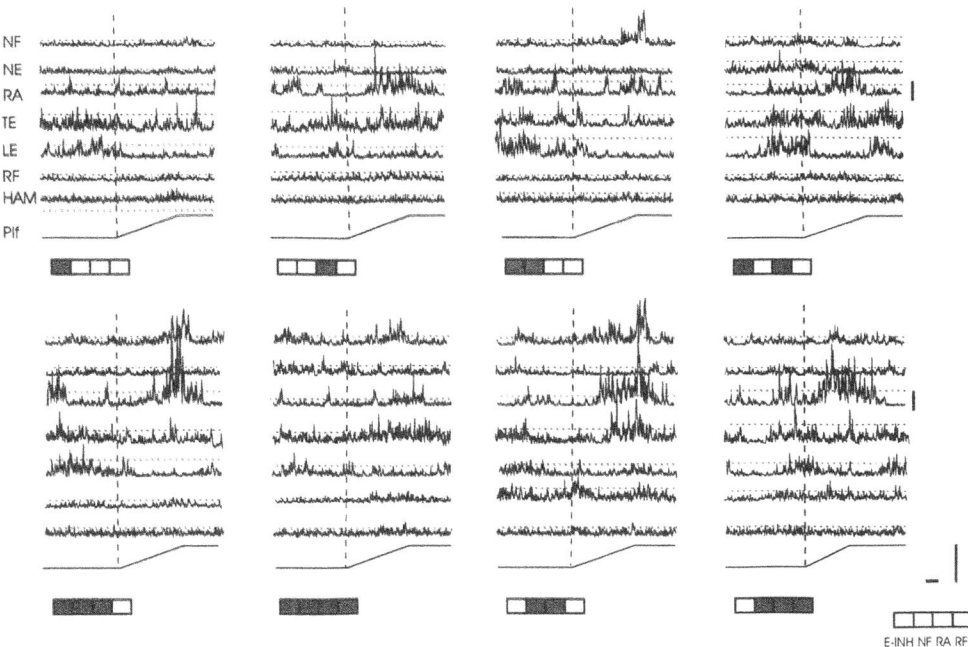

Figure 1. Variation in responses during FW-translation of one infant at 5-6 months of age. PLF = platform signal. Time calibration (horizontal bar) 100 ms, amplitude calibration (vertical bars, RA is the only signal with a separate calibration bar): 0.01 mV.

difference in the development of preference patterns emerged during slow perturbations, most clearly so during FW-translations (Fig. 3). At 7-8 months the trained infants preferred the most complete response patterns, whereas the non-trained infants continued to show a large variation in preference pattern (Fisher p < 0.01). During fast translations this difference was absent.

At 7-8 months all trained infants and 9 of the 11 non-trained infants could sit without help. No relation could be established between the reported moment of independent sitting and the development of the postural responses.

DISCUSSION

The present study showed that complex postural responses were present from the youngest testing age onwards, pointing to their innate and central origin. Already before the age of sitting independently, responses were direction specific. This suggests that from their very beginning onwards postural adjustments appear to be guided by information on direction-specific stability limits stored in an internal body representation, which contains relevant afferent information on postural stability, such as the orientation of the vertical axis and the relation of the center of mass to the margins of the support surface (Gurfinkel et al., 1981; Mittelstaedt, 1983; Massion; 1992, Hirschfeld and Forssberg, 1994).

At 5-6 months of age response patterns varied with any combination of the direction-specific muscles. Variation is a basic characteristic of infant neurological development

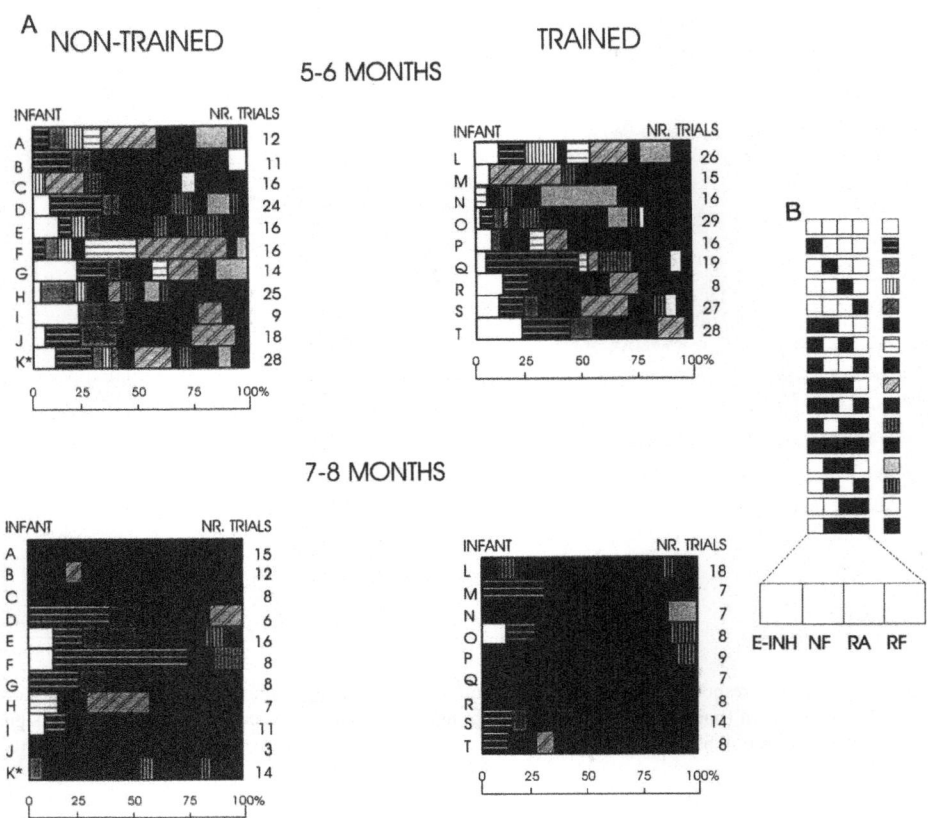

Figure 2. A) Distribution of response patterns during slow FW-translations. Each horizontal bar represents one infant. Infant K is the erroneously non-trained infant. B) Coding of part A.

(Touwen, 1976). Recent EMG-studies on young subjects (*e.g.* Forssberg, 1985; Cazalets et al, 1990; Hadders-Algra et al, 1992) reported that variation decreases with increasing age. The present study showed that selection plays a crucial role in variation reduction and in the development of the limited repertoire of robust muscle activation patterns involved in postural control. Training facilitated the process of selection. Earlier observational studies have suggested that training can induce an acceleration of the development of basic motor skills in infants (McGraw, 1935; Lagerspetz et al., 1971; Zelazo et al., 1972; Zelazo, 1983). Insight into the neural mechanisms governing this acceleration is promoted by the present study. Our findings are in line with Edelman's 'neuronal group selection' theory (Edelman, 1989; Sporns and Edelman, 1993). This theory postulates that development starts with the spontaneous generation of a basic variable movement repertoire. Guided by the afferent output of each movement, those movements are selected which fit best to the task-specific conditions. This selection is mediated by competitive strengthening of neural connections. Animal research on developmental processes in the visual cortex (Wiesel, 1982) and in neuromuscular junctions (Betz et al., 1990) suggested that activity furthers the development of synapses resulting in competent behaviour. Competent connections can arise through selection from an excess of synapses by stabilizing those synapses involved in synchronous firing of the presynaptic terminal and the postsynaptic neuron (Changeux and Danchin, 1976;

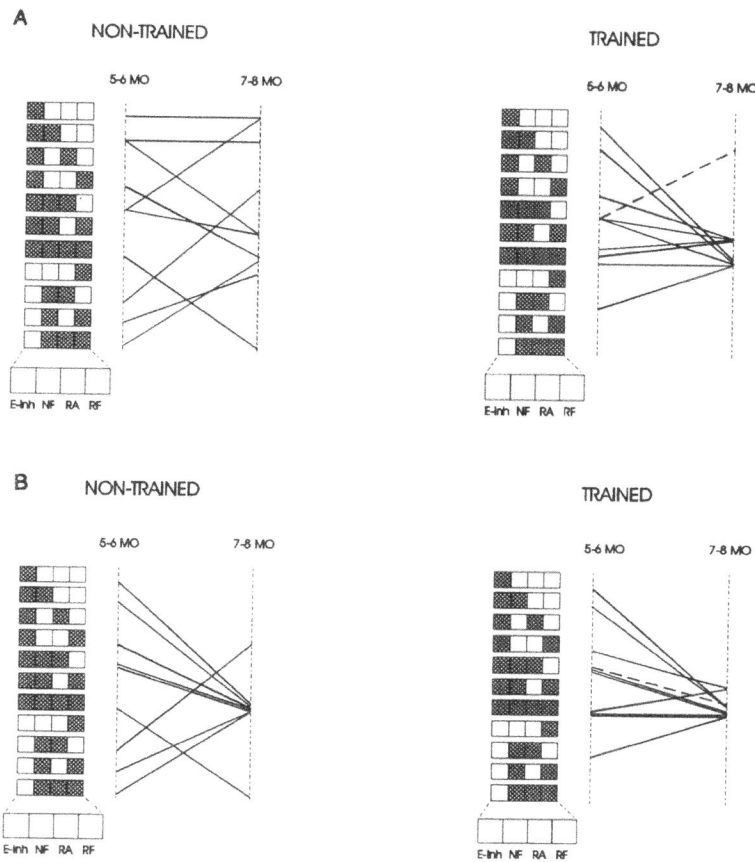

Figure 3. Development of preference patterns during A) slow and B) fast FW-translations. The broken line indicates the erroneously non-trained infant.

Greenough et al., 1987). Another possibility is that conjoint pre- and postsynaptic activity induces growth of the neuropil and stimulates synaptogenesis (Purves, 1994).

The finding that functional sitting behaviour was not related to the maturation of postural adjustments during platform perturbations is remarkable. It underlines the notion that various functional modules in the young nervous system develop autonomously (Touwen, 1976; Hadders-Algra and Prechtl, 1992).

In conclusion: our results suggest that postural responses develop in a predetermined way, starting with a variable repertoire of direction-specific responses. From this repertoire the most appropriate responses, i.e. the most complete patterns which represent the first level of the postulated postural control CPG (Forssberg and Hirschfeld, 1994; Hadders-Algra et al., 1994), are selected. This process can be facilitated by training.

SUMMARY

In order to get insight into the development of the neural circuitry involved in postural control, postural responses during sitting on a moveable platform were assessed in 20 healthy infants. Each infant was tested two times, at 5-6 and 7-8 months of age. After the first session the parents of nine infants trained their child's balance daily (3x5 minutes). During each session multiple surface EMGs were recorded while the infants were exposed to a random series of slow and fast horizontal forward (FW) and backward (BW) displacements of the platform.

In the youngest infants, who were not able to sit without support, FW-translations resulted predominantly in an activation of the neck flexor, the rectus abdominis and rectus femoris muscle, while the neck- , thoracal- and lumbar extensor muscles (NE, TE, LE) and the hamstrings (HAM) showed varying amounts of inhibition. During BW-translations NE, TE, LE and HAM were preferably activated. A large variation in response patterns was present, consisting of activation of any combination of the direction-specific muscles. The results suggest that postural adjustments are epigenetically determined.

During the second examination, when the majority of infants could sit without help, the variation in muscle activation patterns had decreased, resulting in a selection of the most complete pattern of synergist activation. Training facilitated this process of selection. Our results agree with Edelman's 'neuronal group selection' theory, which suggests that development progresses from an epigenetically based variable repertoire of movement patterns, via a process of selection, to those movement patterns which fit best to the environmental constraints. The selection presumably is mediated by competitive strengthening of neural connections, leaving ample opportunity for a facilitating role of training.

ACKNOWLEDGMENTS

We thank Eva Brogren and Ingmarie Apel for technical assistance. The study was supported by the Swedish Medical Research Counsel (4X-5925), Sunnerdahls Handikappfond, Josef och Linnea Karlssons Minnesfond, Norrbacka-Eugenia Stiftelsen and Stiftelsen Sven Jerrings Fond; MHA was supported by a grant from the Ter Meulen Fund, Royal Netherlands Academy of Arts and Sciences.

REFERENCES

Allum, J.H, Honegger, F., and Schicks, H., 1993, Vestibular and proprioceptive modulation of postural synergies in normal subjects, *J. Vest. Res.* 3:59-85.

Betz, W.J., Ribchester, R.R., and Ridge, R.M.A.P., 1990, Competitive mechanisms underlying synapse elimination in the lumbrical muscle of the rat, *J. Neurobiol.* 21:1-17.

Brauth, S.E., Hall, W.S., and Dooling, R.J., 1991, Plasticity of development, MIT Press, Cambridge.

Cazalets, J.R., Menard, I., Crémieux, J., and Clarac, F., 1990, Variability as a characteristic of immature motor systems: an electromyographic study of swimming in the newborn rat, *Behav. Brain Res.* 40:215-225.

Changeux, J-P., and Danchin, A., 1976, Selective stabilization of developing synapses as a mechanism for the specification of neuronal networks, *Nature* 264:705-712.

Diener, H.C., Bootz, F., Dichgans, J., and Bruzek, W., 1983, Variability of postural "reflexes" in humans, *Exp. Brain Res.* 52:423-428.

Diener, H.C., Horak, F.B., and Nashner, L.M., 1988, Influence of stimulus parameters on human postural responses, *J. Neurophysiol.* 59:1888-1905.

Edelman, G.M., 1989, Neural Darwinism. The theory of neuronal group selection. Oxford University Press, Oxford.

Forssberg, H., 1985, Ontogeny of human locomotor control. I. Infant stepping, supported locomotion and transition to independent locomotion, *Exp. Brain Res.* 57:480-493.

Forssberg, H., and Hirschfeld, H., 1994, Postural adjustments in sitting humans following external perturbations: muscle activity and kinematics, *Exp. Brain Res.* 97:515-527.

Gesell, A., 1940, The first five years of life, Harper, London.

Greenough, W.T., Black, J.E., and Wallace, C.S., 1987, Experience and brain development, *Child Development* 58:539-559.

Gurfinkel, V.S., Lipshits, M.I., Mori, S., and Popov, K.E., 1981, Stabilization of body position as the main task of postural regulation *Human Physiology* 7:155-165.

Hadders-Algra, M., and Prechtl, H.F.R., 1992, Developmental course of general movements in early infancy. I. Descriptive analysis of change in form, *Early Hum. Dev.* 28:201-214.

Hadders-Algra, M., Van Eykern, L.A., Klip-Van den Nieuwendijk, A.W.J., and Prechtl, H.F.R., 1992, Developmental course of general movements in early infancy. II. EMG correlates, *Early Hum. Dev.* 28: 231-251.

Hirschfeld, H., and Forssberg, H., 1991, Phase-dependent modulations of anticipatory postural activity during human locomotion, *J. Neurophysiol.* 66:12-19.

Hirschfeld, H., and Forssberg, H., 1994 Epigenetic development of postural responses for sitting during infancy, *Exp. Brain Res.* 97:528-540.

Horak, F.B., and Nashner, L.M., 1986, Central programming of postural movements: adaptation to altered support-surface configurations, *J. Neurophysiol.* 55:1369-1381.

Jacobson, M., 1991, Developmental neurobiology, 3-rd. Ed. Plenum Press, New York, London.

Keshner, E.A., Woollacott, M.H., and Debu, B., 1988, Neck, trunk and limb muscle responses during postural perturbations in humans, *Exp. Brain Res.* 71:455-466.

Lagerspetz, K., Nygård, M., and Strandvik, C., 1971, The effects of training in crawling on the motor and mental development of infants, *Scand. J. Psychol.* 12:192-197.

Macpherson, J., 1988, Strategies that simplify the control of quadrupedal stance. II. Electromyographic activity, *J. Neurophysiol.* 60:218-231.

Macpherson, J.M., 1994, Changes in postural strategy with inter-paw distance, *J. Neurophysiol.* 71:931-940.

Massion, J., 1992, Movement, posture and equilibrium: interaction and coordination, *Progr. Neurobiol.* 38:35-56.

McGraw, M.B., 1935, Growth: a study of Johnny and Jimmy. Appleton Century, New York.

McGraw, M.B., 1943, The neuromuscular maturation of the human infant. Haffner Publishing Company, London, New York. Reprinted in 1969.

Mittelstaedt, H., 1983, A new solution to the problem of the subjective vertical, *Naturwissenschaften* 70:272-281.

Nashner, L.M., 1977, Fixed Patterns of rapid postural responses among leg muscles during stance, *Exp. Brain Res.* 30:13-24.

Purves, D., 1994, Neural activity and the growth of the brain. Cambridge Univ. Press, Cambridge.

Sporns, O., and Edelman, G.M., 1993, Solving Bernstein's problem: a proposal for the development of coordinated movement by selection, *Child Development* 64:960-981.

Thelen, E., 1988, On the nature of developing motor systems and the transition from prenatal to postnatal life. In: Behavior of the fetus. Eds. Smotherman, W.P. and Robinson, S.R., pp. 207-224, The Telford Press, Calwell, New Jersey.

Thelen, E., and Cooke, D.W., 1987, Relationship between newborn stepping and later walking: a new interpretation, *Dev. Med. Child Neurol.* 29:380-393.

Touwen, B.C.L., 1976, Neurological development in infancy. Clinics in Developmental Medicine, Vol., 58, Heinemann Medical Books, London.

Ulrich, B.D., 1989, Development of stepping patterns in human infants: a dynamical systems perspective, *J. Mot. Behav.* 21:392-408.

Waddington, C.H., 1962, New patterns in genetics and development. Columbia University Press, New York.

Wiesel, T.N., 1982, Postnatal development of the visual cortex and the influence of environment, *Nature* 299:583-591.

Zelazo, P.R., 1983, The development of walking: new findings and old assumptions, *J. Mot. Behav.* 2:99-137.

Zelazo, P.R., Zelazo, N.A., and Kolb, S., 1972, Newborn walking, *Science 177:1058-1059.*

14

DEVELOPMENT OF BILATERAL LIMB COORDINATION IN HUMANS

W. Berger and V. Dietz

Neurologische Universitätsklinik
Neurozentrum, Breisacher Str. 64
D-79106 Freiburg
Germany

INTRODUCTION

Upright stance and locomotion in humans differ from those of quadrupedal mammals by the complexity of the system that is needed to regulate body posture and movements in bipeds. Stabilization of balance is achieved by reflex mechanisms depending on signals from the central nervous system, which in turn constantly receives afferent information. The weighting of the different afferent systems is task-dependent (for review, see Dietz, 1992) and undergoes a maturation process (Berger et al., 1984). A major difficulty with the study of developmental changes is to decide to which extent the observed changes during maturation are due to physical or neurodevelopmental factors, or whether both factors are involved. So far only few anatomical structures have been studied in terms of functional normative data during maturation, as, for instance, the somatosensory pathways in terms of conduction velocity. Maturation in this system has been shown to start in the peripheral nerves, followed by spinal pathways, while the supraspinal pathways are the last to mature (Desmedt et al., 1976; Cracco et al., 1979). The authors of the present article have studied age-dependent changes of bilateral coordination over the last years and summarize their main findings in the following.

METHODS

Methods employed were: a treadmill with split belts and with force measuring platforms under each belt; goniometers for the hip-, knee-, and ankle joints; surface electrodes to record electromyograms (EMG) for different muscles; scalp electrodes for recording of mechanically evoked cerebral potentials (CP); highspeed camera and a video analysis system (ELITE) to study body and joint movements.

RESULTS

Newborn stepping as an innate locomotor program in children has been described by Forssberg and Wallberg (1980); it consists in a coactivation of all leg muscles when the feet touch the ground. The first life period considered in our studies was that of a beginning of free standing and of steps induced by a moving treadmill. The typical features of immature gait in children from 1- 2 years of age consisted of (i) a coactivation of antagonistic leg muscles during the stance phase of gait, similar to the finding in newborn stepping, (ii) large, solitary potentials arising with a latency of 20- 30 ms after ground contact and a stretching of gastrocnemius muscle, and (iii) low and tonic activity of gastrocnemius EMG. These early signs of the locomotor pattern are then replaced by a more structured pattern in children from 4-7 years of age, which occurs in parallel to their ability of free and independent gait: (a) a reciprocal activation of antagonistic leg muscles, (b) a suppression of the solitary (monosynaptic) reflexes, and (c) an increase of gastrocnemius EMG activity (Berger et al., 1984). A conclusion was that the group I mediated reflexes are suppressed, and the functionally meaningful group II mediated reflexes are facilitated during development. This notion was confirmed by the finding of compensatory EMG responses following perturbations with stumbling reactions; monosynaptic reflexes were present in early infancy during standing and gait, whereas a suppression during gait took place in older children, together with a reciprocal mode of activation of the antagonistic muscles (Berger et al., 1985).

The use of perturbations during stance and gait allowed us to study not only efferent pathways, but also afferent pathways, by evaluating mechanically evoked potentials and their modulation over age. In children from 6-10 years of age, similar as in adults, a suppression of the monosynaptic stretch reflexes during gait are associated with a suppression of the early part of the evoked potentials. This has been taken to indicate an inhibition of group I afferents at both segmental and supraspinal levels, involving suppression of segmental stretch reflexes as well as of group I volleys to supraspinal levels. In small children up to an age of 2 years, no difference between afferent and efferent reflex behaviour during stance and gait was observed. The conclusion was that the control of afferent information in early infancy still has to be established, and that the maturation of compensatory EMG responses during gait is achieved by a descending inhibition of group I afferents and a facilitation of polysynaptic reflexes via group II afferents.

Developmental changes of bilateral coordination during perturbed stance could be further demonstrated in a study with two groups of children of different age. Figure 1 shows the EMG responses following a unilateral acceleration backward of one leg (A), a unilateral acceleration forward of the other leg (B), and a combination of the two previous perturbations (C), separately in children of 5.5-11 years of age (a) and of 3.5-5.5 years (b). Unilateral displacements were followed by ipsilateral short latency and bilateral longer latency responses, both in the younger and the older children. In the older children the long latency responses were larger, and the short latency responses smaller than in the younger children in all conditions tested. In addition, the bipedal perturbations (C) caused a significant reduction of the long latency EMG responses in both age groups. These findings reflect the fact that already early in childhood a bilateral coordination is established on a spinal level (Berger et al., 1990), similar to adult subjects (Dietz et al., 1989), but still immature with respect to the magnitude of the EMG pattern.

Regarding the physical aspects in the form of body height, body weight and foot length a study was carried out with weight adapted perturbing momenta on the back at the level of the center of gravity in normal subjects from 5-45 years of age and body heights from 113 - 193 cm (Berger et al., 1992). The idea was that keeping balance would be more difficult for a small as compared to a tall (in analogy to the common experience that it is

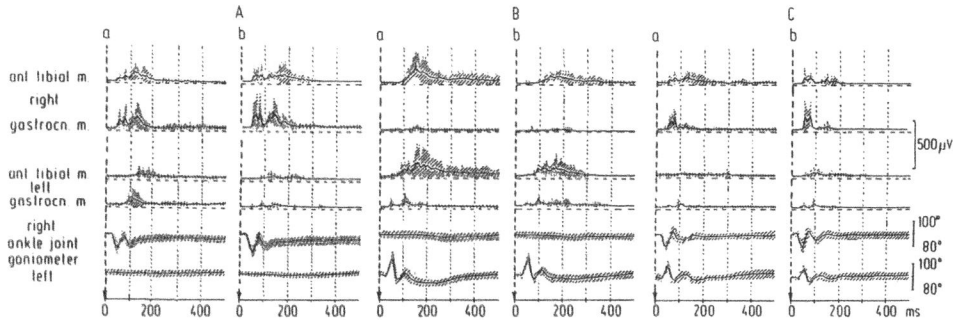

Figure 1. Mean and S.D. values of averaged (n = 20) EMG responses following unilateral backward perturbation of right leg (A), unilateral forward perturbation of left leg (B), and a combination thereof (C). Two groups of children were studied: 5.5-11 years of age (a; n = 12) and 3.5-5.5 years of age (b; n = 19). The arrows indicate onset of treadmill impulse.

easier to balance a stick of 150 cm than one of 15 cm of length; compare Forssberg et al., 1982). We found a close correlation between displacement amplitude at the ankle joint and the height of the subject, with the largest displacements in the smallest subjects. Correspondingly, the compensatory reactions in terms of gastrocnemius response were larger in the small than the tall subjects. Momenta of high strength were associated with large increments of both ankle joint displacement and EMG response. This study could not be performed in children of 1-2 years, but a 'backward speculation' would be that the instability of very small children is largely due to an increment of ankle joint displacement and a lack of compensatory extensor EMG responses typical for this age group.

Video analysis performed in a study with children during unstable stance upon sinusoidal backward and forward movement on a treadmill further confirmed both dependencies: a reduced damping of oscillations from foot to head in small children, and a delayed and more passive mode of EMG reaction to enhanced body sway (Berger et al., in press). A further result of studying this paradigm was an insufficient adjustment of body posture, which is suggested to depend on the immaturity of load receptors, in addition to that of the classical inputs from visual, vestibular, and muscle stretch receptors (Berger et al., 1992a).

Interlimb coordination during split-belt walking in children, however, strongly stressed physical factors for most of the differences seen in the biomechanical parameters such as stride length and frequencies, as compared to adults. The results illustrate that already at an age of 4 years a high degree of flexibility exists in the relative timing of leg movements during walking, thought to be responsible for the rather early maturation of interlimb coordination during locomotion (Zijlstra, W., Prokop, T., and Berger, W., unpublished observations).

CONCLUSIONS

It is generally accepted that also human gait is based upon a number of basic mechanisms such as central pattern generators. A relatively high degree of flexibility of interlimb coordination can already been demonstrated in infants. Developmental changes that take place in the following years are reflected mainly in an increase of the functionally most essential parts of EMG responses, those of longer latency. At the same time, mono- or oligosynaptic reflexes become reduced. An inhibition of group I afferents at both the spinal and the supraspinal levels is suggested. An increase of EMG reflexes with longer latency can be demonstrated in both the ipsilateral and the contralateral leg following unilateral perturbation. In small children the immaturity of proprioceptive processing as well as an immaturity of extensor load receptors are compensated for by enhanced leg flexor activity which is more centrally regulated. A stronger coactivation of anatonistic leg muscles is a typical mode of compensation for perturbing momenta and the pure physical shortcomings in early infancy.

The influence of the vestibular and visual systems appear to be task- and age-dependent. They do not play a major role in the fast compensatory movements reported here.

ACKNOWLEDGMENT

The work was supported by the Deutsche Forschungsgemeinschaft (Be 936/5-1).

REFERENCES

Berger, W., Altenmüller, E., and Dietz, V., 1984, Normal and impaired development of children's gait, *Hum. Neurobiol.* 3:163-170.

Berger, W., Discher, M., Trippel, M., Ibrahim, I.K., and Dietz, V., 1992a, Developmental aspects of stance regulation, compensation and adaptation, *Exp. Brain Res.* 90:610-619.

Berger, W., Horstmann, G.A., and Dietz, V., 1990, Interlimb coordination of stance in children: divergent modulation of spinal reflex responses and cerebral evoked potentials in terms of age, *Neurosci. Lett.* 116:118-122.

Berger, W., Quintern, J., and Dietz, V., 1985, Stance and gait perturbations in children: developmental aspects of compensatory mechanisms, *Electroenceph. Clin. Neurophysiol.* 61:385-395.

Berger, W., Trippel, M., Discher, M., and Dietz, V., 1992b, Influence of subjects' height on the stabilization of posture, *Acta Otolaryngol. (Stockholm),* 112:22-30.

Berger, W., Trippel, M., Assaiante, C., Zijlstra, W., and Dietz, V., 1995, Developmental aspects of equilibrium control during stance: a kinematic and EMG study (Gait and Posture, in press).

Cracco, J., Cracco, R., and Graziani, L., 1979, Spinal evoked potential in man: a maturational study, *Electroenceph. Clin. Neurophysiol.* 46:58-64.

Desmedt., J.E., Brunko, E., and Debecker, J., 1976, Maturation of the somatosensory evoked potentials in normal infants and children, with special reference to the early N_1 component, *Electroenceph. Clin. Neurophysiol.* 40:43-58.

Dietz, V., 1992, *Physiological Reviews,* vol.72, No.1:33-69.

Dietz, V., Horstmann, G.A., and Berger, W., 1989, Perturbations of human posture: influence of impulse modality on EMG responses and cerebral evoked potentials, *J. Mot. Behav.* Vol.21,No.4:357-372.

Forssberg, H., and Nashner, L.M., 1982, Ontogenetic development of postural control in man: adaptation to altered support and visual conditions during stance, *J. Neurosci.,* 5:545-552.

Forssberg, H., and Wallberg, H., 1980, Infant locomotion: a preliminary movement and electromyographic study. In: K. Berg, and B.D. Eriksson (Eds.), Children and Exercise, vol.10, IXth Int. Series Sport Sci., University Park Press, Baltimore, pp. 32-49.

15

VOLITIONAL VERSUS REFLEX CONTROL IN OCULAR PURSUIT

G. R. Barnes

MRC Human Movement and Balance Unit
Institute of Neurology
Queen Square
London WC1N 3BG
United Kingdom

INTRODUCTION

When human subjects track the motion of a visual target the relative velocity of the image of the tracked object on the retina clearly forms an important input to the oculomotor control mechanisms. Indeed, retinal velocity error information provides such a powerful drive to the system that it is difficult to suppress the induced reflex response unless a static fixation cue is also present. However, a consistent finding associated with this type of 'passive' stimulation is that the magnitude of the smooth eye velocity induced is generally less than that generated during 'active' pursuit of the target, in which volitional control is involved. In this article the contribution of volitional and reflex mechanisms to pursuit will be considered.

ACTIVE PURSUIT VERSUS PASSIVE STIMULATION

Although the volitional element of ocular pursuit has long been recognized the precise mechanism by which it operates in conjunction with the feedback of retinal error information to control eye movement has not frequently been addressed. It is well established that the motion of a visual stimulus can excite reflex eye movements without any voluntary participation by the subject. Such passively induced optokinetic eye movements are readily evoked as a nystagmus when an observer views a large moving object such as a passing train. What is less well appreciated is that reflex optokinetic eye movements may even be induced by the type of small targets normally associated with a pursuit task, if no other fixation cues are present (Barnes and Hill, 1984; Pola and Wyatt, 1985). During passive stimulation the subject stares at a point in the vicinity of the target whilst attempting to ignore the target motion. A vigorous nystagmus is generated, in contrast to the mainly smooth eye movement of active pursuit. Although they take such different forms the velocity of smooth movement in both active and passive responses appears to exhibit similar changes with variables such as stimulus frequency (Barnes and Hill, 1984), target eccentricity (Barnes and Crombie, 1985; Pola and Wyatt, 1985) and stimulus predictability (Wyatt and Pola, 1988; Barnes and Ruddock, 1989). However, the velocity induced during passive stimulation is generally only 60-70% of that induced during active pursuit. This evidence suggests that active pursuit may be achieved through a process of volitional potentiation of the basic reflex visual feedback drive to the oculomotor system.

PURSUIT AGAINST A STRUCTURED BACKGROUND

One of the situations in which such a potentiation would be most necessary is during pursuit of a target against a structured background. One of the implications of the finding that passive stimulation can induce eye movement is that any visual motion stimulus, presented anywhere in the visual field, must have some potential to drive the eye. If so, it might be expected that when the subject is presented with a number of such moving stimuli, a conflict could arise. It should be more difficult to pursue a target against a structured background than against a blank background, because the relative motion of the eye across the background should induce an antagonistic optokinetic drive tending to oppose the eye motion. It is possible to demonstrate such an interaction, but the degradation in eye velocity is generally no more than 10-20% in normal circumstances (Collewijn and Tamminga, 1984; Barnes and Crombie, 1985). However, if the strength of the visual feedback for the target is decreased, by placing it in the peripheral visual field (Barnes and Crombie, 1985; Collewijn and Tamminga, 1986) or by randomizing target motion (Worfolk and Barnes, 1992), more degradation is observed and, if the background itself is in motion, the velocity induced by that background increases. This suggests that the background stimulus is not suppressed in these circumstances, but rather, that the potentiation of the feedback from one selected source allows the antagonistic effect of the background to be effectively overcome. This might be expected on the grounds of parsimony, since it must be easier to selectively potentiate a single source than to carry out a blanket suppression of all other motion sources. It also has the advantage that it allows the subject to be constantly aware of the full potential of other moving stimuli in the background and to switch selection to another target if required.

Given that such a potentiation process exists, what form does it take? Is it simply an augmentation of the visual feedback gain that can only be observed in the presence of existing visual feedback, or does it arise from another intrinsic source that can operate independently of, and summate with, visual feedback. In the following evidence will be presented to suggest that it may be either.

EVIDENCE FOR A VELOCITY STORE DURING ACTIVE PURSUIT

Although it is normally not possible to initiate smooth eye movements voluntarily in darkness with any appreciable velocity, experiments have shown that such anticipatory eye movements can be made if there has been some prior exposure to a moving visual stimulus (Becker and Fuchs, 1985; Barnes et al., 1987; Barnes and Asselman, 1991). The stimulus used to elicit such a response is a repeated transient target motion stimulus of the type indicated in the first part of Fig. 1 (closed-loop portion). In this example a small target was made to move horizontally across a semicircular screen in totally dark surroundings. The underlying target motion stimulus was a triangular waveform of frequency 0.32 Hz and velocity 10-50°/s, but the target was exposed for brief periods only (160-640 ms) as it passed through the mid-point. Subjects were initially required to pursue the target during the exposure period in the normal closed-loop mode for four cycles.

The resultant eye movement pattern was composed of both smooth and saccadic eye movements, but after removal of the saccades, an interesting pattern of eye velocity was observed. In the response to the first target presentation there was a latency of 100-150 ms, but with subsequent presentations, the eye velocity became gradually more anticipatory, so that after 3-4 presentations smooth eye movements were being initiated with a velocity that reached as much as 30 °/s prior to target illumination. Such eye velocities could not normally be achieved in darkness, but the prior stimulation allowed this to take place. This type of behaviour cannot be explained by conventional linear feedback theory, but rather, implies the presence of a store, which is charged by the afferent visual feedback information and subsequently released in anticipation of the target appearance. In support of this concept was the finding that, once this steady state response had been achieved, an unexpected change in the motion of the target led to an anticipatory velocity pulse being generated that was correlated with the previous response, but inappropriate for the current stimulus (Barnes and Asselman, 1991).

EVIDENCE FOR A VELOCITY STORE DURING PASSIVE STIMULATION

The experiments described above were carried out with the subject actively pursuing the target, and thus imply some cognitive involvement in the storage process. However, the same type of experiment has also been carried out using passive stimulation techniques with similar results (Barnes and Asselman, 1991). Typically, the subjects were requested to make eye movements in the opposite direction to the target motion during its presentation, a task they could easily do with a little experience. Even though the subjects' active processing was concentrated on this contraversive movement, which resulted in oppositely directed saccades, anticipatory smooth eye movements were still built up over the first 3-4 presentations in the direction of target motion. The implication of these results is that the storage process can be carried out at a non-cognitive level. The results of other experiments (Kowler and Steinman, 1981) have provided similar evidence. As with the experiments described earlier, active responses always had greater velocity than those evoked during passive stimulation. The results of other experiments (Barnes and Hill, 1984; Wyatt and Pola, 1988; Barnes and Ruddock, 1989) have also demonstrated that other aspects of predictive processing appear to function during passive stimulation as well as during active responses.

REPLAYING THE VELOCITY STORE WITHOUT RETINAL ERROR FEEDBACK

In recent experiments (Barnes, Collins and Goodbody, 1994a) we have attempted to verify the role of stored information in the volitional pursuit response by examining the ability of subjects to produce smooth eye movements with a foveally-stabilised image. The initial part of the stimulus was as described above and illustrated in Fig. 1. In the dark period following the fourth cycle the target image was stabilized on the fovea by feeding back the calibrated eye movement signal to drive the target in an open-loop mode. Simultaneously, the external target drive signal was removed, so that any subsequent target movement was driven by the eye alone. However, the target continued to be intermittently illuminated with the same periodicity as before. Subjects were instructed to attempt to generate the same eye movement as they had in the previous closed-loop phase.

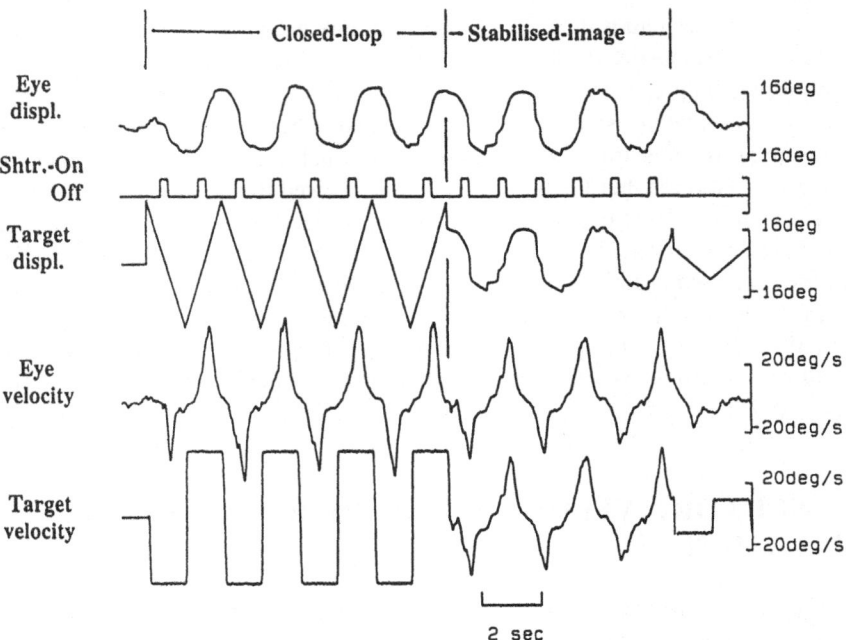

Figure 1. Eye movements generated during intermittent presentation of a target moving with a triangular waveform (frequency 0.325 Hz; peak velocity 40°/s) under closed-loop and stabilized-image conditions. Pulses (second trace) indicate times at which target was illuminated (Pulse duration =320 ms). Saccadic components have been removed from eye velocity trajectory.

It was found that subjects were able to continue to make smooth eye movements both to the left and to the right, as indicated in Fig. 1 (stabilized-image part), by transferring attention from one side to the other of the foveally-stabilised image. Smooth movements could be made with a regular periodicity for at least three further cycles, even though no retinal velocity error feedback was then present. The velocity trajectory was very similar in form to that observed in the closed-loop phase and, in particular, the subjects continued to initiate the smooth movements prior to target illumination. Peak velocity increased with the target velocity of the previous closed-loop phase, but was, on average, 30% lower. These results demonstrate that subjects can continue to generate repeated predictive smooth movements based on stored information if, as with the stabilized image, there is no conflict between the predictive estimate and the visual feedback. Further experiments with continuous illumination of a sinusoidally oscillating target also indicated that the subjects could continue to make quasi-sinusoidal smooth eye movements when the image was suddenly stabilized on the fovea (Barnes et al., 1994a) as demonstrated previously (Cushman et al., 1984; van den Berg and Collewijn, 1987).

INITIATION OF SMOOTH EYE MOVEMENTS BY SHIFTS OF ATTENTION

Although these results indicate that subjects can use stored information to generate smooth eye movements, this is not the only method by which such a response may be elicited with a stabilized image. Kommerell and Taumer (1972) and Grüsser (1986) showed that it is possible to generate smooth eye movements at will with a foveally stabilized image by directing attention to one side of the central image. In recent experiments (Barnes, Collins and Goodbody, 1994b) we have examined this in more detail by presenting the subject with a stabilized display composed of 5 small targets placed at 0°, +/- 2.5° and +/-5° to left and right of center (see inset, Fig. 2). There was no external drive to the display, its movement being controlled simply in response to eye movement. Subjects were instructed to transfer their attention alternately between a pair of targets equidistant from the central (0°) target, which itself was aligned with foveal center. As they did so the whole display moved with the eye so that the subjects could never achieve fixation of the desired target. The timing of attentional shifts was cued by an audio signal.

In response to this volitional shift of attention the subjects were able to generate eye movements composed predominantly of smooth components with some saccadic activity, similar to that shown in Fig. 2A. The direction of smooth eye movement was determined by the direction of attentional shift and the velocity generally increased with increasing eccentricity of attentional shift. At low frequencies (e.g. 0.25 Hz) the smooth eye movement had a velocity profile similar to that shown in the responses of Fig. 1, but at higher frequencies (e.g., 0.8 Hz), the alternating movements merged to give a quasi-sinusoidal waveform.

INTERACTIONS BETWEEN VOLITIONAL CONTROL AND AFFERENT FEEDBACK

Subjects are clearly able to initiate smooth eye movements at will, without recourse to the use of previously stored information, but, if so, what stops them from doing so when there is no moving target present? One possibility is that the retinal error feedback that would be generated by making smooth movements across a stationary target or background effectively inhibits the development of such eye movement. In order to test this, further experiments have been carried out in which the display of Fig. 2 (inset) was only partially stabilized on the retina. This was achieved by feeding back only a proportion (K_f) of the recorded eye movement signal to drive the target. The subjects were instructed to shift attention alternately as before, but as they did so, the display did not move as far or as fast as the eye, thus allowing the subject eventually to fixate the desired target. But during this process the display was naturally swept across the retina in the opposite direction to the prevailing eye movement.

The eye movements generated in this condition were similar in form to those for complete stabilization. However, the velocity of smooth movement became progressively attenuated as the stabilization gain (K_f) was decreased from unity to 0.6, as indicated in Fig. 2. If K_f was decreased below 0.5, negligible smooth eye movement was generated. As K_f was reduced it seemed to require more and more effort to generate the smooth eye movement in opposition to the drag induced by the movement of the display in the opposite direction to intended eye movement.

It seems most likely that this effect was caused by an interaction between the mechanism for volitional generation of eye movement and the visual feedback provided by the relative movement of the display on the retina. Although, the display was relatively small, it would still be capable of providing an effective visual feedback drive as indicated by the responses to passive stimulation described earlier. It is not difficult to see that if the display were stationary it would be difficult to sustain any smooth movement, since this would be equivalent to a value for K_f of 0. In effect, with a partially stabilized image, the subject is attempting to drive the eye at a faster rate than the target. In normal closed-loop pursuit it is

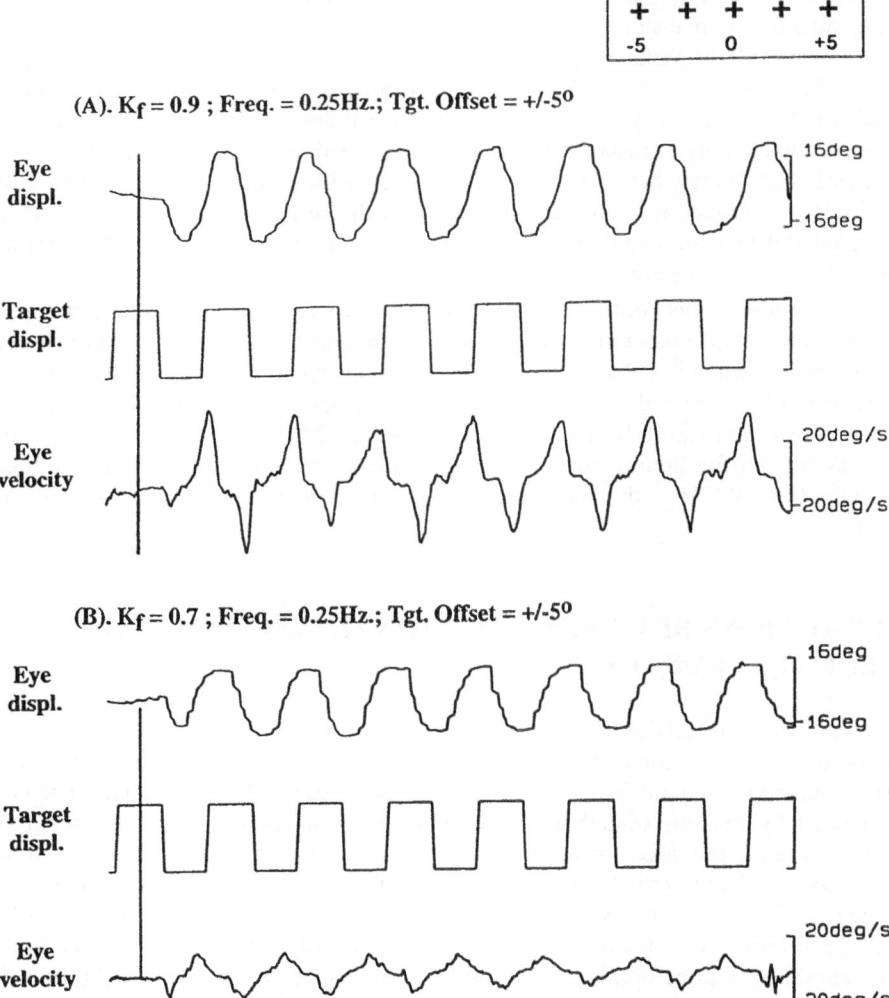

Figure 2. Eye movements generated when viewing a display (shown in the inset figure) which was partially stabilized on the retina by feeding back a proportion (K_f) of the recorded eye movement signal to drive the eye. In (A) $K_f = 0.9$; in (B) K_f=0.7. Subjects transferred attention alternately to targets to the left or right of the central target. The target displacement signal represents the offset of the desired attentional shift which was at +/-2.5° in these examples. Frequency of audio cue signal 0.25 Hz.

also very difficult to drive the eye faster than the target, presumably because, in this more natural situation, the visual feedback effectively acts in opposition to volitional effort and thus maintains average eye velocity close to target velocity.

MODELING THE OCULAR PURSUIT SYSTEM

Taking all these results into account indicates that there must be two primary mechanisms involved in generating active smooth pursuit (Barnes et al., 1987). One is the volitional control mechanism that may be associated with a shift of attention to a para-

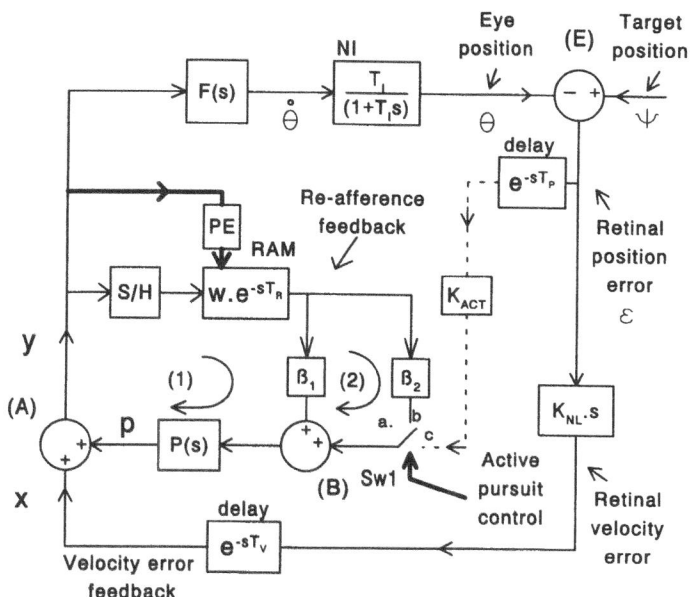

Figure 3. Model of ocular pursuit. S/H - sample/hold module -sampling interval =.05-.25s; RAM - re-afferent memory or self-optimizing variable delay network (= $w.e^{-sT_R}$: for periodic stimuli, $w = -1$ and $T_R = T/2-T_V$ where T= period of stimulus; for non-periodic stimuli, $w =1$ and $T_R = 0$; in darkness, $w = 0$ or -1 and $T_R = 0$); PE - periodicity estimator. Non-linear feedback gain - $K_{NL}= (1+\varepsilon/2)^{-0.5}$, NI - neural integrator. ε = retinal position error. θ = eye position. ψ = target position. $F(s) = (1+2s)/((1+s)(1+0.15s))$. $P(s) = 1/(1+0.1s)$. $T_V = 0.08-0.1s$. $T_P = 0.1s$. $T_1 = 20s$. $K_{ACT} = 1 - 4$. $\beta_1 = 0.5$; $\beta_2 =0.45$.

foveally located target. The other is the reflex velocity error feedback that is associated with optokinesis. Previously, a model of the pursuit mechanisms was developed to explain the predictive behaviour of pursuit (Barnes and Asselman, 1991; Barnes, 1994). The basis of the model (shown in modified form in Fig. 3) is that the pursuit pathways may be split into two components, a retinal velocity error feedback pathway containing a finite time delay (100 ms) and a secondary, re-afferent feedback pathway. The latter is responsible for taking samples of the drive to the oculomotor system, holding them in a short term store (the re-afferent memory; RAM, Fig. 3) and replaying them at an appropriate time so as to boost the visual feedback component. An important aspect of this re-afferent mechanism is that it can be accessed by a periodicity estimator (PE) that, for regularly repeated stimuli, deter-mines the time at which to release the stored predictive estimate from the re-afferent memory. This re-afferent pathway is similar to the efference copy system proposed by a number of authors (Yasui and Young, 1975; Lisberger and Fuchs, 1978; Robinson, 1982). In order to simulate the transient response to an aperiodic stimulus such as a step-ramp, this re-afferent system can operate in its simplest form by setting $w =1$ and $T_R=0$. However, in order to simulate the more complex patterns of behaviour that we have observed during periodic stimulation, it is necessary to represent the re-afferent system in a more complex manner as a short-term memory or variable time delay component controlled by the periodicity estimator (PE) (Barnes and Asselman, 1991; Barnes, 1994). For this purpose an adequate steady state simulation is obtained if $w = -1$ and $T_R = T/2-T_V$, where T is the period of the stimulus and T_V is the visual processing delay of 80-100 ms (Carl and Gellman, 1987). How the changes in the weight (w) and delay (T_R) are achieved is not known, but it is quite possible that they are identified by a single, self-adaptive, neural network, composed of parallel delay lines with variable weighting factors (Widrow and Stearns, 1985).

The re-afferent feedback has been separated into two components (loops (1) and (2)) in order to take account of the difference between active pursuit and passive stimulation. As indicated earlier, anticipatory smooth movements occur even when the subject is not engaged in active pursuit. Thus, this re-afferent mechanism appears capable of operating in an automatic and non-cognitive manner, a function that is served by the inner loop (1) of the re-afferent pathway (Fig. 3; Sw 1 set to position a). The lower gains for smooth eye movement during passive stimulation can be simulated by assuming a value for β_1 of approximately 0.5. During active pursuit, which appears to require cognitive control to augment the passively induced response, this re-afferent gain can be boosted by the addition of volitional effort at junction (B). This may take the form of a reinforcement of the output of the RAM (i.e., switch Sw 1 set to b; $\beta_2 = 0.45$), thus simulating the effects of reproducing stored velocity drive (Fig. 1), or an independent volitional effort (i.e., Sw 1 set to c; $K_{ACT} = 1 - 4$), simulating the oculomotor drive that can be generated by shifting attention as in Fig. 2.

Although this has been represented as two forms of volitional input, in reality it is likely to be one. The hypothesis is that there is an independent mechanism for initiating smooth eye movement, for which there is now strong neurophysiological evidence (Gottlieb et al., 1994). The output from this intrinsic source is essentially uncalibrated, except in a fairly imprecise manner by shifting attention to different eccentricities from the fovea. However, if previous stimulation has already charged the RAM, the stored information can be used by the volitional system as a reference level for driving the eye in the future. Once charged, this internal re-afference mechanism can become self-perpetuating without any further retinal error information (e.g., when the target is stabilized). But if a conflict arises in the form of an antagonistic velocity error feedback, as in the partial stabilization experiments or during normal pursuit when the target velocity is reduced, the attainable eye velocity will also be reduced.

Thus, although evidence suggests that volitional control of pursuit arises from an independent intrinsic source, visual feedback is essential for its proper regulation. The primary advantage of such a system lies in its ability to rapidly adapt to changes in feedback whilst remaining essentially predictive (Barnes, 1994).

REFERENCES

Barnes, G.R., 1994, A model of predictive processes in oculomotor control based on experimental results in humans, in: Information processing underlying gaze control. Ed: Delgado-Garcia, J.M., Pergamon Press, Oxford, pp. 279-290.

Barnes, G.R., Donnelly, S.F., and Eason, R.D., 1987, Predictive velocity estimation in the pursuit reflex response to pseudo-random and step displacement stimuli in man. *J Physiol (Lond)* 389:111-136.

Barnes, G.R., and Asselman, P.T., 1991, The mechanism of prediction in human smooth pursuit eye movements, *J Physiol (Lond)* 439:439-461.

Barnes, G.R., Collins, C.J.S, Goodbody, S.J., 1994a, Perpetuation of ocular smooth pursuit in humans in the absence of a moving target, *J Physiol* 476:21P.

Barnes, G.R., Collins, C.J.S, Goodbody, S.J., 1994b, Volitional regulation of smooth eye velocity in the absence of a moving visual target in humans, *J Physiol* 480: 45P.

Barnes, G.R., and Crombie, J.W., 1985, The interaction of conflicting retinal motion stimuli in oculomotor control, *Exp Brain Res* 59:548-558.

Barnes, G.R., and Hill, T., 1984, The influence of display characteristics on active pursuit and passively induced eye movements, *Exp Brain Res* 56:438-447.

Barnes, G.R., and Ruddock, C.J.S., 1989, Factors affecting the predictability of pseudo-random motion stimuli in the pursuit reflex of man, *J Physiol (Lond)* 408:137-165.

Becker, W., and Fuchs, A.F., 1985, Prediction in the oculomotor system: smooth pursuit during transient disappearance of a visual target, *Exp Brain Res* 57:562-575.

Carl, J.R., and Gellman, R.S., 1987, Human smooth pursuit: stimulus-dependent responses, *J Neurophysiol* 57:1446-1463.

Collewijn, H., and Tamminga, E.P., 1984, Human smooth and saccadic eye movements during voluntary pursuit of different target motions on different backgrounds, *J Physiol (Lond)* 351:217-250.

Collewijn, H., and Tamminga, E.P., 1986, Human fixation and pursuit in normal and open-loop conditions: effects of central and peripheral retinal targets, *J Physiol (Lond)* 379:109-129.

Cushman, W.B., Tangney, J.F., Steinman, R.M., Ferguson, J.L., 1984, Characteristics of smooth eye movements with stabilised targets, *Vision Res* 24:1003-1009.

Gottlieb, J.P., Bruce, C.J., and MacAvoy, M.G., 1993, Smooth Eye Movements Elicited by Microstimulation in the Primate Frontal Eye Field, *J Neurophysiol* 69 (3):786-799.

Grüsser, O-J., 1986, The effect of gaze motor signals and spatially directed attention on eye movements and visual perception, in: Freund, H-J., Buttner, U., Cohen, B., and Noth, J., (eds), Progress in Brain Research. Vol. 64. Elsevier, North Holland, pp 391-404.

Kommerell, G., Taumer, R., 1972, Investigations of the eye tracking system through stabilised retinal images, in: Bizzi, E., (ed), Bibl.ophthal., Karger, Basal, pp. 288-297.

Kowler, E., and Steinman, R.M., 1981, The effect of expectations on slow oculomotor control: III. Guessing unpredictable target displacements, *Vision Res* 21:191-203.

Lisberger, S.G., and Fuchs, A.F., 1978, Role of primate flocculus during rapid behavioural modification of vestibulo-ocular refle. I. Purkinje cell activity during visually guided horizontal smooth-pursuit eye movements and passive head rotation, *J Neurophysiol* 41:733-763.

Pola, J., and Wyatt, H.J., 1985, Active and passive smooth eye movements: Effects of stimulus size and location, *Vision Res* 25:1063-1076.

Robinson, D.A., 1982, A model of cancellation of the vestibulo-ocular reflex, in: Lennerstrand, G., Zee, D.S., Keller, E.L. (eds.), Functional basis of ocular motility disorders. Pergamon Press, Oxford, pp. 5-13.

van den Berg, A., Collewijn, H., 1987, Voluntary smooth eye movements with foveally stabilized targets, *Exp Brain Res* 68:195-204.

Widrow, B., and Stearns, S.D., 1985, Adaptive signal processing. Prentice-Hall Inc, Englewood Cliffs, New Jersey.

Worfolk, R., and Barnes, G.R., 1992, Interaction of active and passive slow eye movement systems, *Exp Brain Res* 90:589-598.

Wyatt, H.J., and Pola, J., 1988, Predictive behaviour of optokinetic eye movements, *Exp Brain Res* 73:615-626.

Yasui, S., and Young, L.R., 1975, Perceived visual motion as effective stimulus to pursuit eye movement system, *Science* 190:906-908.

16

CLASSIFICATION OF VISUAL PROCESSES FOR THE CONTROL OF POSTURE AND LOCOMOTION

George J. Andersen

Department of Psychology
University of California
Riverside, California 92521

INTRODUCTION

Human observers have a multitude of information for the perception of observer motion through the environment. For example, the vestibular system provides information regarding accelerations of the observer through space (Benson, 1990). The somatosensory system provides proprioceptive information that can be used to determine changes in the body position of the observer (e.g., Allum et al., 1993; Dietz et al., 1991). In addition, gravitoreceptors exist which can be used to detect changes in the orientation of the observer with respect to gravity (Mittelstaedt, 1992). However, of all the different sources of information available, the visual system clearly is of central importance.

CLASSIFICATION OF PERCEPTUAL PROCESSES

The use of visual information for the control of posture and the perception of observer motion in the environment has been a topic of extensive research. In this chapter it is proposed that two categories of visual processes exist for the recovery of visual information for postural control and locomotion. One category of processes is concerned with the recovery of optic flow - the perspective transformation of projected light to the retina during motion of the observer. In this chapter it will be argued that optic flow information requires the accurate recovery of velocity (speed and trajectory) information of all visible feature points within the display. Analyses of optic flow typically involve the extraction of local information. For example, Koenderink (1986) presented a mathematical analysis of optic flow based on local differential invariants of divergence, curl, and deformation. Other computational analyses (e.g., Longuet-Higgins and Prazdny, 1981) have proposed that the velocity difference of two feature points - aligned along the same visual direction - can be used to recover relative distance information. These analyses suggest that the processes concerned with the recovery of optic flow information must be capable of accurately recovering local velocity information.

The second proposed category of processes for locomotion and postural control is concerned with the recovery of global velocity - information extracted by the integration of motion (speed and velocity) over large regions of the visual field. Processes concerned with recovering global velocity information would not require a high degree of spatial resolution, but instead would extract information about global speed and movement of an observer through the environment. For example, Dyre and Andersen (1994) have proposed an analysis based on this information, in which the variance of global velocity can be used to determine the direction of self-motion or heading. Recent psychophysical research (Atchley and Andersen, 1992; 1994) has demonstrated that human observers are sensitive to statistical moment properties of global velocity, including variance. These studies also suggest a simple scheme for recovering statistical properties of global velocity using sets or Reichardt detectors. In addition, several studies concerned with circular (e.g., Brandt et al., 1973) and linear vection (e.g., Lestienne et al., 1977) have examined global velocity information.

FACTORS DISTINGUISHING TWO PROCESSES

An important issue is the factors which distinguish these two categories of processes. It is proposed that three factors distinguish these two categories. These factors include the location in the visual field where information is recovered, the spatial resolution required by perceptual processes, and the necessity of recovering relative depth information. While these factors may be viewed as independent of each other they are not. For example, in order to accurately recovery relative depth (the difference in depth of two positions in the visual field) processes must accurately recover the local velocities of feature

points in the visual field. This information can not be recovered unless the mechanisms involved in the extraction of this information have fine spatial resolution.

Location in The Visual Field

According to this classification scheme, processes concerned with extracting optic flow information require the accurate recovery of local velocities. Because the central visual field has the highest spatial resolution for motion detection (Johnson and Scobey, 1980) it is optimally suited for the recovery of optic flow information. Thus, it is proposed that the extraction of optic flow information occurs primarily in the central visual field. In contrast, the processes concerned with the extraction of global velocity would require the integration of motion information over relatively large regions of the visual field. This type of analysis would not be dependent on the accurate recovery of local velocities. The extraction of motion information in the peripheral visual field is performed by motion detectors with large receptive fields (Johnson and Scobey, 1980; Koenderink et al., 1985). This suggests that the peripheral visual field is optimally suited for recovering global velocity information.

Spatial Resolution

A second factor distinguishing these two types of processes is the spatial resolution required. Processes concerned with extracting optic flow must accurately recover local velocity information. Inaccurate recovery of local velocity information would result in large errors in determining heading or the location of obstacles in the path of locomotion. This observation suggests that the mechanisms important for recovering this information must have fine spatial resolution. In contrast, processes concerned with extracting global velocity information do not require fine spatial resolution. Such mechanisms need only integrate motion information (speed and direction) over relatively large regions of the visual field. The model proposed by Dyre and Andersen (1994) involved the integration of speed and direction information for 5 deg diameter regions of the visual field. Their analyses demonstrated that errors in local velocity information had little effect on heading estimates when speed and velocity information was integrated over large regions of the visual field. Bilocal detectors (see Reichardt, 1961; Koenderink et al., 1985) have been proposed which integrate velocity information over relatively large regions of the visual field, and are ideally suited for extracting global velocity information.

Relative Depth Information

A third issue distinguishing these two types of processes is the importance of relative depth information. Previous research has suggested that relative depth information is important for the perception of self-motion from stimulation of the central visual field. In studies by Andersen and Braunstein (1985), subjects were presented with optic flow displays simulating motion of the observer through a 3D volume of random points. The simulated speed of observer motion and visual angle of the display was varied. In one experiment the latency and duration of perceived self-motion was measured. In a second experiment, ratings of perceived depth were measured. They found that those displays which resulted in the longest duration of perceived self-motion were also the displays assigned the highest ratings of depth. They suggested that these results were indirect evidence that induced self-motion from stimulation of the central visual field is dependent on internal depth within the display and that the failure of previous research to find self-motion effects in central vision was due to the lack of optic flow information for internal depth in the stimuli that were investigated.

Further evidence of the importance of internal depth has been found in studies examining heading. Early studies on heading (e.g., Johnston et al., 1973; see also Regan and Beverly, 1982) found poor performance in determining the direction of self-motion. They found that errors in identifying the direction of motion were as large as 8°. However other studies (Warren et al., 1988), which investigated heading accuracy with flow fields containing relative depth information, have found accuracy as low as 1°. Experiments by Rieger and Toet (1985) directly compared heading accuracy for displays which simulated two frontal parallel surfaces separated in depth (and thus contained relative depth information) or a single frontal parallel surface. They found greater accuracy in heading judgments when relative depth information was present.

WHY TWO DIFFERENT PROCESSES?

Before reviewing the research in support of the two types of processes for locomotion and postural stability, an important question is why would the visual system be sensitive to two classes of information. Three arguments can be made in support of this proposition. The first argument concerns the information available to a locomoting observer. Locomotion of an observer requires that the observer be translating through space. Rotations of the observer, such as has been examined in the circular vection literature, provides information for determining the orientation of the observer but not information for determining locomotion of the observer through the environment. During rotation of the observer a uniform field of velocities are present in the visual field whose trajectories are opposite the direction of rotation. The presence of uniform velocities exists throughout the visual field. In contrast, consider an observer looking straight ahead along the path of locomotion during forward motion. The optic flow pattern in the central visual field will consist of a radially expanding or divergence field. However, in the extreme periphery, the feature points translating past the observer will project as approximately parallel trajectories, resulting in a lamellar velocity field. Thus, different types of velocity information will be present in different locations in the visual field during forward motion of the observer. Extracting observer motion information across the visual field would require the presence of two different motion systems - one concerned with recovering divergence optic flow fields, the other concerned with recovering lamellarr flow fields. Koenderink (1977) has proposed a set of simple motion detectors that could be used for extracting optic flow information including divergence. In contrast, Riechardt (1961) has proposed a motion detector which correlates luminance information over space and time, and which could be used for extracting lamellar velocity information.

A second argument concerns the redundant nature of the visual system. It has been well established that considerable redundancy exists in visual processing. For example, the recovery of 3D shape from vision can occur from any one of a multitude of sources including motion, binocular disparity, shading, and texture (see, for example, Gibson, 1966, or Uttal, 1981 for discussions of this issue). From an evolutionary standpoint it is reasonable to expect that the visual system would have evolved with redundant systems. Redundant information could be important for the survival of an organism as it could provide converging evidence of potential threats in the environment or converging evidence of possible locations of food or shelter for an organism. A similar argument can be made that redundant information from different perceptual systems would be important for the control of posture and locomotion. Such redundancy may be important for maintaining balance or accurately maintaining a fixed direction of motion during locomotion. It has already been demonstrated that human observers have information from several different sensory systems for the control of posture and locomotion. In the present chapter, it is argued that redundant processes exist for recovering visual information for the control of posture and locomotion.

A third argument concerns the type of information available from these two types of processes. Global velocity can be used to determine overall velocity (direction and speed) important for locomoting in an environment and maintaining posture. However, because of the low spatial resolution of such mechanisms (e.g., Reichardt detectors) it is poorly suited for detecting small changes in observer motion and posture. In contrast, optic flow information, because of the requirement that local velocity information be accurately recovered, is ideally suited for recovering small changes in observer motion and posture. This system would likely be important for fine-tuned changes in posture and observer motion during locomotion.

PSYCHOPHYSICAL EVIDENCE IN SUPPORT OF TWO PROCESSES

The use of visual information for the control of posture and the perception of observer motion in the environment has been a topic of extensive research. Research on this issue has examined several different aspects of perceptual function and information for locomotion and posture. This research can be conveniently separated into three different categories - postural stability, induced self-motion or vection, and the perception of the direction of observer motion or heading. A brief review of the psychophysical research on these three types of information is presented, with a focus on evidence in support of two types of processes for the control of locomotion and posture.

Self-Motion/Vection

Research using global velocity information for the study of vection have typically involved large visual displays generated by either a rotating drum (e.g., Brandt et al., 1973; Brandt et al., 1975) - for studies concerned with circular vection - or a translating belt or surface (Lestienne et al., 1977) - for studies concerned with linear vection. In general, these studies have used large visual displays presented to the peripheral visual field. For example, the seminal work by Brandt et al. (1973) required subjects to sit with a large vertically oriented drum. The inner wall of the drum consisted of a grating pattern of vertical bars. They found that stimulation of a large area of the visual field (greater than 30 deg), which necessarily included the peripheral visual field, was necessary to produce vection. Based on these results, they concluded that vection was dependent on stimulation of the peripheral visual field. Similar results were also obtained for studies concerned with linear vection (Lestienne et al., 1977).

In contrast, studies on optic flow have used computer generated displays simulating motion of an observer through a 3D environment of random dots. Andersen and Braunstein (1985) presented subjects with optic flow displays in the central visual field and asked subjects to press a button whenever they experienced self-motion. The speed and size of the area of stimulation were varied. That found that strong self-motion was produced with fields of view as small as 15 deg.

The results of these studies, take together, indicate that the perception of self-motion can occur in either the central or peripheral visual field. Global velocity information is effective for large displays presented to areas of the visual field which include the retinal periphery. Optic flow displays are effective when presented to the central visual field. In addition, studies using optic flow displays simulated 3D motion whereas the studies on global velocity information used displays with no relative depth information. Thus, displays found to be effective in the central visual field included relative depth information whereas displays found to be effective in the peripheral visual field did not include relative depth information.

Heading

As discussed above, research on vection/self-motion has primarily used global velocity displays. In contrast, the majority of research on the direction of self-motion (heading) has been concerned with optic flow displays. Gibson (1966) proposed that the focus of expansion - or the point of maximum divergence (Koenderink) - of optic flow could be used to determine the direction of self-motion. Considerable research has examined the usefulness of this information for determining heading. Experiments have studied the usefulness of this information for linear self-motion (Warren, 1976; Warren et al., 1988), curvilinear motion (Warren et al., 1991), the effects of eye movements in determining heading (Warren and Hannon, 1990), and the effectiveness of optic flow for determining heading as a function of the location of the visual field (Warren and Kurtz, 1992). Other studies (Cutting, 1987; Cutting et al., 1992) have examined the effectiveness of differential motion parallax - the variations in local velocity during fixation and tracking of an object in the scene.

However, an inconsistency exists when considering the research on self-motion and the direction of self-motion. Research on self-motion has demonstrated that the optimal stimulus for producing a compelling impressing of self-motion requires stimulation of a large area of the
visual field that usually includes the retinal periphery. However, the research on heading discussed above have involved relatively small displays (less than 10 deg

visual angle) presented to the central visual field. If heading information is extracted solely by mechanisms sensitive to information in the central visual field, then how is the direction of self-motion determined when stimulation is restricted to the peripheral visual field?

This inconsistency led Dyre and Andersen (1994) to consider a third type of information for determining heading - global velocity. To illustrate this information, consider the following three scenarios. Consider the optic flow field produced by forward translation of an observer with the direction of gaze along the path of motion, and the stimulation restricted to the peripheral visual field. Under these conditions the trajectories of image points are lamellar, flowing in the anterior-posterior direction, and the radial structure of the flow field is degraded. Research using this type of display indicate that subjects perceive forward linear vection (Lestienne et al., 1977). Now consider how perceived self-motion changes when the direction of lamellar motion on one side of the visual field is reversed to the posterior-anterior direction, while the other side of the visual field contains anterior-posterior lamellar motion. Previous research using this type of display indicates that subjects perceive circular vection (Brandt et al., 1973). Finally, consider an intermediate case in which anterior-posterior lamellar flow is presented to both sides of the visual field, but one side contains overall faster velocities than the other side. For this condition, subjects report curvilinear self-motion (Dyre and Andersen, 1992; Sauvan and Bonnet, 1993). One source of information that may account for this finding is the overall difference in the speed of global velocity of the two visual fields.

Dyre and Andersen (1994) examined the usefulness of this information by presenting subjects with optic flow displays in which velocity magnitude varied across the display and provided, for some conditions, conflicting heading information relative to the point of maximum divergence. The results indicate that heading discrimination was influenced by differences in global image velocity magnitude, suggesting that the perception of heading was not solely based on velocity information in the central visual field. Recently, Dyre and Andersen (1994) have developed a model for the extraction of heading from global velocity, based on the recovery of variance information for large (5 deg) regions of the visual field.

Postural Stability

Considerable research has examined the importance of global velocity information for postural stability. Lee and Lishman (1975) required subjects to stand within a large room in which the walls of the room were not attached to the floor. When the room was moved backward and forward subjects adjusted their posture in accordance with the room motion. Other studies have found similar results using either a moving room apparatus (Stoffregen, 1985) or projected a moving pattern to the retinal periphery (Lestienne et al., 1977).

In general, studies examining global velocity patterns have presented stimuli to either the entire visual field or to the peripheral visual field. More recently, studies have examined the usefulness of optic flow patterns presented to the central visual field. Andersen and Dyre (1989) presented computer generated displays simulating observer motion through a 3D volume of random points. The simulated motion of the observer was varied according to a sum of sine wave frequencies. Postural stability was measured using a Kistler platform. Their results indicated that subjects adjusted their posture in accordance with the driving frequencies present in the stimulus. These results, consistent with the results for perceived self-motion, indicate that postural adjustment can occur from stimulation of a small area of the central visual field (15 deg) when the display simulates motion of an observer through a 3D scene.

SUMMARY AND CONCLUSIONS

In this chapter a classification scheme has been proposed for perceptual processes concerned with information for locomotion and postural control. According to this scheme, two different types of information

are used by human observers - global velocity information and optic flow. Global velocity information would require processes which integrate velocity information for relatively large regions of the visual field, and would primarily operate in the peripheral visual field. In contrast, optic flow would involve the recovery of accurate speed and trajectory of local information within the visual field, and would operate primarily in the central visual field. It is quite reasonable to expect that the visual system evolved with these two types of processes. There is considerable redundancy in the visual system for processing other types of information (e.g., 3D shape recovery). The extraction of global velocity would be important for determining large or gross changes in the magnitude and/or direction of observer motion. Motion detectors in the peripheral visual field, which involve large receptive fields, are ideally suited for the extraction of this type of information. In contrast, accurate recovery of local velocity information would be critical for the fine tuned control of posture and motion of an observer through the environment. Mechanisms concerned with the recovery of optic flow could extract this type of information. Motion detectors in the central visual field, which involve small receptive fields, are ideally suited for the recovery of this type of information.

ACKNOWLEDGMENTS

The author would like to thank Paul Atchley for comments on an earlier draft.

REFERENCES

Allum, J. H., Honegger, F., and Schicks, H., 1993, Vestibular and proprioceptive modulation of postural synergies in normal subjects, *Journal of Vestibular Research* 3(1):59-85.

Andersen, G. J., 1989, Perception of 3-D structure from optic flow without locally smooth velocity, *Journal of Experimental Psychology: Human Perception and Performance* 15:363-371.

Andersen, G. J., 1990, Segregation of optic flow into object and self motion components: Foundations for a general model, In: R. Warren and A Wertheim (eds), The Perception and control of Self Motion, L. Erlbaum.

Andersen, G. J., 1995, The detection of 3D surfaces from optic flow, *Journal of Experimental Psychology: Human Perception and Performance* in press.

Andersen, G. J., and Atchley, P. A., 1992, Discrimination of velocity fields: Sensitivity to statistical moments, *Investigative Ophthalmology and Visual Science* 33:1141.

Andersen, G. J., and Atchley, P. A., 1995, The discrimination of velocity fields from statistical moments, Vision Research, in press.

Andersen, G. J., and Braunstein, M. L., 1985, Induced self-motion in central vision, *Journal of Experimental Psychology: Human Perception and Performance* 11:122-132.

Andersen, G. J., and Dyre, B. P., 1989, Spatial orientation from optic flow in the central visual field, *Perception and Psychophysics* 45:453-458.

Andersen, G. J., and Dyre, B. P., 1993, The use of statistical moments for the detection of heading, In: Proceedings of the 38th Annual Meeting of the Human Factors Society, Seattle Washington, pp. 240-241.

Benson, A. J., 1990, Sensory functions and limitations of the vestibular system. In: R. Warren and A Wertheim (Eds) Perception and control of self-motion, Lawrence Erlbaum Associates, Inc, Hillsdale, NJ, US. pp. 145-170.

Brandt, T., Dichgans, J., and Koenig, E., 1973, Differential effects of central versus peripheral vision on egocentric and exocentric motion perception, *Experimental Brain Research* 16:476-491.

Brandt, T., Wist, E., and Dichgans, J., 1975, Foreground and background in dynamic spatial orientation, *Perception and Psychophysics* 17:497-503.

Cutting, J.E., 1987, Perception with an eye for motion. Academic Press.

Cutting, J.E., Springer, K., Braren, P.A., and Johnson, S.H., 1992, Wayfinding on foot from information in retinal, not optical, flow, *Journal of Experimental Psychology: General* 121:41-72.

Dichgans, T., and Brandt, J., 1974, The psychophysics of visually-induced perception of self-motion and tilt, In F.O. Schmidt and F.G. Worden (eds) The Neurosciences pp. 123-129, Cambridge MA: MIT Press.

Dietz, V., Trippel, M., and Horstmann, G. A., 1991, Significance of proprioceptive and vestibulo-spinal reflexes in the control of stance and gait, In: A. Patla (Ed) Adaptability of human gait: Implications for the control of locomotion, North-Holland, Amsterdam, Netherlands. 1991. pp. 37-52.

Dyre, B. P., and Andersen, G. J., 1994, The perception of heading from global optical flow, Manuscript under review.

Gibson, J. J, 1966, The senses considered as perceptual systems, Boston, MA: Houghton Mifflin.

Johnson, C. A., and Scobey, R. P., 1980, Foveal and peripheral displacement thresholds as a function of stimulus luminance, line length, and duration of movement, *Vision Research* 20:709-715.

Koenderink, J. J., and van Doorn, A. J., 1977, How an ambulant observer can construct a model of the environment from the geometric structure of the visual inflow, In: G. Hauske and E. Butendant (eds.) Kibernetic, Oldenbourg: Munchen.

Koenderink, J. J., 1986, Optic flow, *Vision Research* 26:161-180.

Koenderink, J. J., van Doorn, A. J., and Van de Grind, W. A., 1985, Spatial and temporal parameters of motion detection in the peripheral visual field, *Journal of the Optical Society of America* A 2:252-259.

Koenderink, J. J., and van Doorn, A. J., 1981, Exterospecific component of the motion parallax field, *Journal of the Optical Society of America* 77:953-957.

Lee, D., and Lishman, J., 1975, Visual proprioceptive control of stance, *Journal of Human Movement Studies* 1:87-95.

Lestienne, F., Soechting, J., and Bertoz, A., 1977, Postural readjustments induced by linear motion of visual scenes, *Experimental Brain Research* 28:363-384.

Longuet-Higgins, H. C., and Prazdny, K., 1980, The interpretation of a moving retinal image, Proceedings of the Royal Society (London) B 208:385-397.

Mittelstaedt, H., 1992, Somatic versus vestibular gravity reception in man, *Annals of the New York Academy of Sciences* 656:124-39.

Reichardt W., 1961, Autocorrelation: a principle of the evaluation of sensory information by the central nervous system, In: Rosenblith, W. A. (Ed.) Sensory communication, New York: Wiley.

Sauvan, X. M., and Bonnet, C., 1993, Properties of curvilinear vection, *Perception and Psychophysics* 53:429-435.

Stoffregen, T.A., 1986, The role of optical velocity in the control of stance, *Perception and Psychophysics* 39:355-360.

von Kries, J., 1962, Notes on the perception of depth, In H. von Helmholtz, Treatise on physiological optics (Vol 3, J. P. C. Southall, Ed. and Trans.). New York: Dover.

von Helmholtz, H., 1962, Treatise on physiological optics (Vol 3, J. P. C. Southall, Ed. and Trans.), New York: Dover. (Original work published in German, 1867, and in English, 1925).

Uttal, W. R., 1981, A taxonomy of visual processes, Hillsdale, NJ: Erlbaum.

Warren, R., 1976, The perception of egomotion, *Journal of Experimental Psychology: Human Perception and Performance* 2:448-456. Warren, W. H., and Hannon, D. J., 1990, Eye movements and optical flow, *Journal of the Optical Society of America* A 7:160-169.

Warren, W.H., and Kurtz, K.J., 1992, The role of central and peripheral vision in perceiving the direction of self-motion, *Perception and Psychophysics* 51:443-454.

Warren, W.H., Mestre, D.R., Blackwell, A.W., and Morris, M.W., (1991), Perception of circular heading from optical flow, *Journal of Experimental Psychology: Human Perception and Performance* 17:28-43.

17

DIFFERENTIAL INFLUENCE OF A VISUAL FLOW PATTERN ON EMG-ACTIVITY OF ANTAGONISTIC LEG MUSCLES DURING UNSTABLE STANCE

V. Dietz,[1*] M. Schubert,[2] and W. Berger[2]

[1] Paraplegic Centre
University Hospital Balgrist
CH-8008 Zürich, Switzerland
[2] Department of Clinical Neurology and Neurophysiology
University of Freiburg
D-7800 Freiburg, Germany

INTRODUCTION

Well-aimed experimental models are required to investigate the human motor system under physiological conditions such as stance and gait. With afferent inputs being necessary to appropriately modulate programmed patterns involved in the regulation of stance and gait, the relative contributions of known receptor systems are still a matter of controversy. These systems include the visual system (Dietz et al., 1979; Nashner and Berthoz, 1978, van Asten et al,. 1988), the vestibular system (Allum and Pfaltz, 1985) and muscle proprioceptive systems (Berger et al., 1984; Dietz, 1992; Dietz et al., 1987).

The aim of the present study was to examine the contributions of visual and muscle proprioceptive afferent input to the programmed EMG pattern. Therefore, stance regulation was investigated under conditions of a continuous instability of body equilibrium which was induced by horizontal sinusoidal oscillations of the support surface and the visual flow pattern both moving with the same frequency. The effect of mismatch between proprioceptive and visual input could be studied by the introduction of various phase-shifts between oscillatory movements of the visual flow pattern and those of the legs. This allowed to study the interaction of proprioceptive and visual cues during continuous regulation of upright stance (Dietz et al., 1992, 1994).

---

[*] Address correspondence to: Dr. Volker Dietz, Schweizerisches Paraplegikerzentrum, Universitätsklinik Balgrist, Forchstrasse 340, CH-8008 Zürich, Switzerland.

METHODS

Subjects stood in an upright position upon a treadmill moving sinusoidally -backward/forward in relation to the subject - with frequencies of 0.25 (or 0.5 Hz) and an amplitude of +/- 8 cm. A visual flow pattern was projected onto a hemispheric screen located in front of the subject's head. The design of the visual flow pattern was chosen to elicit the illusion of viewing a tunnel or a corridor which could be moved, - independently from the subject's supporting surface - along the subject's line of sight (see Fig. 1 A, B). The visual flow pattern was also moving sinusoidally at 0.25 Hz, thus matching the movement characteristics of the treadmill.

Frequency changes in sinusoidal treadmill movement coincided with the projection of the moving visual stimulus (Fig. 1 A, B and C). The superimposed oscillations of treadmill and visual pattern were determined to produce a "retinal image" with various phase-shifts (with respect to the treadmill sinus) of 45, 90, 135, 180, 225, 270, 315 and 360 degrees (Fig. 1 C). Each of the various visual conditions was displayed over a period of five treadmill cycles and each visual condition (consisting of five oscillatory cycles) was averaged over 10 trials for each subject. All conditions were presented randomly during any experimental session. The sinusoidal signals applied to the treadmill and the projector were generated using a two-channel PC-microcomputer-based impulse generator system.

Electromyographic (EMG) recordings were made using surface electrodes from the medial gastrocnemius and tibialis anterior muscles of the right hand side. Ankle- and knee-joint movements were monitored using mechanical goniometers fixed at the lateral aspect of the right foot, leg and thigh (Dietz et al., 1987; Dietz et al., 1989). The treadmill belt was placed over a force measuring system (Kistler). Thus, from the output of 4 Piezo-elements, fixed at the corners of the treadmill, the antero-posterior swaying of the subjects could be measured in terms of the torque exerted by the centre of force (i.e. the body's centre of gravity).

EMG and biomechanical recordings were amplified (FM-microvolt amplifier; time constant 0.15 s, bandwidth 0.1 - 1000 Hz) and, after rectification of the EMG, transferred on-line to a PC-microcomputer system via an A/D-converter sampling at 0.2 kHz (21).

Because the effects of visual stimulation on the leg muscle activation were of major interest, the responses from an additional condition with a phase-shift of 360 deg. were individually subtracted from the recordings obtained in all conditions. The individual EMG patterns of all conditions were normalized with respect to the EMG activity obtained after a change in treadmill frequency from 0.5 to 0.25 Hz during continuing darkness (labeled "relative units"). By dividing each sinusoidal cycle into 50 ms segments the integrated EMG activity of leg muscles during one movement cycle could be studied.

Statistical processing (calculation of means, standard deviation (S.D.), t-tests, correlation coefficients and analysis of variance (repeated measurement design)) was performed upon the individually averaged data using a SPSS/PCTM package.

RESULTS

Fig. 2 shows the effect of two phase-shifts (90 and 180 deg.) between treadmill and visual sinus upon EMG activity in the lower leg muscles (fifth cycle). The different steps of processing of EMG-signals are displayed. In Fig. 2A, the mean EMG activity of 10 trials with S.D. of one individual subject is shown. In order to see the pure

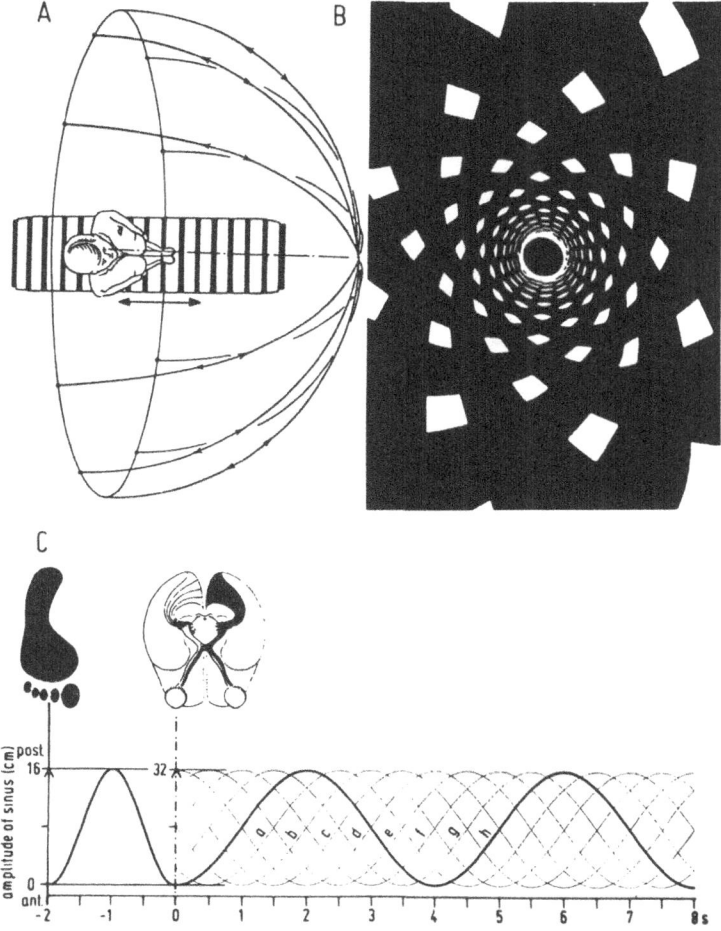

Figure 1. Schematic diagram of the experimental conditions. A, Subject standing on the treadmill within the hemispheric screen. The visual pattern displayed on the screen was moving along the arrows. B, Subject's view of the pattern, which was moved to evoke the illusion of motion along the subjects line of sight. C, Schematic diagram displaying the sinusoidal movements of the treadmill (thick line) and the visual pattern (thin lines) with increasing phase-shifts. The presentation of the visual pattern was coincident with a slowing of treadmill frequency from 0.5 to 0.25 Hz (starting at t = O s).

effects elicited by the visual stimulation, the EMG patterns of the control condition were individually subtracted from those obtained during each experimental condition. All EMG data have been normalized. Then mean values of EMG activity obtained from all subjects were calculated. For comparison the mean EMG traces obtained from all subjects before (Fig. 2B) and after (Fig. 2C) subtraction are shown. The effect of dividing the integrated EMG activity of each sinusoidal cycle into 50 ms segments (mean of all subjects) is shown in Fig. 2D.

Fig. 3A shows the modulated traces of tibialis anterior EMG activity induced by the various phase-shifts. Results were obtained after adaptation. The tibialis anterior EMG was differentially modulated by the various conditions throughout the whole cycle, usually with an activity peak just after the posterior turning point (t = 2 s). Maximal EMG activity of

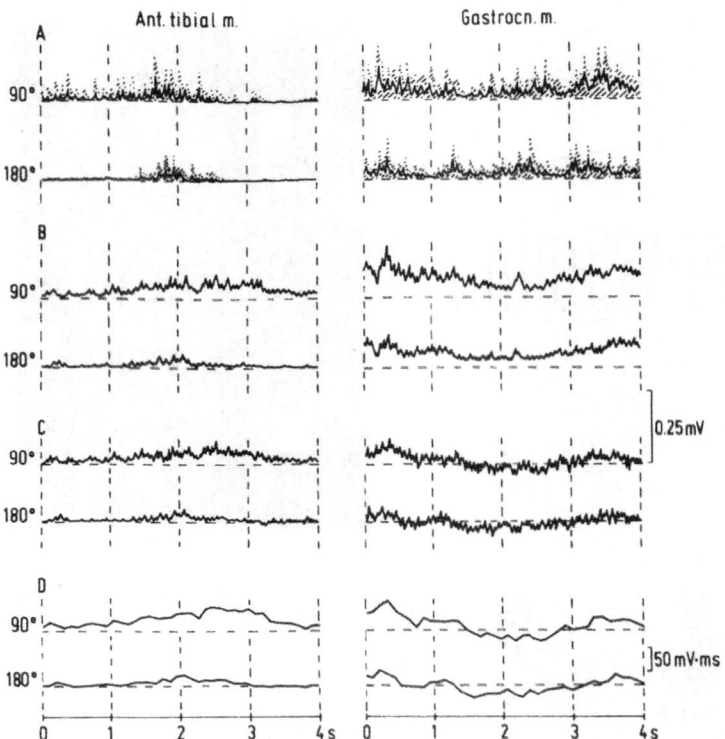

Figure 2. Tibialis anterior (A) and gastrocnemius (B) EMG responses to phase-shifts of 90 and 180 deg. between movement of the treadmill and of the visual pattern. The EMG activity displayed was recorded during the fifth sinusoidal cycle after changing the treadmill frequency with the concomitant presentation of the visual pattern. A, mean and S.D. of the rectified and averaged (n=10) EMG responses of an individual subject. Mean values from all subjects of the (B) unsubtracted and (C) subtracted EMG traces. In the latter case recordings from the control condition were subtracted from the various EMG patterns in order to demonstrate the true effects of visual stimulation on the EMG activity. (D) Mean values from all subjects of the subtracted, normalized, and integrated EMG responses (50 ms integrals). Anterior (t = O and 4 s) and posterior (t = 2 s) turning points.

longest duration occurred during phase shifts of 90 and 135 deg. and of some what smaller amplitudes during phase shifts of 270 and 315 deg. Fig. 3B shows EMG profiles from the gastrocnemius muscle. The main EMG activity phase was more or less restricted to the period after the anterior turning point (from about t = O to 1 s). The largest EMG amplitudes appeared during phase shifts of 90, 135 and 270 deg.

Fig. 4 displays the recorded signals of head acceleration (Fig. 4A) and treadmill torque (Fig. 4B) during the various phase-shifts. The EMG modulation is reflected in these biomechanical signals. The two signals indicate that a forward movement of the body occurred after the posterior turning point (t = 2 s). Around the anterior turning point (t = O s) straightening movements occurred which resulted in the return to a neutral body position ("resetting movements"). The time-relationship between the appearance of the EMG responses and biomechanical signals as well as the movement direction indicate that the movements most probably were the consequence of visually-evoked EMG responses.

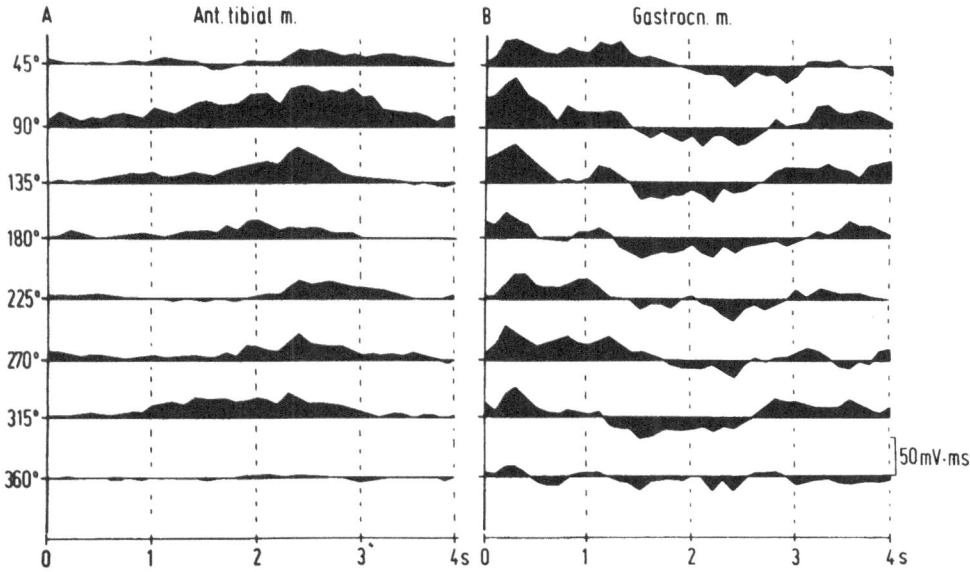

Figure 3. Mean values from all subjects of the subtracted and integrated (A) tibialis anterior and, (B) gastrocnemius EMG activity profiles (50 ms integrals) obtained under the various experimental conditions during the fifth cycle of visual stimulation.

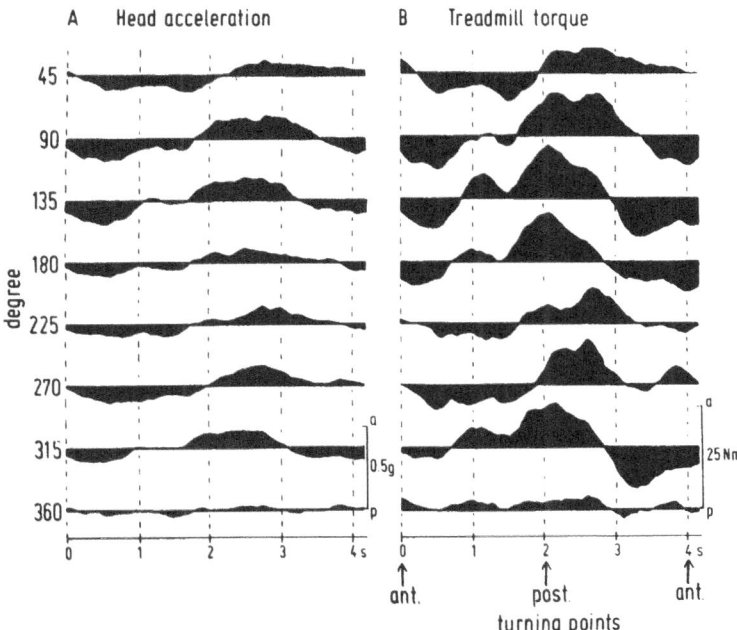

Figure 4. Mean values calculated from all subjects of head acceleration (A) and treadmill torque (B) signals induced by the phase-shifts during the fifth cycle after presentation of the visual pattern (control condition has been subtracted). a: anterior, p: posterior.

DISCUSSION

Postural stabilization during the sinusoidal treadmill movements with congruent visual stimulation is achieved mainly by modulation of leg extensor muscles (Dietz et al., 1993). However, presentation of the incongruous visual information mainly resulted in modulation of EMG activity in leg flexor muscles. At times around the posterior turning point tibialis anterior EMG was modulated in its onset, duration and amplitude. This is the time during which the tibialis anterior muscle is most active during sinusoidal treadmill movements. In comparison gastrocnemius EMG activity was only modulated in amplitude within a short time period restricted to the anterior turning point, the time during which predominant extensor EMG activity occurs.

The difference between the behaviour of the antagonistic leg muscles, where the leg flexors have a higher responsiveness to visual stimuli but the leg extensors to somatosensory input, agrees with observations from experiments in the cat (Beloozerova and Sirota, 1988). In addition, corticospinal projections to lower limb motoneurons were recently shown to be stronger to the tibialis anterior than to the gastrocnemius muscles (Brouwer and Ashby, 1992).

The posterior turning point in the treadmill sinusoid represents a time period during which the tibialis anterior muscle is sensitive towards modulation by a visuo-proprioceptive mismatch. It is also a time period of highest accelerative forces acting upon the body. According to earlier publications (Berthoz et al., 1975) the vestibular system as well as force- and cutaneous pressure receptors will predominantly respond to accelerative forces whereas the visual system is apt to sense the magnitude of velocity. During the experimental conditions with phase shifts of 90/270 degs and 180/360 degs a coincidence was induced of maximal/minimal momentary velocity of the visual flow pattern and, at the anterior/posterior turning point, maximal acceleration of the body. This obviously resulted in maximum and minimum EMG responses in the tibialis anterior muscle.

During quiet stance, direction-specific EMG reactions are usually elicited by visual stimuli (Berthoz et al., 1979). In the experimental condition described here, this effect may be overridden by the programmed pattern of leg muscle activation developing during the treadmill sinus. However, an effect of stimulus direction became apparent when comparing the strength of the visually-evoked EMG activity: The tibialis anterior EMG amplitude was larger and of longer duration when the motion direction of the visual stimulus was opposite to that of the feet than when the direction was the same.

The EMG responses in the gastrocnemius muscle were associated with a straightening of the body and "resetting" of a neutral body position around the anterior turning point: The signals of linear head acceleration and treadmill torque tend to resume lowest values during this phase of the oscillations. The gastrocnemius EMG activity, therefore, seems not directly to depend on visual influences but is rather dependent upon and has to compensate for the strength of the preceding visually-induced tibialis anterior muscle response.

SUMMARY

The influence of visual input upon body stabilization was studied with subjects standing on a treadmill while a visual flow pattern was displayed. Both the treadmill and image were moved sinusoidally - backward/forward - at a frequency of 0.25 Hz. The effect of the presentation of the visual pattern was studied upon leg muscle electromyographic (EMG) activity and corresponding biomechanical signals for various phase-shifts (45 to 360 deg.) between movements of the legs and the visual flow pattern. Around the posterior turning point of sinusoidal treadmill movement, a modulation of the tibialis anterior EMG was observed. The onset, duration and amplitude of the latter were dependent upon the phase-shift between the movements of the legs and the visual pattern. Maximum responses were recorded at phase-shifts with a coincidence of fast velocities of the visual pattern with a phase of maximal body acceleration. At times around the anterior turning point, a modulation only of EMG amplitude occurred in the gastrocnemius muscle, for a "resetting" of the neutral body position. It is suggested that the tibialis anterior is more sensitive than the gastrocnemius muscle to a visual stimulus. The activity of the latter muscle is more subject to proprioceptive input.

ACKNOWLEDGMENTS

This work was supported by the Swiss National Science Foundation (No. 31-33567.92) and by the Deutsche Forschungsgemeinschaft (Be 936/5-1).

REFERENCES

Allum, J. H. J., and Pfaltz, C. R., 1985, Visual and vestibular contributions to pitch sway stabilization in the ankle muscles of normals and patients with bilateral peripheral vestibular deficits. *Exp. Brain Res.* 58: 82-94.

Beloozerova, J. N., and Sirota, M. G., 1988, Role of motor cortex in control of locomotion. In: Gurfinkel, V. S., Joffe, M. E., Massion, J., Roll, J. P., eds., Facts and Concepts. New York: Plenum Press. pp. 163-176.

Berger, W., Dietz, V., and Quintern, J., 1984, Corrective reactions to stumbling in man: Neuronal coordination of bilateral leg muscle activity during gait. *J. Physiol.* (Lond.) 357:109-125.

Berthoz, A., Lacour, M., Soechting, J. F., and Vidal, P. P., 1979, The role of vision in the control of posture during linear motion. In: Granit, R., Pompeiano, O., eds., Progress in Brain Research. Reflex Control of Posture and Movement. Amsterdam: Elsevier, pp 197-209.

Berthoz, A., Pavard, B., and Young, L. R., 1975, Perception of linear horizontal self-motion induced by peripheral vision. *Exp. Brain Res.* 23:471-489.

Brouwer, B., and Ashby, P., 1992, Corticospinal projections to lower limb motoneurons in man. *Exp. Brain Res.* 89:649-654.

Dietz, V., 1992, Human neuronal control of automatic functional movements: Interaction between central programs and afferent input. *Physiol. Rev.* 72:33-69.

Dietz, V., Horstmann, G. A., and Berger, W., 1989, Interlimb coordination of leg muscle activation during perturbation of stance in humans, *J. Neurophysiol.* 62:680-693.

Dietz, V., Quintern, J., and Sillem, M., 1987, Stumbling reactions in man: Significance of proprioceptive and pre-programmed mechanisms. *J. Physiol. (Lond.)* 386:149-163.

Dietz, V., Schubert, M., Trippel, M., 1992, Visually induced destabilization of human stance: neuronal control of leg muscles, *Neuroreport.* 3:449-452.

Dietz, V., Schubert, M., Discher, M., and Trippel, M., 1994, Influence of visuoproprioceptive mismatch on postural adjustments, *Gait and Posture* 2:147-155.

Dietz, V., Trippel, M., Jbrahim, I. K., and Berger, W., 1993, Human stance on a sinusoidally translating platform: balance control by feedforward and feedback mechanisms, *Exp. Brain Res.* 93:352-362.

Nashner, L. M., and Berthoz, A., 1978, Visual contribution to rapid motor responses during postural control, *Brain Res.* 150:403-407.

Van Asten W. N. J. C., Gielen C. C. A. M., and Denier van der Gon J. J., 1988, Postural adjustments induced by simulated motion of differently structured environments, *Exp. Brain Res. 73:371-383.*

18

THE FORMATION OF THE VISUAL AND THE POSTURAL VERTICAL

H. Mittelstaedt

Max-Planck-Institut für Verhaltensphysiologie
D-82319 Seewiesen
Germany

INTRODUCTION

This contribution summarizes the extant facts on the causal processing of the perceived orientation of the visual field and one's own body to the vertical, the conclusions derived from them and the hypotheses evoked by, as well as the theories developed from these findings. Due to limitations of space the evidence which supports them cannot be detailed here, but may be found at especially cited places of the literature.

1. METHODS

1.1. In a fixed but variable roll deviation ρ of their long (Z-) axis from the gravito-inertial vector subjects (Ss) are asked to set a luminous pendulum (LP) in otherwise complete darkness, by remote control, to subjective vertical. The visual vertical (SVV) may then be defined as the mean roll-deviation δ of the LP from the objective vertical. However, since the S, other than the experimenter, cannot observe δ, we better use the variable which determines the afferent input of the visual feedback loop constituted by our experiment, namely, the angle β between the LP and the head's median plane, with $\beta = \rho - \delta$.

1.2. On a tiltable board, then on a sled centrifuge, Ss are asked to change their attitude, also by remote control and in complete darkness, until they feel to be horizontal. The subjective horizontal posture (SHP) of one's body may then be defined as the mean elevation ε of the S's Z-axis from the horizontal plane, or, for ready comparison with the SVV data, as the mean roll deviation ρ of the S's Z-axis from the gravitational vertical in that attitude.

All angles are projections onto the Y-Z plane of the head-fixed right handed system of the insert of Fig.1, and defined with reference to the objective (δ, ρ) or the subjective (β) vertical, that is, in the latter case with the LP as reference, clockwise positive. The head's median (X-Z) plane always coincides with the median plane of the trunk.

Multisensory Control of Posture, Edited by T. Mergner and F. Hlavačka
Plenum Press, New York, 1995

2. THE SUBJECTIVE VISUAL VERTICAL

2.1. Facts

2.1.1. The SVV shows a typical sigmoidal dependency of the angle β on the roll tilt (s. Fig.1), such that β first tends to compensate ($\beta=\rho$) or overcompensate the tilt ($\beta>\rho$: E-Effect) and then undercompensate it ($\beta<\rho$: Aubert- or A-effect), in some Ss even vice versa, until the SVV is veridical again at $\rho=$ 180 deg. Thus during a full 360 deg turn veridicality is reached at least twice, in some Ss four times.

2.1.2. At an increased magnitude G of the gravito-inertial vector the (unsigned) magnitude of β is increased at acute and decreased at obtuse angles of tilt, but by far less than expected if β would depend on Gsin. At or near the four orthogonal attitudes ($\rho =$ 0,±90,180 deg) increased G load has no effect (Schöne and Udo de Haas, 1971).

2.1.3. Centrifugation about an earth vertical axis collinear with the binaural axis (BA) leaves the SVV virtually unchanged (Mittelstaedt, 1985, esp. p 144). If the BA is then shifted craniad or caudad, β changes with respect to the resulting force vector as it does in the control on a tiltable board. The result is the same whether the Z-component of the force vector affects mainly the utricle or mainly the saccule.

2.1.4. Deviations from veridicality analogous to the A-effect are also found for the auditory vertical (Lechner-Steinleitner et al., 1981) as well as for the visual and haptic zenith in pitch (Mittelstaedt, 1983).

2.2. Theory

The deviation δ of the LP from the objective vertical cannot be measured by the Ss' sense organs, hence must be computed from the difference between a measurement of the angle between head and objective vertical and another of the angle between head

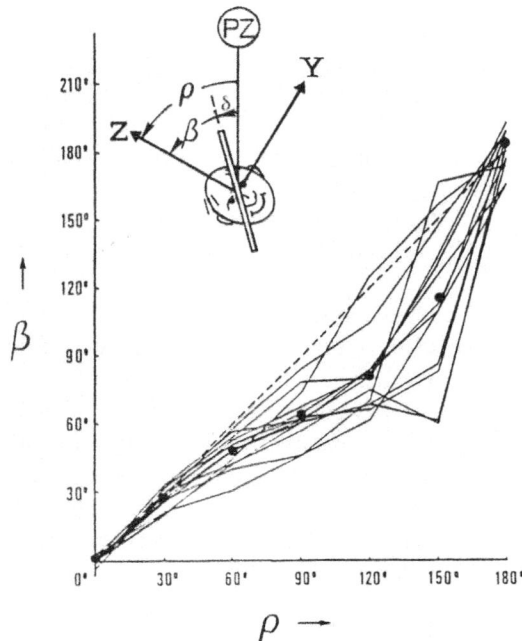

Figure 1. Dependence of the subjective visual vertical upon roll tilt in 13 Ss, as found by Udo de Haas, 1970. δ: angle between a luminous line and the objective vertical (PZ). Abscissa: angular deviation ρ of the head's Z-axis from the physical vertical (X-axis forward, Y- and Z-axis as shown). Ordinate: angle β between the head's Z-axis and the luminous line. All angles are defined with reference to the objective (δ, ρ) or the subjective (β) vertical, as shown by the arrows, clockwise positive. Dots at the means. Note the S whose settings cross the dashed line ($\beta=\rho$), that is, from A- to E-effect, at $\rho \approx 112$ deg.

and LP. Since $\delta=\rho-\beta$, the LP would coincide with the objective vertical if the S rotates the LP until the difference is zero. The theory assumes, however, that the SVV is determined by two tendencies (Mittelstaedt, 1983): The first, controlled by utricles and saccules, tends to rotate the SVV towards $\rho-\beta= 0$, hence towards the objective vertical. The second, controlled by a central nervous agent named "idiotropic vector", tends to rotate the SVV towards $\beta= 0$, hence towards the S's median plane. The first is caused by representations x,y,z of the three force components along the X-,Y-,Z-axes of the labyrinth that are gained by weighting and summating the otolithic afferences (Mittelstaedt, 1975, 1983, 1989). The second is caused by vectorial addition of a constant idiosyncratic quantity (M) of, presumably, central nervous origin. It is essential that the three components are *normalized*, e.g. by division through $N = \sqrt{x^2 + y^2 + z^2}$ or by a feedback loop that leads to the same effect. The idiotropic vector, however, must be unaffected by the normalization, that is, added afterwards. This selective normalization makes the SVV independent of the effect of the non-linearities of the primary otolithic afferences (Fernandez and Goldberg, 1976) at the six orthogonal attitudes, but not in between, where it will show up when the G-load is increased.

The SVV is finally generated by vector multiplication with the respective components of the visual field, in our test of the angle β. Since the eyes are mobile, however, the visual components can only be gained by combining retinal with extraretinal information provided by proprioceptors or efference copies. This process is much more important in the case of pitch tilts (s. Mittelstaedt, 1983, esp. p 278), where large eye deviations occur, than in roll tilt, where the very small ocular counterroll (of less than 10% of ρ) is more or less accounted for (Haustein, 1992). In general, then, the system performs a triple coordinate transformation of the entire visual field from retinal to head and from head to gravito-inertial coordinates, which implies a third transformation from the pitched-up otoliths to the head (Mittelstaedt, 1989).

In order to demonstrate the essentials and the merits of the theory, we shall consider only the case of pure roll, where the non-compensated fraction of the ocular counterroll may be neglected, as well as the otolith-to-head coordinate transformation, which then merely affects the gain of the z-component to some degree. Also, the effect of non-linear deviations of the otolithic primary afference (Fernandez and Goldberg, 1976), which the postulated normalization cannot remove and hence shows up especially at large G-loads, will not be treated here (but s. Mittelstaedt, 1988, esp. p. 235ff).

In pure roll, then, the y-and z-components (y_{L-SV}, z_{L-SV}) of the perceived deviation of the LP from the vertical, under these idealized conditions, are

$$y_{L-SV} = \cos\beta \; y/N - \sin\beta \; (z/N+M) \tag{1a}$$

$$z_{L-SV} = \cos\beta \; (z/N+M) + \sin\beta \; y/N \tag{1b}$$

with $y = U_1\sin\rho$, $z = S_1\cos\rho$, and $N = \sqrt{y^2 + z^2}$ .The subject moves the pendulum until y_{L-SV} is zero. Then

$$\tan\beta = \frac{y/N}{z/N + M} \tag{2}$$

This simple formula can be fairly well adapted to the mean SVV-settings of normal Ss with U_1 set to unity and M determined as the cotangent of β at $\rho= 90$ deg. Thus only one free parameter remains, namely S_1. It turns out to be around 0.6, that is, S_1/U_1 is close to the relation of the number of saccular sense cells to those of the utricle.

2.3. Consequences, Explanations and Predictions

2.3.1. Although this type of coordinate transformation is far from perfect, it nevertheless results in a fairly veridical SVV around the upright attitude where veridicality matters most.

2.3.2. The theory yields a simple explanation for the E- and A- effects. A crossover from over- to undercompensation (from E- to A-effect), or vice versa, should occur at a tilt angle (ρ_{cross}) if and when

$$\tan^2\rho_{cross} = \frac{(S_1/U_1 - 1)^2}{M^2} - \left(\frac{S_1}{U_1}\right)^2 \tag{3}$$

Hence, a crossover can only occur if

$$M^2 < \frac{(S_1/U_1 - 1)^2}{(S_1/U_1)^2} \tag{4}$$

The SVV crosses over from E- to A- effect at acute tilt angles if, as normally, $S_1/U_1 < 1$, but from A- to E- effect at obtuse tilt angles if $S_1/U_1 > 1$. Fig.1 shows such a case, with $\rho_{cross} \approx$ 112 deg, $M \approx 0.28$ and $S_1/U_1 \approx 1.8$, suggesting utricular damage.

2.3.3. The theory yields an explanation for yet another puzzling phenomenon, namely, that the variance of the SVV settings increases monotonically with the tilt, whereas the variance of the otolithic afference in these fixed head postures does not: Due to the idiotropic vector the resulting vector of Eq. 1a, b will be largest at $\rho = 0$ and decrease monotonically with increasing tilt. And the smaller it is, the larger will be the variance, because the sensitivity of the LP settings depends on the relation of the perceived to the objective deviation of the LP from the vertical ($dy_{L-SV}/d\delta$), which decreases accordingly.

2.3.4. A "Ganzfeld" of random dots rotating with constant clockwise (positive) velocity v about the visual (X-) axis causes a clockwise deviation of the SVV (that is, a negative deviation of β), which increases with increasing tilt. When the field rotates in the opposite sense of the head tilt this deviation is larger than when it rotates in the same sense of the head tilt; and this asymmetry also increases with increasing head tilt (Dichgans, Diener, and Brandt 1974).

Exactly this relation is yielded by the theory if representations $K_s v$ and $K_c v$ of the field's velocity v are introduced into Eq. 1 and 2 in the following way (Mittelstaedt, 1991, esp. p. 423ff):

$$\tan\beta = \frac{y/N - K_s vz/N}{z/N + K_c vy/N + M} \tag{5}$$

with $N = \sqrt{y^2 + z^2}$ and with $K_s/K_c \approx 0.6$ (own unpublished results).

The asymmetry is caused by the idiotropic vector, hence should be missing if $M = 0$. This has indeed been found in a patient who suffers from total loss of the myelinated proprioceptive afference below the neck (Yardley, 1990).

2.3.5. The well known effect of structures of the *static* visual field on the SVV (for review s. Howard, 1982, esp. p. 419ff) may also be accommodated by the theory, although in a rather complex way that cannot be treated here (but see Mittelstaedt, 1986).

2.3.6. The theory also yields predictions of the effects of unilateral and bilateral loss of the otoliths (Mittelstaedt, 1995a).

3. THE SUBJECTIVE POSTURAL HORIZONTAL

3.1. Facts

3.1.1. On the tiltable board Ss maintain their subjective horizontal position (SHP) with a mean SD of ±1deg (whereas their SVV at ρ= 90 deg is ±2.5 deg or more), but deviate by an idiosyncratic amount of on average ±4.7 deg from the objective horizontal. This bias of the SHP is fairly constant, even over years. The SHP biases are not correlated with the SVV deviations at that posture (Mittelstaedt, 1983, esp. Fig.3, p 274).

3.1.2. On the sled centrifuge (Fig.2) most Ss do not feel horizontal when the gravito-inertial force vector acting on the otoliths, or its Z-component, is the same as when they feel horizontal on the tiltable board , that is, when the distance c of the centrifuge axis from their binaural axis is equal to the distance $a=-g\omega^{-2}\cotan\rho$ or $b=-g\omega^{-2}\cos$ (where $\omega=0.77$ rad/sec, $g=981$ cm/sec^2 , and ρ the SHP on the tiltable board). In fact, as shown in Fig.3 the differences $\Delta=c-b$ are distributed between zero and 45 cm with a mean around 25-28 cm. Bilaterally neuromectomized Ss, however, show a mean difference Δ of about 45 cm.

3.1.3. On the tiltable board as well as on the centrifuge the SHP is not affected by pressure or tension to the legs; but it depends on leg posture, that is, ρ and Δ increase if the legs are extended (like in standing) from a flexed (like in sitting) posture (Mittelstaedt and Fricke, 1988, esp. Fig.3, p. 29).

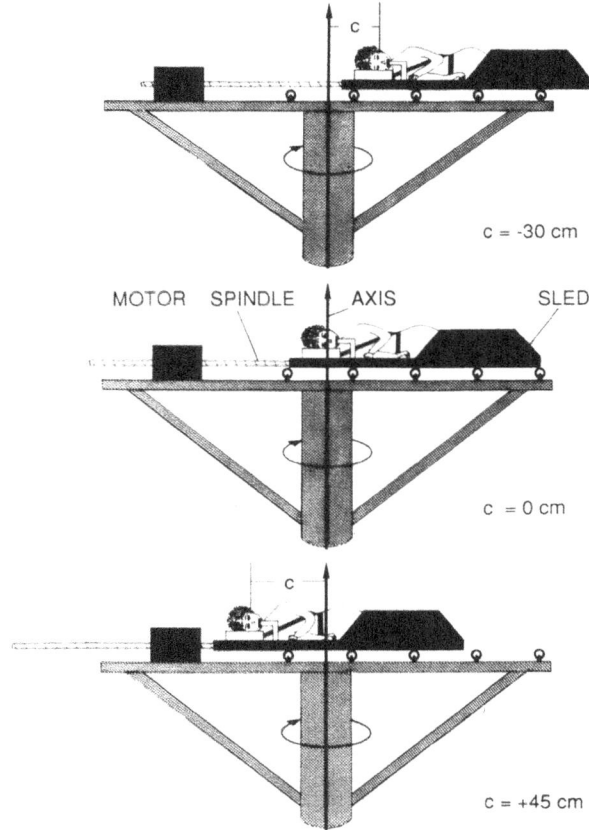

Figure 2. SHP on a rotating sled centrifuge. The S is placed right ear down (as on the preceding control test on a tiltable board) on a sled that can be moved radially along the S's spinal (Z-) axis via remote control by the S (as well as by the experimenter). Due to the additional acceleration, the S feels tilted (as on the board) depending on the distance between the gravity sense organ(s) and the centrifuge axis CA. The SHP is measured as the distance c of the binaural axis from the CA. Under exclusive guidance of the otoliths, the S (assume she had set herself at ρ=90 deg on the tiltable board) would feel tilted upward at negative c, downward at positive c, and horizontal at c=0.

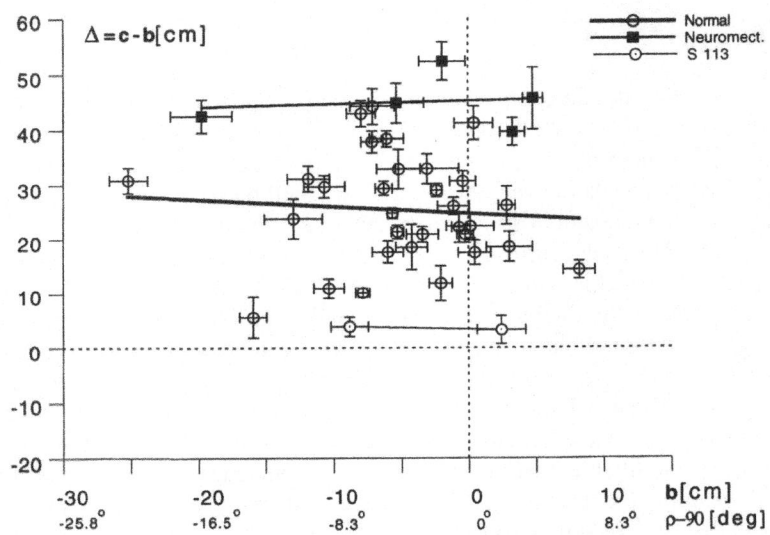

Figure 3. SHP of 32 normal Ss and 5 neuromectomized Ss, right ear down, legs flexed, on a tiltable board and on a sled centrifuge, which produces about 1.5 G per meter radius in the direction of the S's spinal (Z-) axis. Ordinate: difference Δ=c-b between the distance c of the S's binaural from the centrifuge axis where he/she feels horizontal and the respective abscissa value b. Abscissa: distance b computed from the subjectively horizontal tilt angle ρ on the board; b=-gω^{-2}cosρ , that is, the distance where the Ss would set their binaural axis if the otoliths alone would control the SHP. Note that only one S (Vp113) sets herself almost at c-b=0 (in two sessions 4 years apart), 3 Ss behave like Ss without otoliths, and the rest show compromise positions.

3.1.4. The SHP of paraplegics with total bilateral lesion from the 5th lumbar to the 12th thoracic segment depends on leg extension as in normals. Paraplegics with total lesion from the 11th thoracic to the 6th cervical segment (named TC-paraplegics) do not show an effect of leg extension on the tiltable board, yet on the centrifuge, with Δ>>0 at any leg posture. Within both groups of paraplegics no correlation exists between the measured test variables and the height of the lesion (Mittelstaedt, 1992).

3.1.5. Bilaterally nephrectomized Ss behave like TC-paraplegics in all tests (Mittelstaedt, 1992, 1995b).

3.1.6. Positive pressure of 45mm Hg to the legs (PPL) causes a decrease of the absolute distance |c| on the centrifuge, but has no effect in Ss with an SHP of ρ=90deg on the tiltable board (Mittelstaedt, 1995b).

3.2. Consequences, Conclusions and Hypotheses

3.2.1. The SHP is not affected by the idiotropic vector (cf. 3.1.1.).

3.2.2. The SHP depends not only on the otoliths but also (to an idiosyncratic degree) on somatic graviceptors. The centroid of the mass(es) governing the latter lies near the last ribs. The result indicates an additive interaction of the graviceptive systems. Their relative weight in the generation of the subjective postural horizontal may be determined by the compromise distance Δ, e.g. that of the somatic graviceptors as Δ/45 cm, that of the otoliths as (1-D/45cm). Because the SHP is always close to ρ= 90 deg, hence cosρ $\approx$ cotρ and b $\approx$ a, these measures are valid independently of what actually causes the biases (cf. 3.1.2, see also paragraph 4).

3.2.3. Under the conditions of these experiments, mechanoreceptors in the legs do not contribute to the perception of body posture. This does not exclude, of course, that they play a role in the execution and control of the balancing movements released by the graviceptors in everyday life (cf. 3.1.3). Even the effect of leg extension on the tiltable board is not mediated by leg proprioceptors, because it is still present in Th12- and L1-paraplegics who are deprived of them (cf. 3.1.4).

3.2.4. The SHP is affected by inputs from (at least) two distinct sources, channelled by separate afferences entering at two different locations, firstly by the renal nerve, and secondly by the N. vagus or the N. phrenicus. It is very unlikely that many separate sources are distributed all over the length of the trunk, because in paraplegia $\Delta/45$ cm does not decrease with the height of the lesion (cf. 3.1.4).

3.2.5. The result of nephrectomy proves that the kidneys are involved in graviception. Whether they act as statoliths or more indirectly cannot yet be decided. The hypothesis is supported, however, by the existence of a density difference between capsule and kidney, and of pressure receptors around its surface (Niiyima, 1975; Gilmore and Tomomatsu, 1985).

3.2.6. The result of PPL corroborates the hypothesis (Mittelstaedt, 1992, 1995b) that the blood in the large vessels affects graviception through the force f exerted along the Z-axis by its mass m, with

$$f = (1 - \delta_i/\delta_b)mg \cos\rho \qquad (6)$$

(where δ_i and δ_b are the densities of the interstitial fluid and the blood). Hence a craniad shift of the centroid of m, caused by PPL, should indeed lead to a corresponding shift of the centrifuge axis, but have no effect at $\rho = 90$ deg (cf. 3.1.6). The hypothesis also explains the analogous effect of leg extension in TC-paraplegics who are deprived of all afferences from the legs (cf. 3.1.4). It is supported, moreover, by the proven effect of inertial forces on the large vessels (Erickson et al, 1976) and the existence of mechanoreceptors in the structures supporting them against the gravitational load (Kostreva and Pontus, 1993).

4. CONCLUSION AND OPEN QUESTIONS

The results reveal fundamental differences of sensing, processing and performing between the visual and the postural vertical. Whereas the otoliths are involved in both, the somatic graviceptors do not contribute to the generation of the SVV, but may sometimes even have a slight negative (inverting) effect (Mittelstaedt, 1988, esp. Tab.11.1, p.232). Most astoundingly, this is also found after longtime loss of saccular function (Mittelstaedt, 1995a).

Apart from the question of renal graviception, the hypothesis of v. Giercke and Parker (1994) that the abdominal viscera as a whole may have statolith function deserves and needs further scrutiny. The result of PPL shows, at any rate, that the surmised system cannot be the only one.

Finally, it cannot yet be decided whether the SHP biases are caused by vectorial rotation of or by additive superposition onto the setpoint of the SHP. The latter case has consequences for the subjective vertical in weightlessness, because the biases would then persist (Mittelstaedt and Glasauer, 1993). So far the data are in fair agreement with the additive hypothesis, namely, that Ss feel to be horizontal if and when

$$S_0 + V_0 + E_i R_0 + (S_1 + V_1 + E_i R_1)\cos\rho = 0 \qquad (7)$$

where S_0, V_0, R_0 are the biases, S_1, V_1, R_1 the gains of the saccular, the vascular and the renal graviceptors, respectively, and E_i a factor that increases with leg extension.

SUMMARY

Experiments on a tiltable board and a sled centrifuge lead to the conclusion that the visual and the postural vertical are mediated by fundamentally different causal relations.

The visual vertical (SVV) gains its gravity information exclusively from the otoliths, but is also affected by a shift towards the S's Z-axis. A theory developed under this assumption is able to explain the A- and E-effects, the increase of the variance with the tilt angle, the independence of changes in G-load at the six orthogonal attitudes and the peculiar asymmetry of the effect of optokinesis.

The postural horizontal (SHP) does not only depend on the otoliths but also on graviceptors in the trunk. It is experimentally shown

    a. that the centroid of the mass(es) governing the latter is situated near the last ribs,
    b. that their afferences enter the CNS at two distinct locations, the first at the 11th thoracic and the second craniad of the 6th cervical segment.

The effect of the first named input is abolished after bilateral nephrectomy. But whether the kidneys affect graviception like statoliths or more indirectly cannot yet be decided. The second input yields gravity information through the inertia of a fluid. The hypothesis that this fluid may be the blood of the large vessels is corroborated by the effect of shifting the blood craniad by means of positive pressure to the legs.

ACKNOWLEDGMENTS

Thanks are due to Evi Fricke, Willy Jensen and Karl Fischer for their dedicated assistance, and to the directors of the institute, Eberhard Gwinner and Wolfgang Wickler, for logistic support.

REFERENCES

Dichgans, J.M., Diener, H.C., and Brandt, T.H., 1974, Optokinetic-graviceptive interaction in different head positions, *Acta Otolaryng.* 78:391-398.

Erickson, H.H., Sandler, H., and Stone, H.L., 1976, Cardiovascular function during sustained +$G_z$stress, *Av., Space, and Env. Med* 47:250-258.

Fernandez, C., and Goldberg, J. M., 1976, Physiology of peripheral neurons innervating otolith organs of the squirrel monkey. I: Response to static tilts and to long-duration centrifugal force. II: Directional selectivity and force-response relations, *J. Neurophysiol.* 39:970-995.

Giercke, H.V., and Parker, D. E., 1994, Differences in otolithic and abdominal viscera graviceptor dynamics: Implications for motion sickness and perceived body position, *Av. Space Environ. Med* 65:747-751.

Gilmore, J.P., and Tomomatsu, E., 1985, Renal mechanoreceptors in nonhuman primates, Am. J. Physiol, 248: (Regul. Integr. Comp. Physiol, 17), R202-R207.

Haustein, W., 1992, Head-centric visual localization with lateral body tilt, Vision Res. 32: 669-673.

Howard, I.P., 1982, Human visual orientation, Wiley & Sons, New York Brisbane Toronto.

Kostreva, D.R., and Pontus, S.P., 1993, Hepatic vein, hepatic parenchymal, and inferior vena caval mechanoreceptors with phrenic afferents, Am. J. Physiol., 265: (Gastrointest. and Liver Physiol., 28), G15-G20.

Lechner-Steinleitner, S., Schöne, H., and Steinleitner, A., 1981, The auditory subjective vertical as function of body tilt, *Acta Otolaryngol.* 92:71-74.

Mittelstaedt, H., 1975, On the processing of postural information, *Fortschritte der Zoologie* 23:128-141.

Mittelstaedt, H., 1983, A new solution to the problem of the subjective vertical, *Naturwissenschaften* 70:272-81.

Mittelstaedt, H.,1985, Subjective vertical in weightlessness. In: Vestibular and visual control of posture and locomotor equilibrium, Igarashi M, and Black, O., eds., Karger, Basel, pp.139-50.

Mittelstaedt, H., 1986, The subjective vertical as a function of visual and extraretinal cues, *Acta Psychologica* 63:63-85.

Mittelstaedt, H., 1988, The information processing structure of the subjective vertical. A cybernetic bridge between its psychophysics and its neurobiology, In: Marko H, Hauske G, Struppler A, eds., Processing structures for perception and action, Verlag Chemie, Weinheim, pp. 217-263.

Mittelstaedt, H., 1989, The role of the pitched-up orientation of the otoliths in two recent models of the subjective vertical, *Biol. Cybern* 61:405-16.

Mittelstaedt, H., 1991, The role of the otoliths in the perception of the orientation of self and world to the vertical, *Zool. Jb. Physiol.* 95: 419-25.

Mittelstaedt, H., 1992, Somatic versus vestibular gravity reception in man. *Annals of the New York Academy of Science* 656:124-39.

Mittelstaedt, H., 1995a, New diagnostic tests for the function of utricles, saccules and somatic graviceptors, *Acta Otolaryng.* Supplement, (in press).

Mittelstaedt, H., 1995b, Somatic graviception, *Biological Psychology* (in press).

Mittelstaedt, H, and Fricke, E., 1988, The relative effect of saccular and somatosensory information on spatial perception and control, *Adv. Oto-Rhino-Laryng.* 42:24-30.

Mittelstaedt, H., and Glasauer, S., 1993, Illusions of verticality in weightlessness, *The Clinical Investigator* 71:732-739.

Niijima, A., 1975, Observation on the localization of mechanoreceptors in the kidney and afferent nerve fibers in the renal nerves in the rabbit, *J. Physiol* 245:81-90.

Schöne, H., and Udo de Haas, H., 1971, Space orientation in humans with special reference to the interaction of vestibular, somaesthetic and visual inputs, In: Biokybernetik III, Drischel, H. and Tiedt, N., eds., Fischer, Jena, pp. 172-191.

Yardley, L., 1990, Contribution of somatosensory information to perception of the visual vertical with body tilt and rotating visual field, *Perception and Psychophysics* 48:131-134.

19

VISUAL-VESTIBULAR INTERACTION FOR HUMAN EGO-MOTION PERCEPTION

Th. Mergner,[1] G. Schweigart,[1] O. Kolev,[2] F. Hlavačka,[3] and W. Becker[4]

[1] Neurological University Clinic
Neurozentrum, Breisgauer Str. 63
D-79106 Freiburg, Germany
[2] University Hospital of Neurology and Psychiatry
Tzarigradsko Shosse Blvd. 4 Km
1113 Sofia, Bulgaria
[3] Institute of Normal and Pathological Physiology
Sienkiewiczova 1
81371 Bratislava, Slovakia
[4] Sekt. Neurophysiologie
Universität Ulm, O. E.
D-89069 Ulm, Germany

INTRODUCTION

Both vestibular and visual cues are known to contribute to human ego-motion perception in space. While the vestibular system registers actual head motion in space, the visual system records in many situations only the relative head-vs-scene motion; then it signals head motion in space only if the scene is stationary. [*] The use of both cues is probably not merely a matter of redundancy for at least two reasons: (1) Imperfection of the vestibular system; for example, the horizontal canal system fails to register low frequency and constant-velocity head motion. (2) Ambiguity of visual cues; although the visual scene is stationary in most situations of every day life, there are situations in which the visual field, or large parts of it, are filled with moving objects which, when taken as a stationary reference, would create an illusory perception of ego-motion. Therefore, it would be desirable to dismiss visual ego-motion cues whenever the visual scene is moving, and to restrict the use of these cues to conditions where vestibular information deteriorates or is missing. The intriguing problem then is how to detect whether the visual scene is stationary or not. If detection of scene motion is based on vestibular cues, what happens if, for example, the vestibular system fails to distinguish between body stationarity and body motion of constant or low speed?

---

[*] In many situations only global velocity information is picked up by the visual system. With respect to the term global velocity information, see Andersen, this Volume.

Multisensory Control of Posture, Edited by T. Mergner and F. Hlavačka
Plenum Press, New York, 1995

Yet, it is a common experience that our brains solve these problems rather well. There are few exceptions only. An illusory ego-motion perception may arise, for instance, when we are housed within a potentially moving vehicle and look through a window at a large visual object slowly moving past the window. A frequently cited example is the illusion of ones own train leaving the station, while actually the train on the neighboring track is moving. The illusion dissolves once we cognitively realize that our train cannot be moving. This experience nicely illustrates that auxiliary sensory information (lack of movement related wagon vibrations), cognition (departure time of own train is later), expectation (direction of illusion counter that of expected train motion), and memory (same illusion detected on earlier occasions) is permanently used to check, and correct if necessary, the possibly ambiguous information conveyed by the vestibular and visual channels. It is these latter factors which make it difficult for neurophysiologists to gain access to the physiological mechanisms that underlie visual-vestibular interaction for ego-motion perception (e.g., Mergner and Becker, 1990).

Neurophysiologists have tried to create experimental conditions which circumvent these factors and which would allow to study the visual effect upon ego-motion perception in a consistent manner. Horizontal constant velocity rotation of an optokinetic drum about the stationary subject has been frequently used to this effect in the past (see Dichgans and Brandt, 1978). When subjects are suddenly presented with a drum rotating about the stationary body, they initially perceive it as rotating in space and experience themselves as stationary. However, some seconds later, the drum appears to slow down and the body is perceived as accelerating until a state is reached where the body appears to spin inside a subjectively stationary drum, a state that has come to be known as circular vection (CV).

The CV in the above situation is cogent, but may be interrupted by 'drop outs' of unknown reasons. Its value for the understanding of visual-vestibular interaction is limited in that the vestibular cues are ambiguous in this condition (the activity of horizontal canal receptors would not allow to differentiate between constant velocity body rotation and the resting condition). The slow build up of CV originally was related to low pass characteristics of the visual input, but this notion did not fit the findings obtained with sinusoidal optokinetic stimulation in later studies (Büttner and Henn, 1981; Mergner and Becker, 1990). A better explanation might be the concept of a visually evoked 'expected vestibular signal' with dynamics mimicking those of the vestibular signal that would arise if the visual stimulation was due exclusively to a motion of the subject in a stationary environment; the expected vestibular signal would suppress CV as long as it is not canceled (confirmed) by an actual vestibular signal (compare Zacharias and Young, 1981).

When we began to study visual-vestibular interaction several years ago by means of sinusoidal rotations, we soon came to realize that we could obtain consistent data only if we controlled our subjects' perceptual state by giving specific instructions and/or creating particular conditions (Mergner and Becker, 1990). Fig. 1 demonstrates how the response curves obtained for ego-motion perception during isolated visual stimulation (subject stationary, drum rotating) could be modified by instruction. When subjects were told to attend to their 'bodily feelings', they felt stationary during the visual stimulation (curve A). However, when they were attending a stationary visual fixation point and its relative motion with respect to the rotating optokinetic pattern, they were irreversibly entrained into a state of CV (curve B). Up to frequencies of 0.8 Hz CV was 'full' ('gain' approximately unity). Beyond 0.8 Hz 'gain' decreased rather abruptly. Interestingly, in a more recent investigation three patients with chronic bilateral loss of vestibular function yielded very similar gain-vs-frequency characteristics as in curve B, without being given specific instructions though. Thus, these studies revealed little in the way of a visual-vestibular interaction; the subjects based their ego-motion perception either predominantly on vestibular cues (A) or on visual cues (B). Accordingly, we have outlined a concept of visual-vestibular cue rivalry in ego-motion perception as one of several aspects of this phenomenon (Mergner and Becker, 1990).

The present study represents a further attempt to identify the various factors contributing to ego-motion perception. We succeeded to obtain quite consistent CV estimates that ranged in between the two extremes described above and, hence, were suggestive of "true" visual-vestibular interaction. Since we knew from previous studies (Mergner et al., 1991, 1992) that detection thresholds may considerably shape a subject's response in the range of stimulus intensities we are using, we included an evaluation of thresholds before trying to establish a formal description of the results in terms of a dynamic model.

METHODS

Two experimental series were performed in seven normal adult subjects. In the first series, subjects' ego-motion perception was evaluated in terms of gain and phase using suprathreshold stimuli. In the second series, detection thresholds for ego-motion perception were measured.

For stimulation the subjects were seated on a rotation chair. Their heads were attached to the chair with a 15° nose-down inclination by means of a bite board. Chair rotation (head rotation in space, HS) represented the vestibular stimulus (VEST.). For optokinetic stimulation, a black-and-white random patch pattern was projected onto a cylindrical screen which surrounded the chair (r = 0.9 m). The pattern was rotated about the same vertical axis as the chair. It covered almost the entire visual field. Rotation of the visual pattern relative to the head (VH) represented the visual stimulus (VIS.). Isolated visual stimulation was obtained by rotating the pattern and keeping the head stationary.

Rotations were sinusoidal about the primary position at 0.025, 0.05, 0.1, 0.2, and 0.4 Hz. Chair and pattern could be rotated independently of each other, but only in-phase and counter-phase combinations were used with one stimulus being held at a constant amplitude while the amplitude of the second one was varied (0° to ±16°). Auditory cues were minimized by plugging the subjects' ears.

Special care was taken to avoid that subjects saw parts of their bodies or body-linked objects (like the pointer). Seeing relative motion between body parts and the visual stimulus is known to greatly facilitate the occurrence of CV (Mergner and Becker, 1990; see Henn et al., 1980). The illumination of the pattern projection was turned down to skotopic levels restricting the subjects' visual field such that they no longer saw the rim of their orbits or their noses. We ascertained that they still experienced CV during constant-velocity visual stimulation at this luminance level (compare Leibowitz et al., 1979). Furthermore, subjects were instructed not to fixate or track with their eyes any item of the visual pattern, but rather to stare through it. Pilot experiments had suggested that pursuing the visual pattern can prevent the occurrence of CV unless the stimulus excursions are clearly larger than 16° and the subjects' ocular pursuit movements became disrupted by nystagmus (compare Fischer and Kornmüller, 1930). The pilot experiments also indicated that handling a pointer for indication of perceived ego-motion can reduce the frequency of occurrence of CV, as compared to a verbal estimation procedure. However, the use of pointer indications is advantageous in that these provide a measure of the temporal stimulus-response relationship, in addition to that of response magnitude.

Subjects indicated perceived body motion in space by using a pivotable pointer that was fixed to the chair (pointer and chair axes colinear). They tried to keep the pointer's spatial orientation constant by counter-rotating it in response to perceived body excursion ("concurrent pointer indication"; for details, see Mergner et al., 1991). Vision of the pointer was occluded by a black circular board. Readings of pointer, chair and pattern positions (potentiometers) were recorded with the help of a laboratory computer. Pointer responses were referred to chair rotation (to pattern rotation if the chair remained stationary) and analyzed in terms of gain and phase (G=1 and φ= 0°, if pointer was kept truly stationary in space; G=0, if subjects made no pointer excursion, so that the pointer turned with the chair).

For the evaluation of the detection threshold of ego-motion perception we used the methods of lower and upper limits. In order to obtain a clear relation between perception and stimulus at threshold levels, subjects indicated perceived ego-motion by means of the pointer as described above (for details, cf. Mergner et al., 1991).

The data presented in the following are given in terms of median values (n= 14; 7 subjects, 2 values each).

RESULTS

Suprathreshold Stimulation

Figure 1 shows our subjects' ego-motion perception obtained when the visual pattern was rotated about their stationary bodies (CV during pure visual stimulation). 'Gain' and phase of the pointer indications are plotted as a function of stimulus frequency (dotted lines interconnecting median values and their 95% confidence intervals). Also shown are results from previous work (Mergner and Becker, 1991; magnitude estimates) in which normal subjects experienced either no CV (A) or a "full vection" (B) as did patients with bilateral vestibular loss (C; cf. Introduction).

CV obtained under the present conditions was clearly different from these earlier results. It showed a median gain close to zero at 0.4 Hz, but increased monotonically with decreasing frequency, reaching 0.42 at 0.025 Hz. Its phase was essentially counter to pattern rotation (approximately -180°) and was almost independent of stimulus frequency. We attribute the difference between present and previous findings mainly to the fact that our subjects stared through the pattern rather than trying to track it (see Discussion).

Fig. 2A shows the results obtained when a constant vestibular stimulus (8° peak velocity) was combined with various visual stimuli of same or opposite direction (see right insets for physical situations used to create these vestibular-visual stimulus combinations). Gain is plotted with respect to the vestibular stimulus; thus, a gain of unity and a phase of

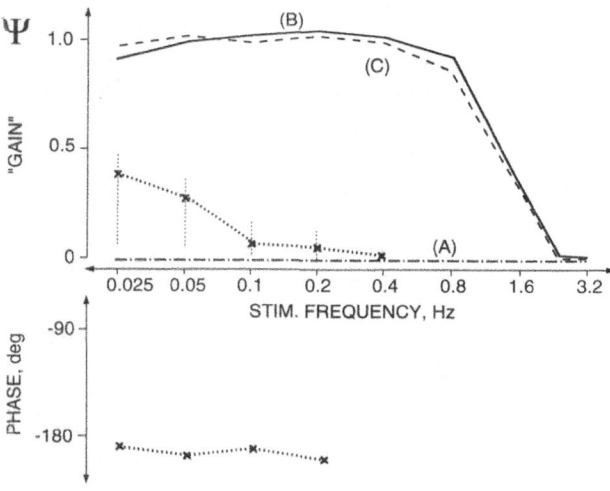

Figure 1. Ego-motion perception during pure visual stimulation (visual pattern rotated about stationary subject). Median 'gain' and phase values are plotted as a function of stimulus frequency. Dotted curve, results of present study (median pointer indications, 95% confidence intervals). Dash-dot, dashed and continuous curves give data from previous work in normal subjects experiencing either no CV (A) or "full vection" (B) and in patients with bilateral vestibular loss (C), respectively (pattern displacement, ±8°; magnitude estimates, cf. Introduction).

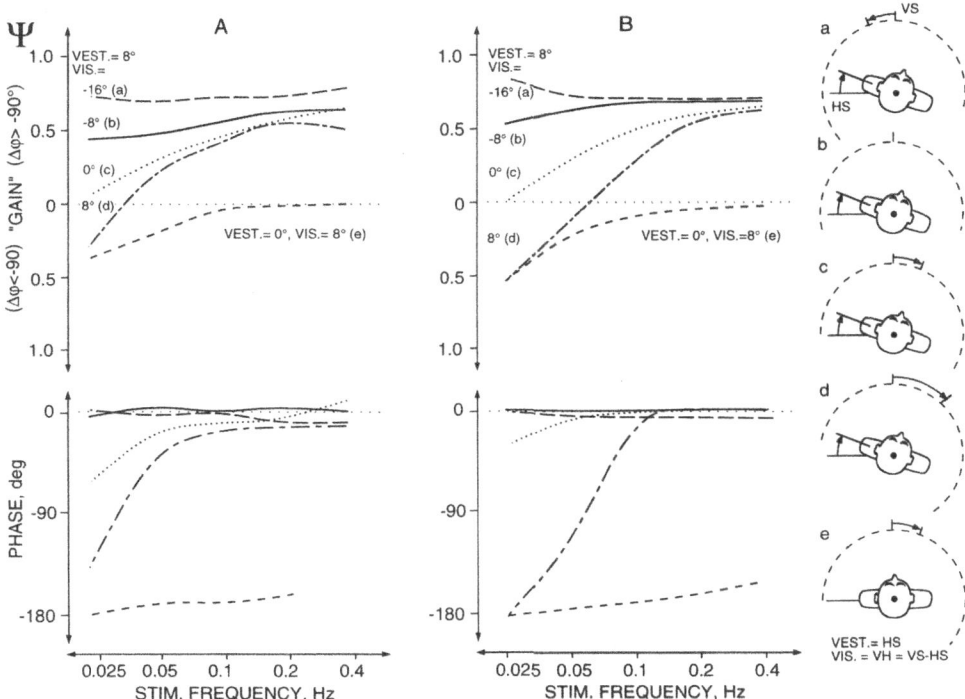

Figure 2. Ego-motion perception during visual-vestibular stimulus combinations. Smoothed median gain and phase values are plotted as a function of stimulus frequency. A vestibular stimulus (peak displacement, 8° = constant) was combined with inphase and counter-phase visual stimuli of different amplitudes as indicated (see also insets on the right side for physical stimulus condition). Curve e is a repetition from Fig. 1 with the gain values mirrored about the x-axis to be consistent with the gain conventions of Fig. 2 (see text). A Experimental data. B Simulation using model in Fig. 4A.

zero indicate that the subjects' responses were veridical (pointer stabilized in space by excursions of same magnitude and opposite direction as actual head rotation). Phase values of 180° indicate that subjects felt rotated counter to actual head rotation; this illusion can be evoked by visual stimuli that have the same direction as the vestibular rotation (e.g. curve d at low frequencies; curve d corresponds to drum rotation in phase with chair at twice its amplitude, creating a visual stimulus of 8° in phase with head rotation; cf. inset d). To ease reading of the figure, we have mirrored gain values of counter-phase responses about the x-axis.

The stimulus combinations yielded a family of response curves (smoothed median curves) which showed similar gain and phase values at high stimulus frequencies, but which clearly diverged at low frequencies. Consider first the response curve for pure vestibular stimulation (curve c, VEST.= 8°, VIS.= 0°; visual pattern rotating with head). Gain decreased from a value of about 0.6 at 0.4 Hz to almost zero (0.06) at 0.025 Hz. Phase was close to zero at frequencies above 0.05 Hz and exhibited some lag at lower frequencies. With regard to gain these frequency characteristics were similar to those of a previous study with pure vestibular stimulation in the dark (Mergner et al., 1991), whereas the phase in this earlier study exhibited a lead at low frequencies instead of a lag. The gain attenuation at low frequencies became considerably less when the vestibular stimulus was combined with a counter-phase visual stimulus of same magnitude (curve b, VEST.= 8°, VIS.= 8°; pattern stationary in space). Gain was further enhanced when the counter-phase visual stimulus was doubled in amplitude (a). On the other hand, when an inphase visual stimulus was added (d), gain was decreased and phase became reversed at low frequencies; at 0.025 Hz the response to this stimulus combination appeared to be dominated by the visual cue and became similar to that evoked by pure visual stimulation (curve e, replotted from Fig. 1 with gain values below x-axis to be consistent with the conventions of Fig. 2).

Threshold Measurements

In Figure 3 the detection thresholds of ego-motion perception are plotted as a function of stimulus frequency for four different stimulus combinations. The data are presented in terms of (median) values of peak angular velocity (compare Mergner et al., 1991). Occasionally, some of these combinations failed to elicit an ego-motion perception even at large stimulus intensities. Fig. 3 therefore indicates also the percent frequency of the occurrence of ego-motion perception. Also shown are threshold values obtained from simulation of a model (bold curves) which will be considered in the Discussion.

As shown in Fig. 3A, pure vestibular stimulation (visual pattern fixed with respect to head) always evoked an ego-motion perception (100% incidence). Threshold values were on the order of 1°/s; they exhibited little dependence on stimulus frequency and resembled those obtained previously with pure vestibular stimulation in the dark (Mergner et al., 1991).

On the other hand, with pure visual stimulation (Fig. 3B) subjects often perceived the visual pattern veridically as rotating about their stationary bodies. An (illusory) perception of ego-motion (body rotation relative to an apparently stationary visual pattern) occurred in 23% of the trials at 0.4 Hz and reached a frequency of 78% at 0.05 Hz and 0.025 Hz. Yet, the thresholds observed in this condition were essentially independent of stimulus frequency. Thus, like the vestibular cue, the visual effect upon ego-motion appears to have a velocity threshold which is by a factor of 3 lower, though (0.3°/s).

When a vestibular stimulus was combined with a counter phase visual stimulus (head rotating, visual pattern stationary), ego-motion occurred again in 100% of the trials and thresholds were clearly below 1°/s (Fig. 3C). At low frequencies the thresholds exhibited similar values as during pure visual stimulation; at higher frequencies they increased without reaching those obtained with pure vestibular stimulation, though.

Finally when the visual pattern rotated together with the head at twice its amplitude, two different thresholds could be observed. When increasing stimulus intensity from clearly subthreshold values, subjects often first experienced ego-motion counter to pattern and body rotation (obviously visually determined). With a further increase of intensity this sensation first became vague or disappeared and eventually an ego-motion perception in opposite direction could arise (in the direction of body and pattern rotation; obviously vestibularly determined). The probability of experiencing the visual and/or the vestibular ego-motion perception clearly depended on stimulus frequency. The incidence of the visual ego-motion perception was similar to that during pure visual stimulation (dash-dot curve in Fig 3D). In contrast, the vestibular ego-motion perception (dotted) occurred less frequently than during pure vestibular stimulation at most frequencies except at the highest ones where it reached 100%. The thresholds for visual ego-motion (dashed) were similar to those obtained previously during pure visual stimulation (compare Fig 3B; the slightly higher values in Fig. 3D result from the fact that in the latter experiment thresholds were determined only with increasing stimulus intensities yielding upper limits). Those for vestibular ego-motion (continuous) were quite large at low frequencies and reached values similar to those obtained with pure vestibular stimulation only at high frequencies (0.4 Hz).

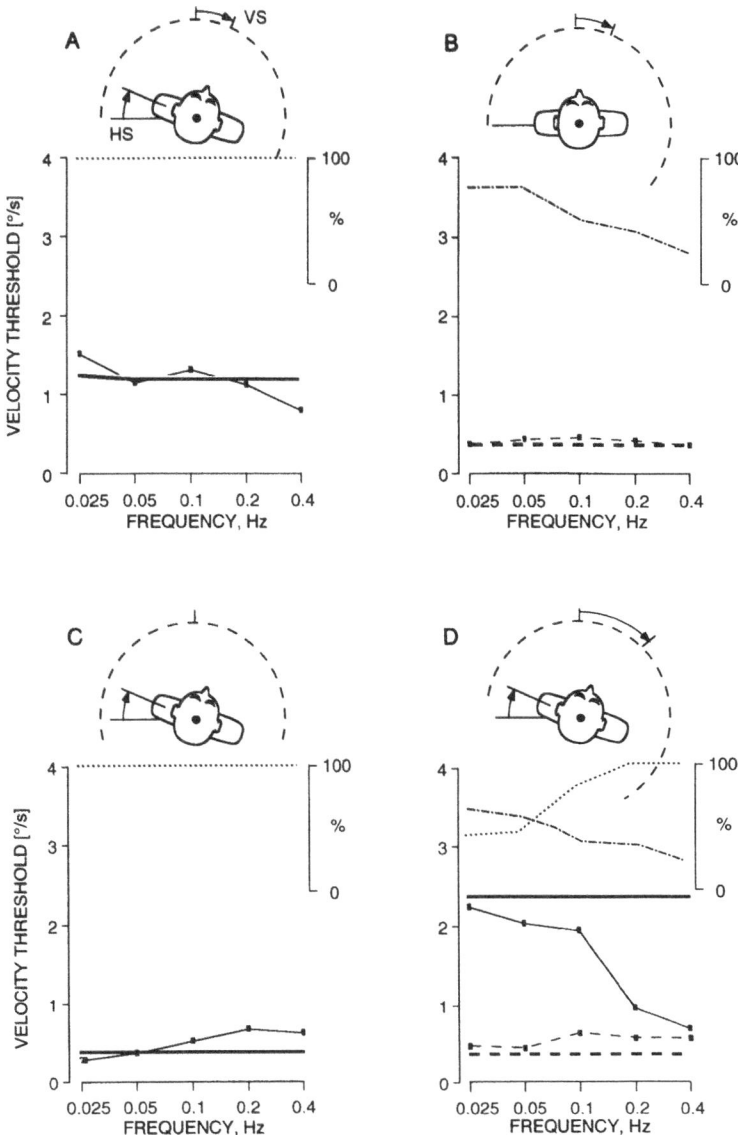

Figure 3. Detection thresholds of ego-motion perception for four different visual-vestibular stimulus combinations. The thresholds are plotted in terms of median peak velocity values over stimulus frequency. Upper parts of graphs give in addition the percent frequency of the occurrence of ego-motion perception. Bold curves show simulated thresholds (model in Fig. 4A). A Body rotation with head-aligned pattern (VEST.= 8°, VIS.= 0°). B Pattern rotation about stationary body (VEST.= 0°, VIS.= 8°). C Body rotation relative to stationary pattern (VEST.= 8°, VIS.= -8°). D Body rotation, inphase pattern rotation with twice the amplitude (VEST.= 8°, VIS.= 8°). In D two threshold curves were obtained (see text).

DISCUSSION

The present investigation elaborates on a previous series of experiments on visual and vestibular mechanisms of ego-motion perception (Mergner and Becker, 1991) which had suggested a dualistic view of ego-motion: Perception appeared to be determined either by vestibular or by visual signals, but rarely by a true interaction of the two cues (such as a weighted average). Using a somewhat different experimental approach, we now demonstrate (cf. Fig. 2A) that there are also conditions in which visual input can modify the vestibular evoked ego-motion perception in a graded way.

The methods used in the present experiments differed from those in our previous series in two aspects: (1) subjects tried to stare through the pattern rather than fixating at the pattern or at a small target, that moved independently of the pattern, and (2) subjects used a pointer for concurrent indication of their perception rather than giving verbal estimates. As mentioned above, staring at the pattern tended to facilitate the occurrence of CV; the reasons for this facilitation are unknown at present. Handling the pointer tended to suppress CV to some extend, but it did not prevent its occurrence.

While at high frequencies of rotation (>0.2 Hz) the visual effect upon ego-motion perception was fairly small, a conspicuous visual modulation of self-motion perception was evident at low frequencies. Since the vestibular cues deteriorate considerably at low frequencies due to their high-pass characteristics and, not to neglect, due to a relatively large velocity threshold, the visual effect can be interpreted as helping to make ego-motion perception independent of stimulus frequency. It must be kept in mind, however, that this applies only to the "standard" situation where the subject rotates with respect to a stationary surround (curve b in Fig. 2A). In other situations like the rotation of the surround in phase with the body at twice its amplitude, the effect becomes detrimental rather than beneficial (curve d in Fig 2A).

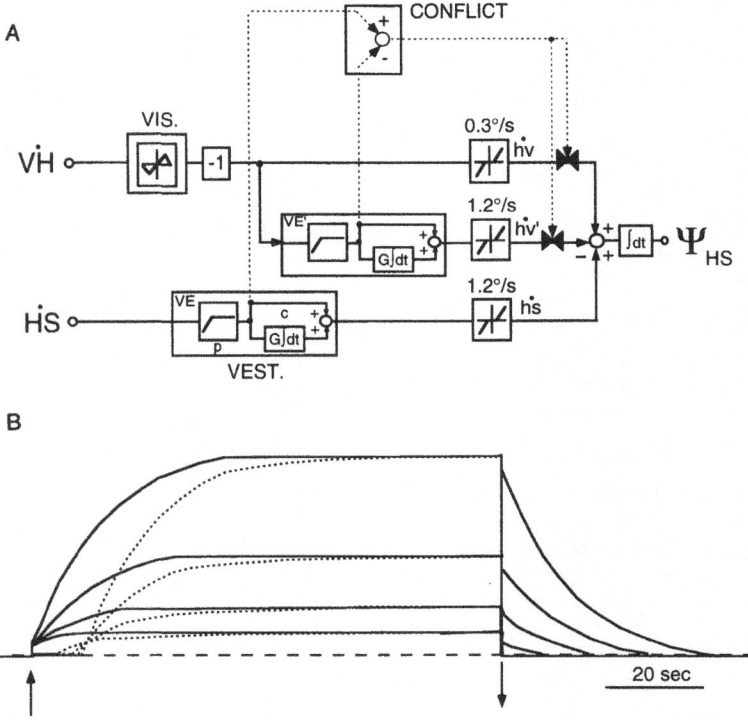

Figure 4. A Dynamic model of visual-vestibular interaction for human ego-motion perception. $H^{\cdot}S$, head-in-space velocity; $V^{\cdot}H$, visual pattern-relative-to-head velocity; Ψ_{HS}, perception of head-in-space displacement. See text for further details. Results of simulation of experimental data are given in Figs. 2B, 3. B Simulation of ego-motion perception during pure visual stimulation (CV) with constant-velocity steps of 2, 4, 8, and 16°/s (stimulus onset and end indicated by arrows). Continuous profiles give results obtained without, and dotted profiles with conflict mechanism.

On a first glance, the gain-vs-frequency characteristics of the visual effect obtained with pure visual stimulation (curve e in Fig. 2A) - which is reminiscent to the slow build-up of CV with constant velocity rotation (see Introduction) - would seem to indicate a low pass behavior. However, the fact that the phase of the visually induced ego-motion illusion remained essentially constant across all frequencies is incompatible with the notion of a linear systems low-pass mechanism. We had encountered a similar incompatibility, in terms of linear systems theory, in a previous study on vestibular-proprioceptive interaction for which we had proposed a non-linear model (Mergner et al., 1991). We now have adapted this earlier model so as to provide also a reasonable description of the present data. The adapted model is shown in Fig 4A. Basically, perception of head rotation, Ψ_{HS} (identical to the perception of body rotation if no head-vs-trunk rotation is allowed), results from a summation of the vestibular cue about head-in-space velocity (h·s; pathway VEST.) with a "direct" visual signal (h·v; pathway VIS.) and an "indirect" one (h·v') about the relative velocity between head and visual scene. To convert these velocity coded cues into a perception of position (by using the pointer, subjects indicated positions), the output of the summing point is passed through an integrator. The vestibular cue is shaped by peripheral (stage p) and central (stage c) mechanisms. Stage p represents the well known high pass characteristics of the vestibular organ (here we assume a time constant of 5 s corresponding to a cut off frequency of 0.032 Hz), whereas stage c reflects our hypothesis that processing at various central levels results in a partial compensation for the low frequency cut-off in stage p (G=0.11, corresponding to a frequency of 0.0175 Hz). Finally, errors which can easily accumulate as a result of the integration in c are suppressed by the 1.2°/s threshold element. This threshold element is instrumental in creating the fading of vestibular motion perception at low frequencies. Experimental evidence for this role of the threshold comes from the observation that the frequency below which perception begins to decrease depends on the amplitude of stimulation (the higher the amplitude the lower the critical frequency; Mergner et al., 1991).

The visual cues are subject to a saturation and gain reduction for velocities exceeding 33°/s (pathway VIS.), which explains the decrease in gain of CV at high frequencies (curves B and C in Fig 1; note that the cut-off frequency will again depend on the amplitude of stimulation). The direct visual pathway has been given a threshold of 0.3°/s corresponding to the detection threshold for CV during pure visual stimulation. The indirect visual pathway is basically a replica (VE') of the vestibular pathway VEST. Since its signal is being subtracted at the final summing point, it will cancel the vestibular signal during rotation of the subject in a stationary surround so that perception will be dominated by the direct visual pathway.* This is not true for high frequencies, though; when V·H exceeds the linear range of VIS. at high stimulus frequency, both visual cues decline and, moreover, cancel each other so that perception becomes determined by the vestibular signal.

Using the model in Fig. 4A, we have simulated the perceptual response to the various stimulus combinations used in the present experiments. The results of these simulations, which are shown in Fig. 2B, were qualitatively quite similar to the responses of our subjects depicted in Fig. 2A. In particular, the model predicts correctly the graded effect of the visual input upon ego-motion perception; indeed, notwithstanding the model's many nonlinear

---

*If the model is drawn with a common summation point for all internal signals at the output site, as in Fig. 4A, an aspect of 'distributed processing' emerges, which illustrates that a verbal description of signal flow is problematic. Imagine an observer A who considers a fast body rotation relative to a stationary scene towards a given side (e.g., 20°/s; then H·S= 20°/s and V·H=-20°/s) and starts the interpretation of the model by summing the internal signals from above: the sum of the direct visual signal and the indirect one yields approximately zero, so that he would be inclined to state that then Ψ_{HS} is determined mainly by the vestibular signal. An observer B, starting from below, might say that the sum of the vestibular and the indirect visual signal yields approximately zero, so that then Ψ_{HS} is determined essentially by the direct visual signal.

elements, visual-vestibular interaction occurs basically by linear summation. However, as outlined in the Introduction, earlier work by a number of groups including our own, had indicated that ego-motion perception is often dominated in an apparently non-linear way either by visual or by vestibular cues. To explain this behavior, we suggest that there is an internal monitor which compares the indirect visual cue to the vestibular cue and which derives a measure of conflict if the two cues are unequal. If this measure (conflict) exceeds a certain threshold it would "choke" the visual pathways so that perception would be determined exclusively by the vestibular cue. [*]

To illustrate this mechanism, conceive of a stationary observer who is suddenly exposed to an environment rotating at constant velocity. The vestibular replica VE' will faithfully transmit the sudden onset of V'H much as the vestibular system would if there was a sudden onset of head rotation. However, with the head being actually stationary, a conflict is created which prevents the visual cue from influencing ego-motion perception. Therefore, as shown by the simulation in Fig 4B (dashed profiles), the S initially experiences no CV. As time progresses the signal about scene-vs-head rotation becomes progressively smaller because of the high-pass characteristics of VE', and so does the conflict; this terminates the suppression of the visual cues allowing their difference to create an ego-motion illusion. The time it takes the conflict to decline to an uncritical value depends on the magnitude of V'H; therefore, as evident from Fig 4B, the higher the velocity of rotation, the later the onset of CV, which is in good agreement with experimental observations (compare Henn et al., 1980). Also, suppressing the visual cues contingent on conflict is useful in avoiding aftereffects when the pattern movement stops. As depicted by the continuous profiles in Fig 4B, without suppression CV would considerably outlast the end of rotation, owing to the low pass character of the difference between the two visual cues.

Although the topology of the model in Fig. 4A is very similar to our earlier model of vestibular-neck proprioceptive interaction, the functional similarity between the two models is restricted to the creation of frequency independent phases and to the "preference" for the most commonly occurring combination of stimuli, that is rotation of the head on the stationary trunk in the case of vestibular-neck interaction, and rotation of the subject within a stationary environment in the present study. On the other hand, whereas the hypothesis of an indirect neck signal shaped by a replica of the vestibular system helps to explain the transformation from a trunk based to a space based coordinate system in agreement with experimental results, the indirect visual pathway provides a means to alert to discrepancies between visual and vestibular messages. The use of this "expected vestibular signal" to derive a measure of conflict is similar to that in the model of Zacharias and Young (1981). Their model differs from the present one mainly in how conflict is used to control the weighting of the vestibular and visual channels, respectively.

Although the model in Fig. 4A provides an accurate description of a number of observed phenomena, it is far from being complete. For example, it lacks a formal description of the observation of different probabilities of CV-occurrence and of their frequency dependence. Also, it does not predict the phenomenon of "full" CV in which the gain of the illusion becomes almost unity (Fig. 1, curve B). Yet, we consider it an important piece necessary to solve the complex puzzle of ego-motion perception.

ACKNOWLEDGMENTS

This work was supported by the Deutsche Forschungsgemeinschaft (SFB 325/7). O. Kolev was supported by DFG 436 BUL 17/4/93 and F. Hlavacka by DFG 436 CSR 113/27/17.

---

[*] The chokes in the model were realized by subtracting from the visual signals (direction separation by rectification, absolute values) the conflict value (absolute value).

REFERENCES

Büttner, U., and Henn, V., 1981, Circular vection: Psychophysics and single-unit recordings in the monkey, Annals of the New York Academy of Science, 374:274-283.

Dichgans, J., and Brandt, T., 1978, Visual-vestibular interaction: Effects on self-motion perception and postural control. In: Held, R., Leibowitz, H., and Teuber, H.L. (eds) Handbook of sensory physiology, Vol VIII. Springer New York, pp755-804.

Fischer, M.H., and Kornmüller, A.E., 1930, Optokinetisch ausgelöste Bewegungswahrnehmung und optokinetischer Nystagmus, J Psychol Neurol, 41:273-308.

Henn, V., Cohen, B., and Young, L.R., 1980, Visual vestibular interaction in motion perception and the generation of nystagmus, Neurosci. Res. Program Bull., 18:459-651.

Leibowitz, H.W., Shupert, C., and Dichgans, J., 1979, The independence of dynamic spatial orientation from luminance and refraction error, Perception and Psychophysics, 25:75-79.

Mergner, T., and Becker, W., 1990, Perception of horizontal self-rotation: Multisensory and cognitive aspects. In: Warren, R., Wertheim, A.H. (eds) Perception and control of self-motion. Lawrence Erlbaum, Hillsdale London, pp219-263.

Mergner, T., Siebold, C., Schweigart, G., and Becker W., 1991, Human perception of horizontal head and trunk rotations in space during vestibular and neck stimulation, Exp Brain Res, 85:389-404.

Mergner, T., Rottler, G., Kimmig, H., and Becker, W., 1992, Role of vestibular and neck inputs for the perception of object motion in space, Exp Brain Res 89:655-668.

Zacharias, G.L., and Young, L.R., 1981, Influence of combined visual and vestibular cues on human perception and control of horizontal rotation, Exp. Brain Res., 41:159-171.

20

MODIFICATION OF THE GALVANIC SWAY RESPONSE BY VISUAL CONDITIONS

B. L. Day and C. Bonato

MRC Human Movement and Balance Unit
Queen Square
London WC1N 3BG
United Kingdom

INTRODUCTION

If vestibular information is used to help control upright stance, then it is probable that vestibular signals are acted upon in the light of other sources of information about body motion and orientation. This is because vestibular receptors signal motion and orientation of the head in space but, in isolation, do not directly inform on motion of the body relative to the support surface. The rules used by postural centers for interpreting the total pattern of afferent input from diverse sources in order to generate appropriate postural adjustments, are largely unknown.

Low-intensity, galvanic stimulation over the mastoid processes of a subject standing in the dark produces a relatively consistent postural response. The response, which consists of an increase in body sway in a direction towards the anodal ear irrespective of the position (yaw) of the head relative to the feet (Lund and Broberg, 1983), is most likely caused by a change in vestibular afferent firing. EMG recordings have shown that the latency of this postural adjustment is around 120 ms in leg muscles (Britton et al., 1994). Presumably, the response is organized to compensate for the apparent disturbance of the body signaled by the artificial vestibular input. Therefore, during vestibular stimulation, the availability of visual signals that give information about body motion relative to the support surface, would provide additional sensory information that conflicts with the apparent disturbance. Other studies have shown that visual information is capable of modifying the postural response to galvanic stimulation (Britton et al., 1993; Fitzpatrick et al., 1994; Smetanin et al., 1990) although there is not agreement regarding the underlying mechanism. In the present experiments we explore whether, and in which way, a small change in the amount of available visual information is able to modify the motion of the body induced in standing subjects by a change in vestibular input.

METHODS

Seven healthy subjects were studied with ethical committee approval. Subjects stood with feet together looking straight ahead in front of a vertical screen placed at head height at a distance of 80 cm. A disc of light (2.5 cm dia.) was projected onto the screen in an otherwise blacked-out room. In 50% of trials the disc of light would either remain illuminated or be extinguished during the 5s data collection period. In 67% of trials a long duration, low intensity (4 s at 0.5 mA) galvanic stimulus was applied across the mastoid processes 1s after the start of data collection. The availability of visual information, the presence of galvanic stimulation and the polarity of stimulation (anode on the right (R+) or anode on the left (L+)), were intermixed such that subjects underwent 10 trials each of 6 randomized stimulation conditions. Measurements were made from averages computed from all trials within a condition for each subject. An infra-red emitting diode was fixed to the back of the subject over C7 and its three-dimensional position was measured at 20 ms intervals using a Selspot II system. The components of velocity along the three principal axes were obtained by digital differention of the position record. The speed of the marker was taken as the amplitude of the vector sum of these three components of velocity.

RESULTS

Galvanic stimulation produced a body sway to the right or left depending upon the polarity of the stimulus. With the subject in total darkness, the stimulus induced a movement of the body towards the anodal ear such that lateral velocity increased steadily, reached a peak and then declined. The movement tended to reach a maximum lateral excursion and then stopped or recovered slightly even though the stimulus was maintained (Fig. 1). When the disc of light was visible, the response had a similar form but the maximum lateral displacement of the body was significantly less. The lateral component of velocity, when averaged across subjects and plotted against time, indicated that the presence of the disc of light had little effect on the early part of the response (Fig. 2). The initial acceleration of the body appeared similar for the two visual conditions. The effect of vision was to induce a

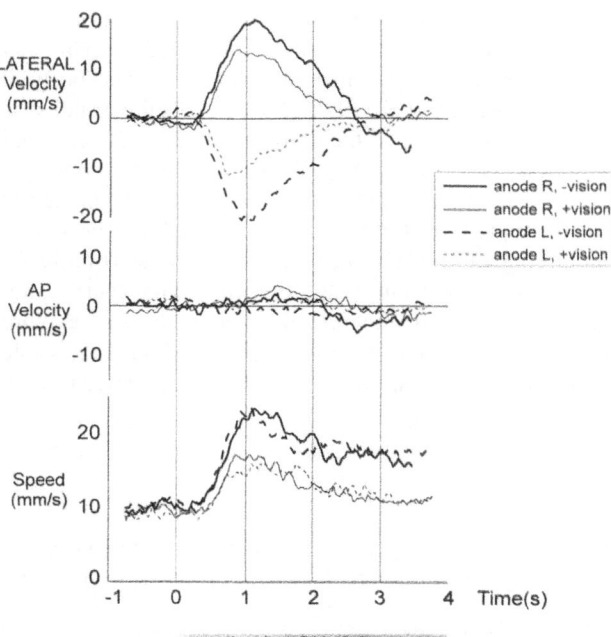

Figure 1. Mean movement path of the body at the level of C7 when viewed from above (plan view) is shown in (A). Each trace represents the average displacement from all 7 subjects obtained during galvanic stimulation. (B) shows a bar graph representing the maximum lateral displacement of the body measured from each subject individually. Data obtained using the two polarities have been combined.

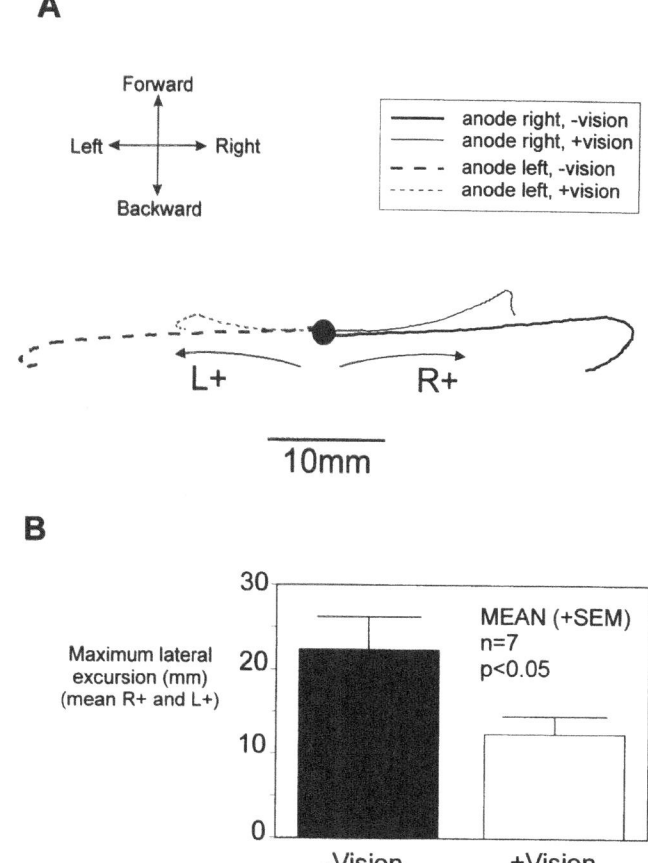

Figure 2. Mean velocity profile of body at level of C7 averaged across all 7 subjects. Shown are the lateral and anteroposterior components of velocity together with tangential velocity (speed) for four conditions of stimulation.

deceleration of the body at an earlier latency compared to no-vision. Inspection of responses from individual subjects show that this behaviour was quite consistent. However, there were some differences between subjects. Only three subjects showed clear bilateral effects of vision. The four remaining subjects showed unilateral effects; for three subjects, the effect was seen only when postural responses were induced to the left, and for one subject only for movements to the right.

CONCLUSIONS

These results indicate that an increase in sensitivity to body motion along visual channels can act to modify vestibular-induced automatic postural responses. A small amount of visual information was found to produce a reduction in the amplitude of the galvanic body sway response. The velocity profile, however, showed that the initial part of the response was relatively unaffected by the visual scene suggesting that the restricted visual information used in these experiments was insufficient to lead to a change in gain of the vestibular-induced response. Instead, the net response was made smaller by decelerating the body at an earlier time. We suggest that information from the visual system modifies vestibular-based postural responses by acting in a feedback capacity, by detecting the induced body motion and arresting it part way through its progression.

REFERENCES

Britton, T.C., Day, B.L., Brown, P., Rothwell, J.C., Thompson, P.D., and Marsden, C.D., 1993, Postural electromyographic responses in the arm and leg following galvanic vestibular stimulation in man, *Exp. Brain Res.* 94:143-151.

Fitzpatrick, R., Burke, D., and Gandevia, S.C., 1994, Task-dependent reflex responses and movement illusions evoked by galvanic vestibular stimulation in standing humans, *J. Physiol.* 478:363-372.

Lund, S., and Broberg, C., 1983, Effects of different head positions on postural sway in man induced by a reproducible vestibular error signal, *Acta. Physiol. Scand.* 117:307-309.

Smetanin, B.N., Popov, K.E., and Shlykov, V.Yu., 1990, Changes in vestibular postural response determined by information content of visual feedback, *Neurofiziologiya* 22:80-87.

21

INTERACTION OF OPTOKINETIC REFLEX AND VESTIBULO-OCULAR REFLEX DURING ACTIVE AND PASSIVE HEAD ROTATION

G. Schweigart,[1] Th. Mergner,[2] S. Morand,[2] and I. Evdokimidis[2]

[1] Ruhr-University Bochum
Department Neurophysiology MA4
44780 Bochum
Germany
[2] Neurologische Universitätsklink
Neurozentrum
79106 Freiburg
Germany

INTRODUCTION

For accurate vision, we have to stabilize the visual images in our eyes on the fovea. During head movements, this goal is mainly achieved by the combined actions of the vestibulo-ocular reflex (VOR) and the optokinetic reflex (OKR). The VOR is evoked by head movements in space and gives rise to a compensatory counter-rotation of the eyes in the orbits, with the aim to stabilize the eyes in space. Accordingly, the VOR contributes to clear vision only if the visual scene is stationary during self-motion. The OKR, on the other hand, tries to zero relative motion between visual scene and retina, thereby improving vision independently of whether the scene, the head or both are moving. Therefore, the OKR would be better apt than the VOR to guarantee accurate vision. However, the OKR suffers from limited band width at high temporal frequencies (see Barnes, 1993).

A generally accepted notion in the past was that the main function of the OKR is to augment the eye stabilization brought about by the VOR, by providing a signal about head velocity in the low frequency range where the vestibular canals no longer function properly (Henn et al. 1980, p. 582). However, when recently studying the interaction of the OKR and the VOR in macaque monkeys during various stimulus combinations (Schweigart et al., 1995), we found that the eyes were primarily stabilized with respect to the optokinetic stimulus rather than to space, a response that depended mainly on the OKR. The VOR appeared to be effective only at high rotational frequencies where the gain of the OKR declined, and it contributed considerably to visual image stabilization only in situations where the visual scene was stationary. We described our results by a simple model assuming linear summation of the two inputs. The model (see Fig. 3C) is similar to those of Raphan et al. (1979) and Robinson (1977), apart from its rather simple and straight forward topology, a substantial delay time and a velocity non-linearity.

In the present experiments we studied VOR-OKR interaction in healthy adult human subjects. In particular, we were interested to know whether our previous notion of OKR-VOR interaction (Schweigart et al. 1995) can be extended to humans and whether it applies similarly to active as compared to passive head movements in human subjects.

METHODS

OKR-VOR interaction was studied in seven healthy human subjects. Their age ranged from 24-47 years. The subjects were seated on a Bárány rotation chair which allowed whole body rotation in the horizontal plane. Subjects' heads were attached to the chair by means of a head holder with a 15° nose-down inclination. The intersection of the inter-aural and naso-occipital lines was centered at the vertical rotation axis of the chair. Head rotation in space (HS) represented the vestibular stimulus. For horizontal optokinetic stimulation, a black and white random patch pattern was projected onto a cylindrical screen (radius, 0.9 m) which surrounded the chair and which was rotated about the same vertical axis as the chair. It always covered the whole visual field. The rotation of the visual pattern relative to the head (VH) represented the optokinetic stimulus.

In the passive condition, isolated vestibular stimulation was obtained by head (chair) rotation in the dark. Isolated optokinetic stimulation was obtained by rotating the pattern while keeping the head stationary. The rotations had a sinusoidal wave form, symmetrically about the primary position. Chair and pattern could be rotated in either the same direction or in opposite directions (in-phase or counter-phase) and with different amplitudes (0°, ±8° and ±16°), such as to produce various vestibular-optokinetic stimulus combinations. During pure vestibular and pure optokinetic stimulation, we used five different stimulus frequencies (0.05, 0.1, 0.2, 0.4, 0.8 Hz). During the stimulus combinations we restricted the frequencies to 0.05 Hz and 0.8 Hz. Auditory cues were minimized by occluding the subjects' ears. Subjects were instructed to look passively at or through the pattern, rather than to fix the eyes on an item of the pattern, and to perform pre-instructed mental arithmetics.

In the active condition, the rotation chair always was stationary. The head holder was made pivotable to allow active head rotation in the horizontal plane. Subjects were instructed to move their heads in a sinusoidal way symmetrically about the primary position at different frequencies and velocities. This represented our vestibular stimulus. The optokinetic stimulus was varied by feeding a head position signal (potentiometer reading from the head holder) into the input of the optokinetic rotation device, and by modifying gain and sign of the signal. In this way similar vestibular-optokinetic stimulus combinations as in the passive condition were obtained. Subjects were instructed to look passively at the pattern, similar as in the previous experiment. However, they were unable to perform the pre-instructed arithmetics, since their attention was absorbed by the performance of the head movements.

Horizontal eye movements were recorded with conventional EOG (DC-recording). Position readings of the stimuli and of the eyes were recorded using a sampling rate of 200 Hz and stored in a laboratory computer. Data analysis was off-line. The actual frequency of each cycle was verified by Fourier analysis. Smooth (slow) and saccadic components of the horizontal eye position were separated using an interactive computer program. The slow eye-in-head responses (EH; see Fig. 1) were characterized in terms of gain and phase by using the fundamental waves of the Fourier analysis. Eye-in-space displacement (ES) was obtained by vector summation of the eye-in-head response with the head-in-space signal (chair or head holder position signal, respectively). The results to be presented give the average across all subjects (values per stimulus condition and subject, n= 4-30).

For calibration of the eye movements, the subjects tracked a moving red light spot (diameter, 0.5° of visual angle), which was projected onto the screen and rotated in the horizontal plane. With the rotation axes of spot and pattern projection being collinear with the axis of head rotation, we had to take into account that subjects' eyes were located in front of the rotation axes and that, therefore, the visual angle of spot rotation was approximately 10% larger than the corresponding head rotation angle. In the following we characterize the physical stimuli by giving the rotation angles of head or pattern movement in space. The eye movements were referred to the visual angles; for instance, with the head stationary, ideal eye tracking of the rotating pattern was defined as having unity gain.

RESULTS

Figure 1 shows an original recording of VOR-OKR interaction of a subject performing active head rotations. The visual pattern was rotated in phase with, but with twice the amplitude of the head (active condition). The subject tried to move the head sinusoidally, first at a high frequency (approximately 0.5 Hz) and then at low frequency (0.1 Hz). Note that, at the high frequency, the smooth component of the eye-in-head movement (EH) is rather small and is in counter-phase with respect to the visual pattern-in-space (VS) and head-in-space (HS) rotations. At the low frequency, in contrast, the EH response is considerably larger and in phase with VS and HS. The insets at the bottom of the figure repeat this response behavior in a schematic form, displaying in addition the eye-in-space (ES) movement calculated from EH and HS (vector summation). The ES response has the same direction as VS (and HS), the amplitude being similar to that of VS at the low frequency and less than half of VS at the high frequency. In the following we relate in most of the depicted data the ES response to the VS stimulus, apart from conditions where the visual pattern was

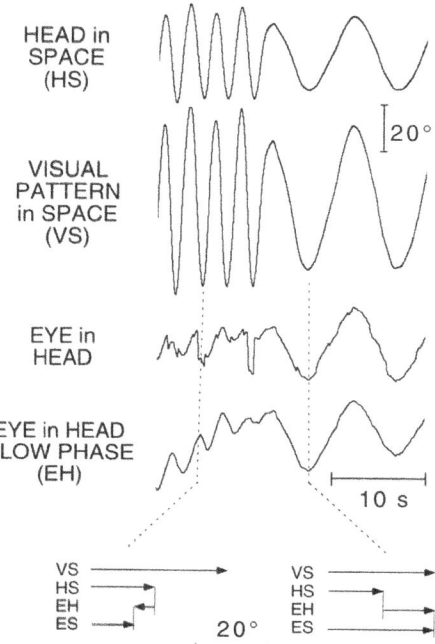

Figure 1. Examples of VOR-OKR interaction (active condition). The subject performed sine-like head movements in space (HS). The HS position reading was used to rotate the optokinetic pattern in space (VS) in the same direction, but with twice the amplitude of the head. Eye-in-head slow phase position (EH) was extracted from the eye-in-head curve by removing the quick phases. From this data the evoked eye-in-space response (ES) was calculated (as shown schematically in insets).

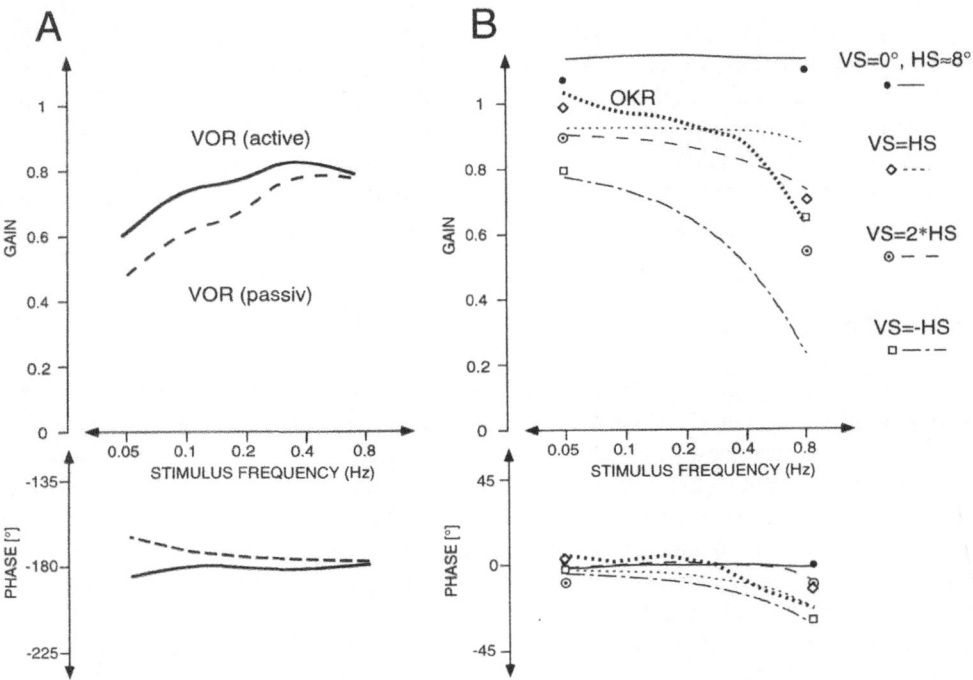

Figure 2. (A) VOR in the dark in the active condition (continuous curve) and in the passive condition (dashed curve). (B) OKR (thick dotted curve; passive condition) and eye-in-space responses during the various stimulus combinations as indicated on the right side. Data obtained with active head movements, thin lines, and with passive movements, symbols. The plots in A and B show mean gain and phase values as a function of stimulus frequency.

extinguished (VOR alone) or where VS was zero (head rotation with pattern stationary; EH related to VH, the calculated displacement of the visual pattern relative to the head).

Figure 2A gives the frequency response of our subjects' VOR (in the dark) in the active as compared to the passive condition. Mean gain and phase values are plotted as a function of stimulus frequency. At the highest frequency (0.8 Hz) gain and phase are similar in the two conditions, amounting to approximately 0.75 and 0 deg, respectively. With decreasing frequency the gain attenuates, and does so more in the passive condition as compared to the active condition, reaching 0.47 and 0.57, respectively, at 0.05 Hz. The differences between the two gain curves were statistically not significant, however, due to a large variability of the individual data. The phase developed a slight lag in the active condition (-7° at 0.05 Hz) and a slight lead in the passive condition (14° at 0.05 Hz).

The OKR response is depicted in Fig. 2B (stimulus amplitude, ±8°; passive condition only). The gain was close to unity at 0.05 Hz and 0.1 Hz, it decreased slightly at 0.2 Hz and 0.4 Hz, and it dropped to a value of 0.67 at 0.8 Hz. The OKR phase was near to ideal (0°) up to 0.2 Hz, but developed a lag at 0.4 and 0.8 Hz (-24° at 0.8 Hz).

Figure 2B gives, in addition, the results obtained with the stimulus combinations. In these combinations the vestibular stimulus was constant (precisely ±8° in the passive condition and close to ±8° in the active condition) and the optokinetic stimulus was varied. The results of the following combinations are shown: VS=-HS; peak displacement of visual pattern-in-space VS counter to that of head-in-space HS (active condition, mean value of

HS=9.95°±1.51°). VS=0°, HS≈8°; pattern stationary in space (HS=9.65°±1.05°). VS=HS; pattern head-stationary (HS=7.94°±1.74°). VS=2*HS; pattern in-phase with, but having twice the amplitude of the head (HS=8.19°±1.51°). In the active condition stimulus frequency varied to a considerable degree, depending on the subjects' performance of the head movements. Therefore, rather than giving mean gain and phase values, we present the best fitting regression lines to the individual data (polynomial, second order). As a reminder, in the passive condition only two stimulus frequencies were used (0.05 Hz and 0.8 Hz). Note that a gain of unity and a phase of 0° in this plot would indicate an ideal stabilization of the eyes on the pattern (compare above).

The gain of the combination responses attenuated with increasing stimulus frequency, while the phase developed some lag at high frequency, similar to the OKR. This applied to both the active and the passive condition. A clear exception was the combination 'VS=0°, HS≈8°' (visual scene stationary), in which gain and phase remained essentially constant across all frequencies. Remarkably, the phase at mid- to low frequencies was near to ideal (0°) in all combinations. Thus, eye and pattern movements were rather well synchronized, independently of the stimulus combination. The gain values, in contrast, showed a considerable variation in relation to the stimulus combination.

In order to better understand the relation between gain and stimulus combination, the data were replotted in Fig. 3A. The plot shows peak displacement values of head (H), visual pattern (V) and eyes (symbols and regression lines for passive and active conditions, respectively) relative to space (S) as a function of the stimulus combination. The regression lines used for the active condition gave a good description of the eye displacement values at a given frequency, indicating that vestibular and optokinetic effects summed linearly. Note that the regression lines for 0.05 Hz almost coincide with the lines representing visual pattern displacement (V), indicating that the eyes were rather effectively locked on the pattern at this frequency. With increasing frequency the slopes become smaller and the regression lines progressively rotate towards the horizontal lines S, which correspond to stationarity with respect to space. Thus, with increasing frequency the effect of the OKR became smaller, so that the VOR shifts the eyes towards a stabilization in space. An additional interesting feature in the plot is that the regression lines cross each other at a point close to the intersection of the V and S lines (visual pattern stationary, only the head rotates; VS=0°, HS≈8°). At this point gaze stabilization is largely independent of stimulus frequency. Furthermore, the results obtained in the passive condition (symbols) are similar to those just described for the active condition.

The results shown in Fig. 3A are summarized in Fig. 3B in the form that the slopes of the regression lines are plotted as a function of stimulus frequency. In this graph, a value of 1 would indicate that the regression line is parallel to the V line, i.e., that the eye displacement varies in perfect concert with the visual pattern displacement, whereas a slope of 0 would indicate that the eyes exhibit no response to the pattern motion. The slope values are close to unity at 0.05 Hz, but decrease with increasing frequency. They depend to some extent on the stimulus condition; the slope curve of the active condition (thick continuous curve) is somewhat higher than that of the passive condition (thick dashed curve).

DISCUSSION

Using sinusoidal rotations in the horizontal plane, we assessed VOR-OKR interaction in humans in the frequency range of 0.05 - 0.8 Hz. It is known that head movements during active behavior like walking and running contain rather high frequencies. For instance, the predominant frequency of horizontal head oscillations during walking in man is about 0.8 Hz, and during running the frequencies in the pitch plane may reach even 5 Hz (King et al.,

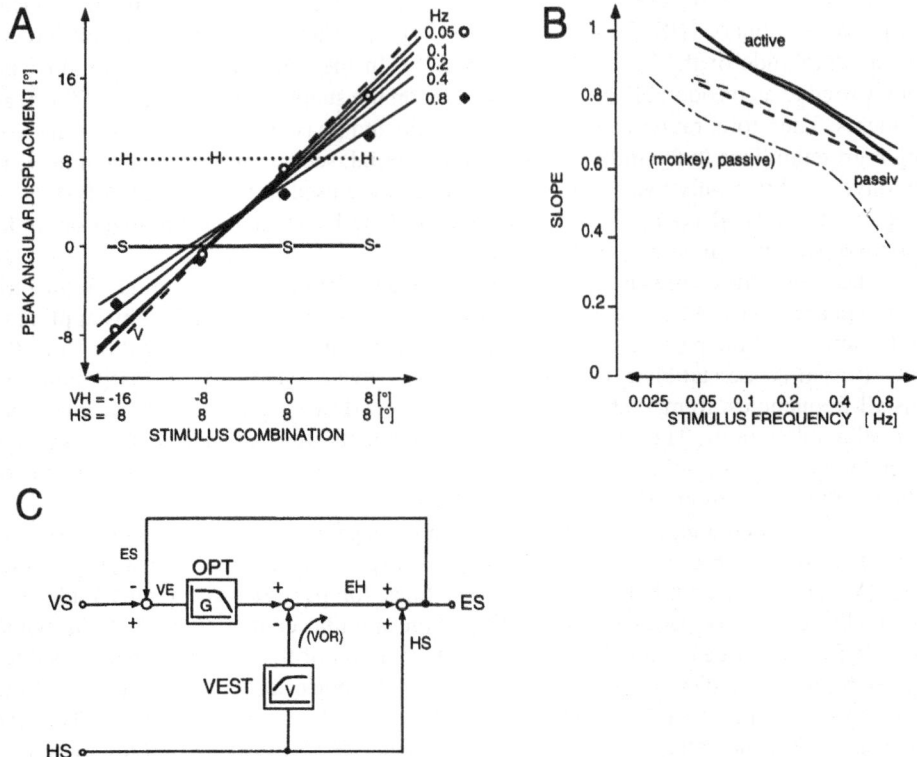

Figure 3. (A) Peak angular displacement values of eye, visual pattern (V) and head (H) relative to space (S) as a function of stimulus combination. The abscissa gives peak displacement values of visual pattern relative to head, VH (optokinetic stimulus), and head in space, HS (vestibular stimulus), with equal signs indicating in-phase combinations, opposite signs counter-phase combinations. Eye displacements are given by thin oblique (regression) lines for the active condition and by symbols for the passive condition, separately for different stimulus frequencies as indicated. (B) Slopes of regression lines plotted as a function of stimulus frequency. Thick solid curve, active condition, and thick dashed curve, passive condition. The superimposed thin curves give the corresponding data simulated with the help of the model shown in C. Thin dashed-dotted curve gives the experimental data of macaque monkeys, for comparison (Schweigart et al., 1995). (D) Model of OKR-VOR interaction.

1992). Furthermore, orienting head movements to auditory or visual targets may be very rapid ('saccade like'), containing high frequencies. Thus, it would have been desirable to study a broader range of frequencies. However, there were a number of obstacles. For instance, most subjects were unable to perform sine-like active head movements at frequencies below 0.05 Hz, and stimuli and measurements became rather inaccurate above 0.8 Hz. Furthermore, we believe that the investigated range of frequencies is of functional relevance; most of our time we are not running or performing jerky head movements, but rather are involved in activities that entail predominantly slow head movements or sways as, for instance, during upright stance, sitting, reaching movements, reading or passive body motions when using vehicles. Naturally occurring motions of the visual scene, on the other hand, usually are of low frequency, as for instance when watching traffic, wind agitated trees or drifting clouds. Finally, our findings in monkey (Schweigart et al., 1995) suggested that non-linearities in the VOR-OKR interaction become pronounced at high stimulus frequencies, a fact that poses major problems to the interpretation of the data.

A major result of the present study is that VOR and VOR-OKR interaction are basically similar in the active as compared to the passive condition, despite minor differences across the two conditions. It appeared from our data that, as a rule, VOR gain is somewhat larger in the active than in the passive (cf. Fig. 2A). This finding would be in line with earlier observations (see Barnes, 1993). Furthermore, during the active head rotations of our subjects with the pattern stationary, eye displacement gains even exceeded unity slightly, unlike with passive head rotation. Gain values above unity in comparable conditions (i.e., when actively rotating the head and viewing a target) have been observed earlier, at least in selected subjects (Collewijn et al., 1983). As a possible source of the gain increase in the active condition, one would have to consider neck afferent input (the subjects moved their heads relative to their trunks). Neck stimulation alone evokes a small and variable cervico-ocular reflex (COR) and in combination with the vestibular input increases the gain of the eye response slightly (Jürgens and Mergner, 1989). Alternatively, the effect could result from a general increase in arousal during the volitional activity.

Another important result of the study is that VOR-OKR interaction in man can be described in terms of a simple model (Fig. 3C) which has been developed earlier to describe this VOR-OKR interaction in monkeys (Schweigart et al., 1995). Similar as previous models with other topology, it assumes a linear interaction. It resembles in many respects the one suggested by Lau et al. (1978) except that it considers gaze position in space instead of eye-in-head position, and in that it relies on the results of a larger variety of stimulus frequencies and stimulus combinations.

Our model comprises: (i) A closed loop with negative feedback representing the optokinetic system which tries to zero the visual pattern-on-eye motion (VE, retinal error), i.e., the difference between visual pattern-in-space (VS) and eye-in-space (ES) motions. (ii) A summing junction (ES= EH + HS), where "noise" under the form of head movements is injected, reflecting the kinematic situation of the head as a generally unstable platform for the eyes, and a feed forward compensation of this noise by means of the VOR. The transfer function (V) of the vestibular system (VEST) is assumed to be a high pass with a time constant of 16 s (behavioral time constant).

The transfer function (G) of the optokinetic system (OPT) was modeled empirically, such that the model simulated the experimentally observed transfer function of the OKR (Fig. 2B). Basically, the box "OPT" consists of: (a) A "direct" pathway comprising a delay time of 75 ms (the shortest visuo-oculomotor latency in man; Gellman et al. 1990), a low pass of 1.5 Hz corner frequency, and a non-linearity in the form of a velocity saturation at 7°/s. (b) An "indirect" pathway integrating the signals of the direct pathway at frequencies above 0.015 Hz. Others have circumvented the problems associated with a substantial delay time being included in the closed loop of the optokinetic system (e.g., by assuming a central reconstruction of visual pattern-in-space movement through positive feedback of an efference copy (Raphan et al., 1979) without addressing the question of how the efference copy could be temporally matched with the delayed eye-in-space signal.

With G being the transfer function of the open loop optokinetic system (box OPT) and V that of the VOR (box VEST), eye displacement in space then will be

$$ES = (G/(1+G))*VS + ((1-V)/(1+G))*HS. \qquad (1)$$

According to equation (1) the eye-in-space movement is mainly determined by the visual pattern-in-space (VS) movement weighted by the closed loop gain (G/(1+G)) of the OKR, whereas the head-in-space (HS) movement contributes only to the extent that the VOR gain is different from unity.

When we consider the optokinetic stimulus (visual pattern motion relative to head, VH) instead of visual pattern-in-space motion, equation (1) becomes

$$ES = (G/(1+G))*VH + (1-(V/(1+G)))*HS, \tag{2}$$

which predicts that eye displacement in space is a linear function of the visual pattern-to-head displacement (VH prediction) and of the head-in-space displacement (HS prediction). The VH prediction is clearly born out by the data in Fig. 3A. Preliminary data (not presented here) support the HS prediction, at least in the passive condition, similar as in monkey (Schweigart et al., 1995). Note that, since G is a function of stimulus frequency, equation (2) defines a different regression line for each frequency.

If the vestibular stimulus is kept constant (Fig. 3A), equation (2) predicts a slope of

$$dES/dVH = G/(1+G). \tag{3}$$

Inserting the empirically determined transfer function G into equation (3), one obtains predictions for the slope of ES versus VH (thin curves in Fig. 3B). As evident from Fig. 3B, these predictions closely match the values determined from the regression lines in Fig. 3A (thick lines in Fig. 3B) over the frequency range tested. Also shown in Fig. 3B are the results of our previous data on the macaque monkey (Schweigart et al., 1995). The somewhat lower slope values in monkeys appears to result from the lower gain of their OKR; otherwise the results parallel those found in man, although man and monkey differ with respect to their VOR transfer characteristics.

Our findings of a linear interaction of VOR and OKR are in line with the earlier observations of Raphan et al. (1979) in the time domain and of Paige (1983) in the frequency domain. For interaction, Paige (1983) used the conditions of passive head rotation relative to a stationary pattern (VS=0°, HS=x°; 'VOR enhancement condition') and of head rotation with head-stationary pattern (VS=HS; 'VOR suppression condition'). One could argue that these two conditions behaviorally are particular in that they are cognitively well defined, whereas other stimulus combinations are rare in everyday life and might create unnatural sensations (yielding possibly a somewhat different VOR-OKR interaction). However, our results clearly demonstrate that VOR-OKR interaction is essentially the same (linear) in all stimulus combinations.

In view of the task to be accomplished, namely stabilization of the visual environment on the retina, and of the present findings, we suggest that the OKR should be considered the primordial system in VOR-OKR interaction. Owing to its negative feedback character, the OKR acts to reduce retinal slip whatever causes the slip, be it ego-motion, motion of the visual scene, or both. At low frequencies, the high gain of the OKR even overrules contributions of the VOR. However, the ability of the OKR to stabilize the eyes on the scene deteriorates at high frequencies. It is at these high frequencies where the VOR "helps" the OKR. However, this help is directed only at reducing the "noise" introduced by head movements performed in a stationary visual environment, whereas it becomes a nuisance in situations in which the visual scene, or large parts of it, are moving.

SUMMARY

The interaction of the vestibulo-ocular reflex (VOR) and the optokinetic reflex (OKR) was studied in normal human subjects using various vestibular-optokinetic stimulus combinations. The vestibular stimulus was generated by sinusoidal head movements in the horizontal plane, both during passive (rotation chair) and active (self-generated) head movements. The optokinetic stimulus was generated by a pivotable random patch pattern. At low stimulus frequencies (< 0.2 Hz in most optokinetic-vestibular stimulus combinations), the eyes tended to be stabilized on the optokinetic pattern, independently of whether the head, the pattern, or both were rotated. At higher frequencies, the OKR gain attenuated, and in the stimulus combinations the eyes became increasingly stabilized in space. The results were similar during active and passive head movements. We conclude that eye stabilization on the visual scene, a prerequisite for accurate vision, is primarily performed by the OKR. The VOR becomes a useful addition only at high stimulus frequencies/velocities during head rotation in a stationary visual environment, compensating for the limited bandwidth of the OKR in case of rapid head movements. We present a model which describes OKR-VOR interaction both in monkey and man, and both for passive and active head rotations.

ACKNOWLEDGMENT

Supported by DFG SFB 325

REFERENCES

Barnes, G.R., 1993, Visual-vestibular interaction in the control of head and eye movement: the role of visual feedback and predictive mechanisms. *Prog. Neurobiol.* 41:435-72.

Collewijn, H., Martins, A.J., and Steinman, R.M., 1983, Compensatory eye movements during active and passive head movements: fast adaptation to changes in visual magnification, *J. Physiol.* 340:259-286.

Gellman, R.S., Carl, J.R., and Miles, F.A., 1990, Short-latency ocular-following responses in man. *Vis. Neurosci.* 5:107-122.

Henn, V., Cohen, B., and Young, L.R., 1980, Visual-vestibular interaction in motion perception and the generation of nystagmus. *Neurosci. Res. Prog. Bull.* 18:457-651.

Jürgens, R., and Mergner, T., 1989, Interaction between cervico-ocular and vestibulo-ocular reflexes in normal adults. *Exp. Brain Res.* 77:381-390.

King O.S., Seidman S.H., and Leigh R.J., 1992, Control of head stability and gaze during locomotion in normal subjects and patients with deficient vestibular functions. In: Berhoz A., Graf W., Vidal P.P., eds. The head-neck sensory motor system. New York, Oxford: Oxford University Press, pp. 568-570.

Lau, C.G., Honrubia, V., Jenkins, H.A., Baloh, R.W., and Yee, R.D., 1978, Linear model for visual-vestibular interaction. *Aviat. Space Environ. Med.* 49:880-885.

Paige, G.D., 1983, Vestibuloocular reflex and its interaction with visual following mechanisms in the squirrel monkey. I. Response characteristics in normal animals. *J. Neurophysiol.* 49:134-151.

Raphan, T., Matsuo, V., and Cohen, B., 1979, Velocity storage in the vestibulo-ocular reflex arc (VOR). *Exp. Brain Res.* 53:229-248.

Robinson, D.A., 1977, Vestibular and optokinetic symbiosis: an example of explaining by modelling. In: Baker R., Berthoz A., eds. Control of gaze by brain stem neurons. Amsterdam: Elsevier, pp. 49-58.

Schweigart, G., Mergner, T., and Becker, W.,1995, Eye stabilization by vestibulo-ocular reflex (VOR) and optokinetic reflex (OKR) in macaque monkey: which helps which?, *Acta otolaryngol. (Stockh.)* 115:19-25.

22

INERTIAL REPRESENTATION OF VISUAL AND VESTIBULAR SELF-MOTION SIGNALS

Bernhard J. M. Hess and Dora E. Angelaki

Department of Neurology
University Hospital
CH-8091 Zürich
Switzerland
Department of Surgery (Otolaryngology)
University of Mississippi Medical Center
2500 North State Street
Jackson, Mississippi 39216-4505

INTRODUCTION

Recently, we have investigated the behavioral consequences of central interactions of semicircular canal-born head velocity signals and otolith signals coding absolute head position relative to gravity (Angelaki and Hess, 1994, 1995). In these studies, we determined the spatial orientation of head angular velocity following step-like head tilts applied in a plane orthogonal to a previous earth-horizontal head rotation. The basic finding is that the vestibular system of rhesus monkeys behaves like a gyroscope keeping head angular velocity signals invariant relative to gravity. Thus, for example, a transient head tilt can induce a rather precise spatial transformation of a vestibulo-ocular response which compensates for the amount of head tilt in direction and angle. As a result, the overall response continues to reflect faithfully the spatial head movement. As vestibular motion signals are usually accompanied by visual information related to the associated optic flow pattern, one would expect that both vestibular and visual inputs interact centrally in a coherent way in order to represent the state of head motion in space. In the following paragraphs, we present evidence supporting the hypothesis that both vestibular and optokinetic signals are processed in a network which represents the state of head motion in a common inertial (gravity-related) reference frame.

STATIC VERSUS DYNAMIC TESTING OF INERTIAL PROPERTIES OF THE VESTIBULAR SYSTEM

Head orientation relative to gravity has a profound influence on the spatio-temporal response characteristics of vestibular and optokinetic ocular nystagmus. With regard to the optokinetic system, visual self-motion information can be easily dissociated experimentally from the normal vestibular motion information, in particular from a dynamic otolith input as it is normally associated with head movements. For example, optokinetic full field stimulation about the head yaw axis can be performed in different head positions relative to gravity. Under such somewhat artificial conditions of visual full-field stimulation in tilted head positions (vs. gravity) there is a reorientation of an initially primarily horizontal optokinetic afternystagmus relative to the head. This reorientation of the eye angular velocity vector of the principal response component is generated by build-up of an orthogonal response component in the plane of head tilt (Raphan and Cohen, 1988; Dai et al., 1991). Interestingly, no such reorientation of eye angular velocity occurs in static tilt positions during torsional and vertical optokinetic afternystagmus following a constant velocity rotation of the visual surround about the head roll or pitch axis. Based on these premises, Raphan and Sturm (1991) proposed that the underlying vestibular mechanism is best described by a spatial transformation that keeps the yaw axis eigenvector close to the spatial vertical while the roll and pitch axis eigenvectors remain head fixed. Such a system would not care about reorienting the direction of torsional or vertical head velocity signals as it is the case during optokinetic afternystagmus in static tilt positions nor would it keep such signals space-invariant following a reorientation of the head by a transient head tilt relative to gravity. In mathematical terms, such an operator that keeps only the yaw eigenvector space-fixed when the head is tilted or reoriented relative to gravity can be expressed in form of an upper triangular matrix with respect to head coordinates (i.e., all elements below the diagonal are zero):

$$\Omega_{space} = \begin{pmatrix} tor\leftarrow tor & tor\leftarrow ver & tor\leftarrow hor \\ ver\leftarrow tor & ver\leftarrow ver & ver\leftarrow hor \\ hor\leftarrow tor & hor\leftarrow ver & hor\leftarrow hor \end{pmatrix} \Omega_{head}$$

$$= \begin{pmatrix} tor\leftarrow tor & 0 & tor\leftarrow hor \\ 0 & ver\leftarrow ver & ver\leftarrow hor \\ 0 & 0 & hor\leftarrow hor \end{pmatrix} \begin{pmatrix} \Omega_{tor} \\ \Omega_{ver} \\ \Omega_{hor} \end{pmatrix}_{head}$$

As it is reasonable to assume that optokinetic and vestibular signals share the same central vestibular mechanisms, one would expect that the same spatial transformation principles apply to vestibular as well as optokinetic signals. The experimental proof of such assumption is not straightforward, at least not with regard to static tilt experiments since there is no easy way to dissociate the direction of stimulation of the semicircular canals and the direction of gravity while keeping at the same time head orientation constant relative to gravity. We therefore adopted a different experimental approach to investigate the basic spatio-temporal properties of the vestibular system using dynamic head tilts. Similar to our earlier study on the vestibulo-ocular reflex, we measured three-dimensional optokinetic nystagmus and afternystagmus in the rhesus monkey in a paradigm aimed at dissociating the head-fixed semicircular canal-related reference and the inertial reference. Both reference frames coincided during the optokinetic nystagmus (and the initial afternystagmus) which was generated by rotating the visual surround about the earth-vertical axis. The dissociation paradigm consisted of applying fast transient head tilts about an earth-horizontal axis through various tilt angles (pulse of 180 °/s peak velocity and 180°/s² acceleration) and in different head planes after stop of optokinetic stimulation (i.e., light off inside the rotating optokinetic sphere which surrounded the animal followed by stop of drum rotation). In our earlier study with an analogous test paradigm , we found that the vestibular system tended to keep angular eye velocity constant in space not only for the horizontal but in a similar way also for the torsional and vertical vestibulo-ocular reflex

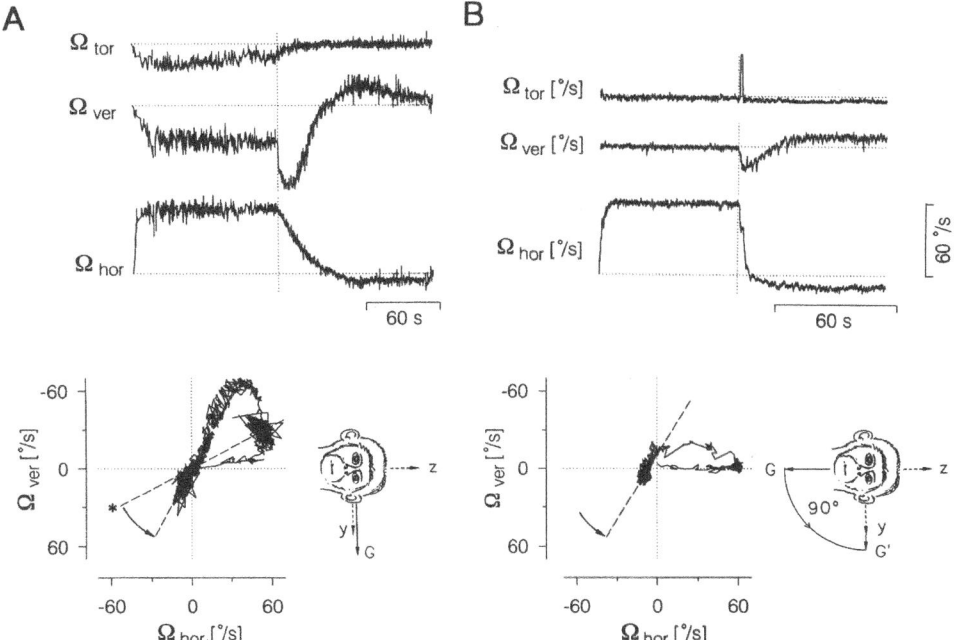

Figure 1. Spatial reorientation of optokinetic afternystagmus in the head roll plane. A: Horizontal optokinetic nystagmus and afternystagmus following stimulation at 60 °/s about the head yaw axis in left-ear down position. B: Similar as in A but optokinetic stimulation in upright position was followed by a rapid roll tilt to left-ear down position immediately after light-off. *Upper panels:* Time course of torsional (Ω_{tor}), vertical (Ω_{ver}) and horizontal (Ω_{hor}) eye angular velocity. Dotted vertical lines indicate stop of stimulation (light off). Dotted horizontal lines indicate zero base lines. Peak velocity of torsional vestibulo-ocular reflex in B is cut. *Lower panels:* Vertical versus horizontal eye velocity depicting the eye velocity trajectory in the head roll plane. Dashed lines show the best fitted lines through the eye angular velocity trajectory after the vertical response component had reached its peak value during optokinetic afternystagmus (A: tilt angle = -62.3°, B: tilt angle = -59.3°). Average tilt of eye velocity vector during steady-state optokinetic response is -28.4° in A (dashed line marked with asterisk) and zero in B.

(Angelaki and Hess, 1994). These findings were obviously not in line with the results of Dai et al. (1991) obtained during static testing of the optokinetic afternystagmus. At first sight, this might suggest that there are basic differences in the central representation of vestibular and optokinetic motion signals. Before addressing this issue, we like to compare the dynamic and the static test paradigm in the case of optokinetic nystagmus and afternystagmus. To this end, we studied in the same animal the spatial reorientation of horizontal optokinetic afterresponses in different head tilt positions and compared the results with those obtained by applying a head tilt from upright through the same angle to the same final position (Fig 1). As this example shows, there was little difference in the final spatial orientation of the eye angular velocity vector when comparing a static with a dynamic roll tilt during the afterresponse of a horizontal optokinetic nystagmus.

The major difference in the response characteristics elicited in the two experimental paradigms consisted in the shape of the eye velocity trajectory: While the eye velocity vector followed typically a large loop in the static tilt situation, this loop was much smaller following a dynamic reorientation of the head. The loop area enclosed by the velocity trajectory mainly depends on how fast the principal response component decays in relation

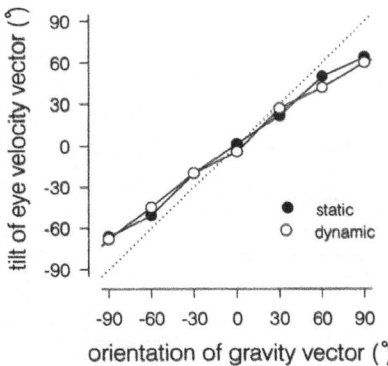

Figure 2. Relation between tilt of eye angular velocity and gravity vector during horizontal optokinetic afternystagmus. Results obtained during static roll tilt (solid symbols) are compared with those obtained following dynamic roll tilts from upright through corresponding tilt angles (open symbols). Upright (0°), left ear-down (90°), right ear-down (-90°).

to the simultaneous build-up of the orthogonal response component (given similar gains of the principal response). Thus, the observed relative reduction in loop area due to the dynamic head tilt reflects a stronger dumping of the principal response component. This enhanced dumping allows the system to build-up a correspondingly smaller orthogonal response component with still the same result in terms of final orientation of the eye velocity vector. This most conspicuous difference in the trajectory loop was little due to a difference in the habituation of the animal. The time constant of the horizontal optokinetic afternystagmus was $T_l = 82.2$s and $T_r = 30$ s in this animal for leftward and rightward optokinetic stimulation in upright position, respectively. The static head tilt reduced this time constant by a factor of 3.1 to $T_l = 26.1$s. At the time of the dynamic tilt experiments the animal exhibited time constants of $T_l = 28.3$ s and $T_r = 23.5$ s in upright position. The dynamic head tilt reduced the time constant by a factor of 8.1 to $T_l = 3.5$ s. The large vertical eye velocity component in the static tilt situation reflects a bias in the orientation of eye velocity in the roll plane. This bias was more pronounced for horizontal nystagmus with slow phase velocity directed towards gravity (leftward in left-ear down, rightward in right-ear down position) which was always associated with a downbeat nystagmus (slow phase up). Similar results have been reported earlier by Raphan and Cohen (1988). There was also a small torsional bias velocity during yaw axis optokinetic stimulation in the head tilted position which was not present in head upright position. Its slow phase velocity was opposite in direction to the static counterroll in torsional eye position (not shown). Nevertheless, the difference in the final spatial orientation of the eye velocity vector of optokinetic afternystagmus in the static or dynamic head tilt situation was minor (Fig 2). In both paradigms, there was the typical undershoot in final orientation of eye angular velocity at larger tilt angles as described earlier for both the optokinetic afternystagmus in static head tilt positions (Dai et al., 1991) and postrotatory vestibulo-ocular responses following dynamic head tilts (Angelaki and Hess, 1994).

Spatio-Temporal Properties of Inertial Processing of Vestibular and Optokinetic Signals

The spatio-temporal transformations leading to a reorientation of head velocity signals exhibit different characteristics for signals originating from the different semicircular canals: Velocity signals originating from the lateral semicircular canals are transformed according to a rotation rule while those originating from the vertical semicircular canals follow rather a projection rule (Angelaki and Hess, 1994). The projection rule is best characterized by the fact that orthogonal response components change signs for tilts through

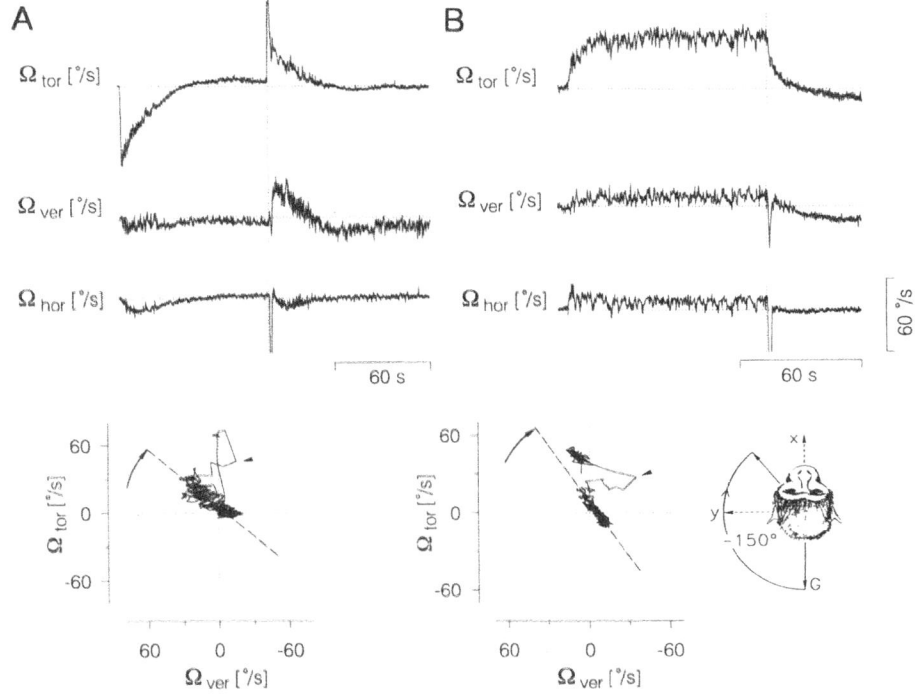

Figure 3. Spatial reorientation of postrotatory torsional vestibulo-ocular and optokinetic afternystagmus after yaw tilt through an obtuse angle. A: Per- and postrotatory torsional vestibulo-ocular nystagmus elicited by constant-velocity rotation at 90 °/s about the head roll axis in supine position. B: Torsional optokinetic nystagmus elicited by optokinetic stimulation about the head roll axis in supine position. In A and B, the head was tilted through -150° in the yaw plane immediately following onset of postrotatory response and optokinetic afternystagmus. Peak velocity of transient horizontal vestibulo-ocular reflex is cut. *Upper panels:* Time course of torsional (Ω_{tor}), vertical (Ω_{ver}) and horizontal (Ω_{hor}) eye angular velocity. Dotted vertical lines indicate stop of stimulation (light off). Dotted horizontal lines indicate zero base lines. *Lower panels:* Torsional versus vertical eye velocity depicting the eye velocity trajectory in the head yaw plane. Dashed lines show the best fitted lines through the eye angular velocity trajectory after the vertical response component had reached its peak value during postrotatory response or during optokinetic afternystagmus (A: tilt angle = -130.4°, B: tilt angle = -145.2°).

obtuse angles compared to tilts through acute angles. To facilitate synergistic interaction of optokinetic and vestibular self-motion signals one would expect that the optokinetic input associated for example with a head roll or head pitch movement would be transformed through a similar projection mechanism as the respective semicircular canal signals. This is indeed the case as illustrated in Fig. 3 for the spatial modification of a torsional optokinetic afternystagmus by a yaw tilt through an obtuse angle (-150°). In this example, we compare the spatial transformation of a torsional optokinetic and a vestibulo-ocular response obtained from two different animals for the same dynamic tilt. Typically, the trajectory of eye velocity followed for a certain time the same direction as the tilt (see arrow heads in Fig 3). After this initial deviation the trajectory changed direction to follow the shortest path towards alignment with the new direction of gravity.

The similarity in the response characteristics shown in this example suggests that optokinetic and vestibular signals are processed through the same basic mechanism, at least when there is a dynamic reorientation of the head relative to gravity. It is apparent that in both situations, not only velocity signals of the vertical semicircular canal system are processed through the projection mechanism but also the respective optokinetic head velocity signals. Clearly, any quantitative explanation of the underlying spatial transformation mechanism must account for this parallel processing of optokinetic and vestibular signals during dynamic head tilts. The Raphan-Sturm model of visual-vestibular interaction falls short in explaining the spatial reorientation illustrated in Figure 3. This spatial transformation of central head velocity signals requires a non-zero matrix element "ver←tor" (see transformation matrix above) which transforms a torsional vector component into a vertical vector component.

Are The Gyroscopic Vestibular Properties Adequately Described by an Upper Triangular Operator?

Our recent study on the spatio-temporal properties of the vestibular system using dynamic tilt experiments clearly revealed that head velocity signals in the vestibulo-ocular reflex are kept space-constant (Angelaki and Hess, 1994). The important consequence of this finding is that the vestibular system exhibits gyroscopic properties. These gyroscopic properties are feeding through a three-dimensional low pass filter before being coupled to the oculomotor output via a gain element. We assume that the same or similar head velocity signals are fed through a different gain element into the vestibulo-spinal system to facilitate control of head and body posture (Angelaki and Hess, 1995). The properties of this system could be summarized as an inertial guidance system to provide orientation in absolute space. Our data in both the vestibular and optokinetic protocol suggest that all matrix elements of the underlying spatial transformation operator are non-zero. Alternatively, it has been argued that the "pitch-roll" transitions in eye velocity signals which we described following dynamic yaw tilts in the vestibulo-ocular reflex (see for example Fig 3 A) charge the yaw state of the velocity storage integrator with a small time constant which then discharges along the roll or pitch axis with a longer time constant when the animal is supine/prone (Raphan, Wearne and Cohen, 1994). Such an argument is inadequate to explain our findings. The short-lasting transient head tilts lead to a similarly short on-off response of eye velocity in the same movement plane (see examples Fig. 1, 3 and 4). If there is any charge of the velocity storage during the on-phase of the tilt stimulus it is discharged during the off-phase. Furthermore, inactivation of the semicircular canals in the plane of the head tilt does not abolish the spatial reorientation of vestibular nystagmus suggesting that the transformation of head velocity does not depend on an appropriate semicircular canal signal in the tilt plane. In fact, it appears to depend entirely on otolith input signals (Angelaki and Hess, 1995). Finally, not only dynamic head tilts about yaw but also about pitch and roll result in similar reorientation of eye velocity in the vestibulo-ocular reflex as well as in the optokinetic (after)nystagmus. In all these cases, reorientation of eye angular velocity relative to the head can be explained by a simple rotation or projection of the principal response without recurring to a charge of the velocity storage integrator along the axis of head tilt.

An example of a transition from vertical to horizontal eye velocity during optokinetic afternystagmus is illustrated in Figure 4. This example demonstrates that the principle of spatial constancy of steady-state head velocity signals holds true not only for yaw but also for pitch axis optokinetic stimulation when applying a head roll tilt during the afternystagmus. Thus, the spatial transformation matrix must exhibit also a non-zero element "hor←ver" below the diagonal, contrary to the assumption of an upper triangular matrix.

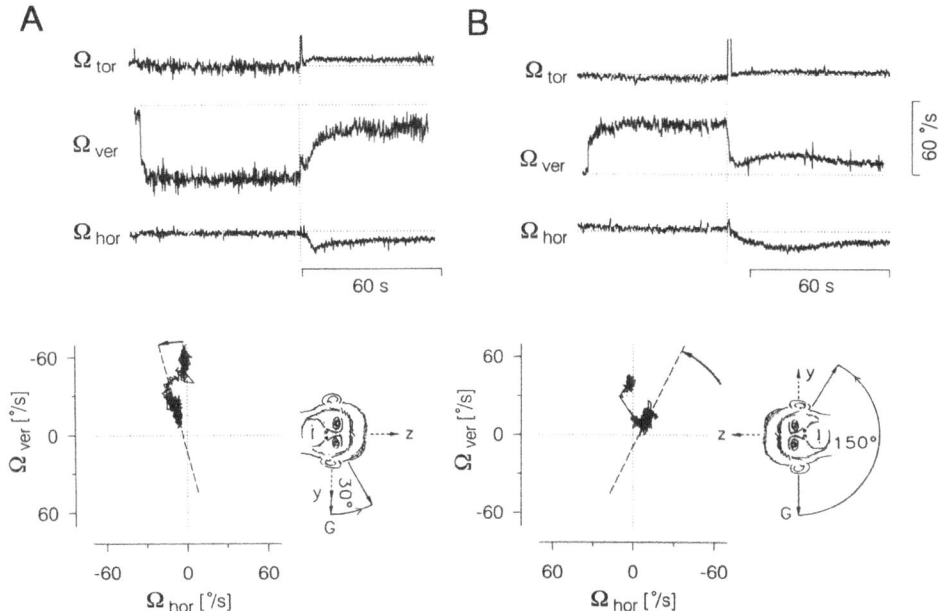

Figure 4. Spatial reorientation of vertical optokinetic nystagmus after a rapid tilt in the roll plane during the afternystagmus. A: Vertical optokinetic nystagmus and afternystagmus to optokinetic stimulation at 60 °/s about the head pitch axis in left-ear down position. Onset of optokinetic afternystagmus was followed by a rapid roll tilt through 30° towards upside-down. B: Similar as in A but optokinetic stimulation in right-ear down position. Onset of optokinetic afternystagmus was followed by a rapid roll tilt through 150° towards left-ear down. *Upper panels:* Time course of torsional (Ω_{tor}), vertical (Ω_{ver}) and horizontal (Ω_{hor}) eye angular velocity. Dotted vertical lines indicate stop of stimulation (light off). Dotted horizontal lines indicate zero base lines. Transient torsional vestibulo-ocular reflex is cut in B. *Lower panels:* Vertical versus horizontal eye velocity depicting the eye velocity trajectory in the head roll plane. Dashed lines show the best fitted lines through the eye angular velocity trajectory after the horizontal response component had reached its peak value during optokinetic afternystagmus (A: tilt angle = 14.3°, B: tilt angle = 154.1°).

CONCLUSION

Our results strongly suggest that vestibular as well as optokinetic signals are processed through the same spatio-temporal transformation mechanism during dynamic head tilts. Thus, it appears that all 9 matrix elements of the operator describing the neural transformation in the vestibular system are non-zero. How can we reconcile this finding with the experimental observation that there is no spatial reorientation of eye angular velocity during torsional or vertical optokinetic afternystagmus in static tilt positions? This peculiar feature of the optokinetic reflex could be related to the fact that retinal image slip signals must undergo some additional processing before feeding into the vestibular spatial transformation mechanism. It has been proposed that the optokinetic head velocity signals are computed as the sum of a retinal slip signal and an efference copy of eye velocity (Robinson, 1977). A channel-specific gating of this summing process could alter the overall spatial transformation process in the appropriate way. Thus it is conceivable that optokinetic stimulation in static head positions is associated with a gating of the visual input to the inertial vestibular system in such a way that only yaw velocity signals are reoriented towards gravity. In summary, vestibular and optokinetic signals are processed centrally in a network exhibiting gyroscopic properties: it represents faithfully the state of self-motion in inertial space. While in the dynamic motion situation, optokinetic and vestibular signals appear to be processed in a similar way, transformation characteristics differ for optokinetic signals in static tilt situations, presumable due to non-linear gating of the visual feedback signals. Inertial coding of head motion signals appears to be instrumental for optimal control and coordination of head, gaze and body in space.

ACKNOWLEDGMENT

This work was supported by grants from the Swiss National Science Foundation (31-32484.91) and the National Institute of Deafness and other Communication Disorders (F32 DC-00092).

REFERENCES

Angelaki, D.E., and Hess, B.J.M., 1994, Inertial representation of angular motion in the vestibular system of rhesus monkeys. I. Vestibulo-ocular reflex, *J. Neurophysiol.* 71:1222-1249.

Angelaki, D.E., and Hess, B.J.M., 1995, Inertial representation of angular motion in the vestibular system of rhesus monkeys. II. An otolith-controlled coordinate transformation that depends on an intact cerebellum, *J. Neurophysiol,* 73:1729-1751.

Dai, M., Raphan, T, and Cohen, B., 1991, Spatial orientation of the vestibular system: dependence of optokinetic after-nystagmus on gravity, *J. Neurophysiol.* 66:1422-1439.

Raphan, T., and Cohen, B., 1988, Organizational principles of velocity storage in three dimensions. The effect of gravity on cross-coupling of optokinetic afternystagmus. *Ann. N. Y. Acad. Sci.* 545:74 - 92,

Raphan, T., Wearne, S., and Cohen, B., 1995, Static and dynamic effects of gravito-inertial acceleration (gia) on spatial orientation of velocity storage, *Soc. Neurosci. Abstr.* 20:1195,

Robinson, D.A., 1977, Vestibular and optokinetic symbiosis: an example of explaining by modelling. In: Control of gaze by brain stem neurons, edited by R. Baker and A. Berthoz, Amsterdam: Elsevier/North Holland, p. 49-58.

23

THE VISUAL GUIDANCE OF BALLISTIC ARM MOVEMENTS

Jeroen B. J. Smeets[*] and Eli Brenner

Vakgroep Fysiologie
Erasmus Universiteit Rotterdam
Postbus 1738
NL-3000 DR Rotterdam
The Netherlands

INTRODUCTION

To hit a moving object, you must reach some position at the same time as the object. The position at which you will hit the object depends on the object's speed and direction of motion, as well as on the timing of your own movement. We investigated the information used to extrapolate the position of a moving object during both the planning and the execution of fast (ballistic) arm movements.

To make goal-directed movements, the nervous system has to transform sensory information into activations of various muscles. If a target is moving, or if one is moving oneself, the sensory information is continuously changing. The best way to control ones action is to take these changes into account to predict the future position of the target. How does the nervous system use information about target motion in the control of goal-directed action?

Spatial information is only meaningful when it is defined with respect to a frame of reference. Two frames of reference are important when considering visuo-motor co-ordination. Arm movements are defined with respect to an egocentric frame of reference (Paillard, 1991), which is also used to perceive positions. For the perception of motion, however, an allocentric frame of reference is also used: the (stable) visual surrounding (Brenner and Smeets, 1994a; Smeets and Brenner, 1994). This dissociation between the perception of position and motion is the basis for our first question: does the nervous system use (allocentrically perceived) motion to predict a future target position for goal-directed arm movements? To answer this question we studied goal-directed movements to targets which were moving over a structured background. We separated the contributions of position and motion by moving the background on some of the trials.

---
[*] e-mail: smeets@fys1.fgg.eur.nl. Phone: +31-10-4087565; fax: +31-10-4367594.

Spatial information is not only used to plan fast goal-directed arm movements. It is also used to adjust the movements (Prablanc, Pélisson and Goodale, 1986). To study the control of ongoing movements, we studied arm movements towards targets which started moving when the hand started moving. The duration of the arm-movements was about the same as the reaction time. Assuming that correcting ongoing movements is based on the same spatial information processing as the planning of these movements, it would take longer than the duration of these movements to react to changes in the environment. For this reason, such fast movements are in general named 'ballistic'. It is possible, however, that more limited spatial information is used for faster feedback. By studying 'ballistic' movements, we ensure that we are dealing with on-line control of the movements, rather than with corrective (sub)movements superimposed on the ongoing movements.

METHODS

We used a graphical workstation to present the stimuli, and active infrared markers to measure the subjects' movements (see Fig. 1). Targets moved at five different velocities or in five different directions. Allocentric information was selectively manipulated by moving the background in some conditions. In all experiments, trials with and without background motion, and with various directions and speeds, were randomly intermixed. In all experiments, subjects were completely free to move their eyes and heads. An extensive description of the set-up and experimental procedure is given elsewhere (Smeets and Brenner, 1995).

To examine which spatial information is used for planning arm movements, two experiments were carried out: one to compare the use of position and speed, the other to compare the use of position and direction of motion. Each experiment consisted of two parts. In the first part, we investigated how subjects perceived the targets' motion. In the second part, we studied how subjects used this perceived information to control their goal-directed action. It has been argued that visual information is treated separately for perception and action (Bridgeman et al, 1981; Goodale and Milner, 1992). However, we have shown that both egocentric and allocentric information is used in both perception and action (Brenner

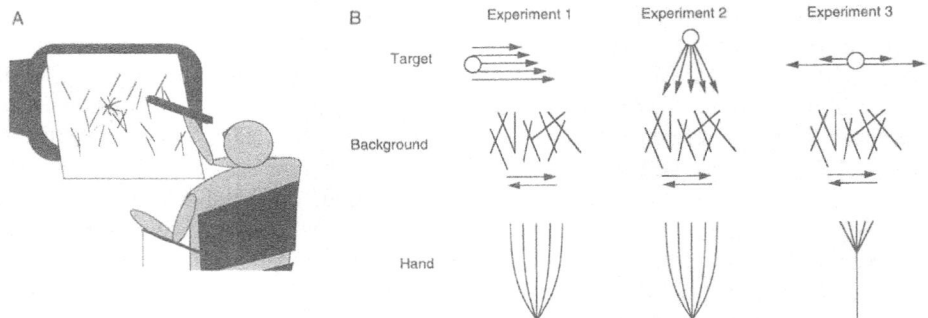

Figure 1. (A) The experimental set-up. Subjects sat in front of a screen onto which a target and a background were projected. (B) Both target and background could move (shown schematically in front-view), which influenced the hand's trajectory (shown schematically in top-view). In experiment 1, we varied the apparent speed of the target, by varying the target's motion, and by moving the background. In experiment 2, we varied the apparent direction of motion of the target by varying the target's motion and by moving the background. In experiment 3 the target and the background remained static until the hand started moving. Thereafter, either the target, or the background could start to move.

and Smeets, 1994; Smeets and Brenner, 1995). The distinction appears to be between position and motion, rather than between perception and action.

In the first part of experiments 1 and 2, we asked the subjects to indicate the perceived position, speed and direction of motion of the target with the computer mouse. In the second part of these experiments, we asked the subjects to hit the moving target as quickly as possible with a rod. We presented the same stimuli as in the first part of the experiments. Subjects received visual feedback about their performance after each trial.

To investigate whether spatial information influences ongoing ballistic movements, a third experiment was carried out. In this experiment, the target and background remained static until the hand started moving. Thereafter, either the background or the target could start moving, either to the left, or to the right. If the background moved to the left, the target appeared to move to the right, and vice versa.

All figures show results averaged over all subjects and trials. These were 6-12 subjects who each performed 5-15 trials for each condition (numbers depend on the experiment).

RESULTS

The perceived target motion in experiments 1 and 2 is shown in Fig. 2. The perceived position is independent of motion of the background, whereas the perceived speed and the perceived direction of motion are influenced by motion of the background. This is in line with our expectation: position is perceived egocentrically; motion is perceived (partially) allocentrically. Having a good perceptual characterization of our stimulus, we can now describe how this perceptual information is used in motor control.

Our subjects started to move their hand about 300 ms after the target appeared on the screen. They accelerated their hand almost continuously towards the screen. After about 300 ms, their hand was decelerated by the screen when they hit it. The direction in which the hand moved at motion onset depended on the position of the target, but was independent of its perceived speed (Fig. 3A). However, the perceived speed did influence motor control: the maximum velocity of the hand (and thus the movement time) depended on the perceived target speed (Fig. 3B).

One could conclude from experiment 1 that only egocentric information is used to direct the hand towards the target. To examine whether this is so, the role of another allocentric source of information was evaluated in experiment 2: the direction of target motion. Fig. 3C shows that the direction in which the hand moved at movement onset depended on the (allocentrically perceived) direction of target motion. We therefore reject the hypothesis that only egocentric information is used to direct the hand.

In experiment 3, the target was stationary when it appeared on the screen. Once the subjects moved their hands, it started to move. Subjects always adjusted the movements of their hands (Fig. 4). If the background moved instead of the target, the response of the hand was later, smaller, and in the opposite direction than the target appeared to move. Thus, the allocentrically perceived direction of motion is not used for the on-line control of fast movements. The latency of the response to target motion was always less than 150 ms; examination of the lateral acceleration showed that the latency of response to a change in target position was 110 ms (Brenner and Smeets, 1995). The maximum speed of the arm movements was independent of whether the target started to move. The control of an ongoing movement is therefore based on purely egocentric information on the target's position.

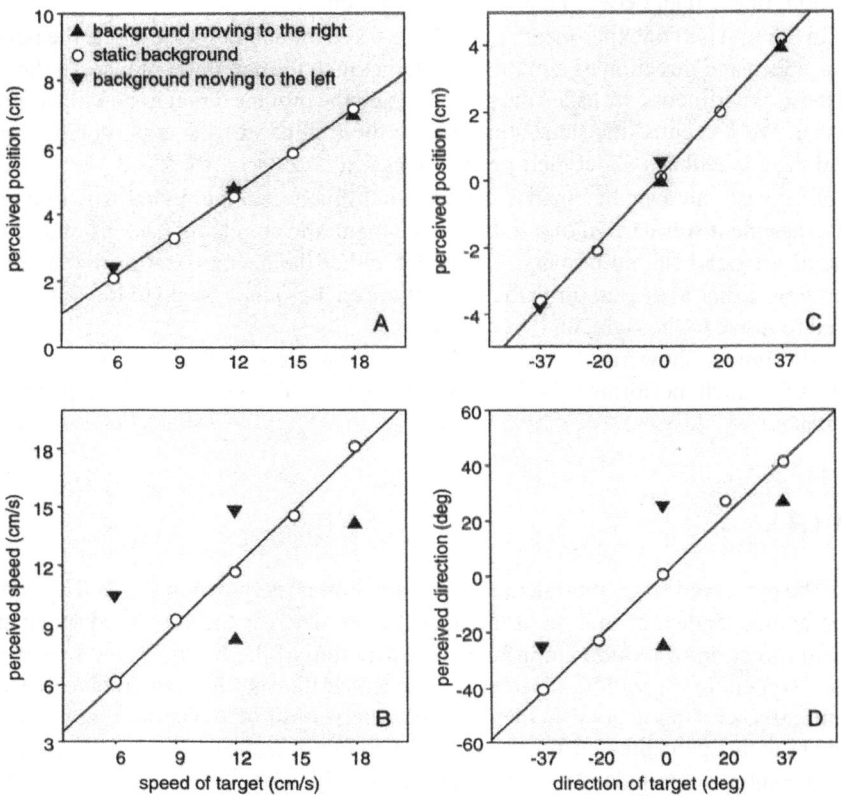

Figure 2. Perception of spatial information. (A) Subjects were asked to indicate the positions at which targets moving at different speeds disappeared (experiment 1). These positions were perceived correctly, irrespective of background motion. (B) Subjects were asked to match the targets' speed (experiment 1). Background motion induced large errors in the perceived speed. (C) Subjects were asked to indicate the positions at which targets moving in different directions disappeared (experiment 2). The different positions at which the targets disappeared were again perceived correctly, irrespective of background motion. (D) Subjects were asked to indicate the directions in which targets moved (experiment 2). Background motion induced large errors in the perceived direction.

DISCUSSION

We started this work with the idea that the distinction between allocentric and egocentric information would be useful to describe visuo-motor behavior. Our work shows that these sources of information are used differently during the planning and the execution of movements. During the control of ongoing movements, only egocentric information (perceived position) was used. In the planning stage, our subjects used both allocentric and egocentric information. Perceived direction (allocentric) and perceived position (egocentric) were used to predict where the target would be hit; perceived speed was used to control when the target was hit. The planning stage (reaction time) takes about 250 ms longer than the time needed to adjust ongoing movements (110 ms).

On-line control of movements can only be useful if the delay in using the sensory information is short. It therefore seems to be a good strategy to only use information which

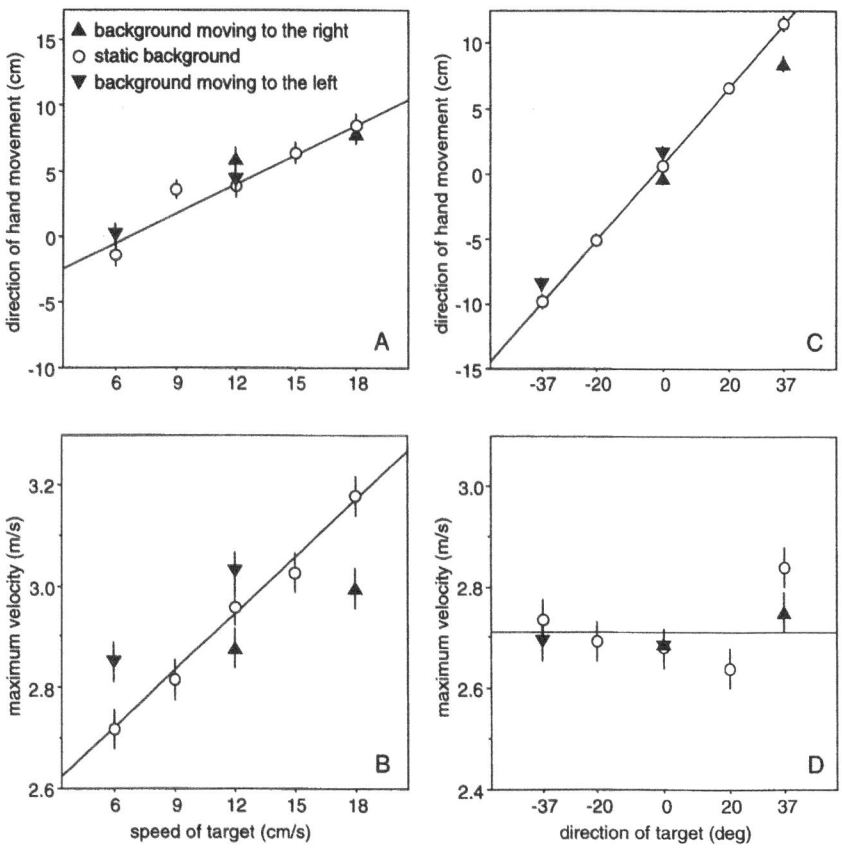

Figure 3. The use of spatial information for planning goal-directed arm movements. The direction of the movement of the hand is expressed as the intersection-point between the tangent to the trajectory and the surface on which the target moved. The maximum velocity is the maximum of the velocity-component orthogonal to this surface. Error bars indicate the average of the individual subjects' standard error of the mean. (A) When the speed of the target was varied (experiment 1), the direction in which the hand started to move was independent of motion of the background, and thus independent of the perceived speed. The direction of the hand movement did depend on the target's position at reaction time and thus indirectly on the target's speed (see Fig. 2A). (B) The maximum speed of the hand during the movement depended on the background motion in the same way as the perceived speed did (see Fig. 2B). (C) When the target's direction of motion was varied (experiment 2), the direction of the hand at movement onset did too. Note that the range of directions of hand movement was much larger than in the first experiment (Fig. 3A). This is because the direction of the hand at movement onset is influenced both by the position of the target at reaction time (see Fig. 2C) and by the extrapolation of its position using the perceived direction of motion (see Fig. 2D). The latter is evident from the fact that the direction of the hand movement depended on the motion of the background. (D) The maximum speed of the hand during the movement does not depend on the direction of target motion.

is in the same frame of reference as the motor-commands (Paillard, 1991). This saves the time needed for the nervous system to transform information. During the preparation, more time is available. Why we do not use all the available information to estimate where the target will be hit is not that easy to explain. It is important to realize that the target can only be hit at a certain position, if it is hit at the right instant. The strategy our subjects chose was to determine the position on the basis of position and direction, whereas the timing was based on the perceived speed. In this way, space and time can be determined independently (Brenner and Smeets, 1994b).

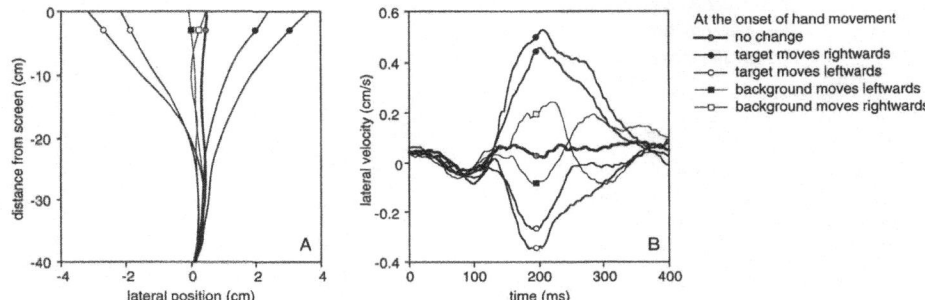

Figure 4. The use of spatial information for adjusting ongoing movements. Open symbols: target appears to move rightwards; closed symbols: target appears to move leftwards. (A) Average trajectories of the hand towards the screen. The target started moving when the hand was about 38 cm from the screen. (B) The lateral velocity as a function of time after the target started moving. The effect of background motion is small and opposite to the effect one would expect on the basis of the apparent motion of the target.

Our results question the notion of ballistic movements. Some authors claim that some kinds (or some parts) of movements cannot be changed by sensory information (reviewed by Jeannerod, 1988). It has been argued that fast goal-directed arm movements must be ballistic because the first 100 ms of an EMG-pattern of such a movement is not influenced by proprioceptive information about that movement (Wadman et al, 1979). However, proprioceptive information can change fast goal-directed movements at any instant during the movement (Smeets et al., 1990, 1995). The only reason that some (parts of) movements seem to be ballistic is that the sensory-motor loop (both visual and proprioceptive) has a delay of more than 100 ms. Recently, Pratt and Abrams (1994) claimed that the acceleration phase of a goal-directed movement is ballistic. The movements in our experiments only consisted of an acceleration phase, but were modifiable by changes in visual information during the movement. We conclude therefore that the title of this chapter is inadequate: goal-directed movements are never ballistic.

REFERENCES

Brenner, E. and Smeets, J.B.J. (1994a) Different frames of reference for position and motion. *Naturwissenschaften* 81:30-32.

Brenner, E. and Smeets J.B.J. (1994b) Why we hit slow targets more gently. *Journal of Physiology* 479P:53-54.

Brenner, E. and Smeets, J.B.J. (1995) Fast responses of the human hand to changes in target position. Submitted for publication.

Bridgeman, B., Kirch, M. and Sperling, A. (1981) Segregation of cognitive and motor aspects of visual function using induced motion. *Perception and Psychophysics* 29:336-342.

Goodale, M.A. and Milner, A.D. (1992) Separate visual pathways for perception and action. *Trends in Neuroscience* 15:20-25.

Jeannerod, M. (1988) The neural and behavioral organization of goal-directed movements. Clarendon Press, Oxford.

Paillard, J. (1991) Knowing where and knowing how to get there. *In:* Brain and space, ed. J. Paillard, Oxford University Press

Pratt, J. and Abrams, R.A., (1994) Action-centered inhibition: effects of distractors on movement planning and execution. *Human Movement Science* 13:245-254.

Prablanc, C., Pélisson, D. and Goodale, M.A. (1986) Visual control of reaching movements without vision of the limb. I. Role of retinal feedback of target position in guiding the hand. *Experimental Brain Research* 62:293-302.

Smeets, J.B.J. and Brenner, E. (1994) The difference between the perception of absolute and relative motion: a reaction time study. *Vision Research* 34:191-195.

Smeets, J.B.J. and Brenner, E. (1995) Perception and action based on the same visual information: distinction between position and velocity. *Journal of Experimental Psychology: Human Perception and Performance* 21:19-31.

Smeets, J.B.J., Erkelens, C.J. and Denier van der Gon, J.J. (1990) Adjustments of fast goal-directed arm movements in response to an unexpected inertial load *Experimental Brain Research* 81:303-312.

Smeets, J.B.J., Erkelens, C.J. and Denier van der Gon, J.J. (1995) Perturbations of fast goal-directed arm movements: different behaviour of early and late EMG-responses. *Journal of Motor Behavior* 27:77-88.

Wadman, W.J., Denier van der Gon, J.J., Geuze, R.H. and Mol, C.R. (1979) Control of fast goal-directed arm movements. *Journal of Human Movement Studies* 5:3-17.

24

SELF-CONTROLLED REPRODUCTION OF PASSIVE LINEAR DISPLACEMENT

Distance, Duration, and Velocity

P. Georges-François, R. Grasso, A. Berthoz, and I. Israël

Laboratoire de Physiologie de la Perception et de l'Action
CNRS - Collège de France
Paris, France

INTRODUCTION

Idiothetic signals (Mittelstaedt and Mittelstaedt, 1973) are the sensory signals generated by the movement of a subject: optic flow, proprioception, efference copy, and inertial signals. Mittelstaedt and Mittelstaedt (1980, 1982) formulated the use of these idiothetic signals for spatial orientation in the "path integration" hypothesis. With path integration, a subject should continuously be aware of his own position with respect to the starting point, during self-motion. Recent studies with humans (Thomson, 1983; Israël and Berthoz, 1989; Klatzky et al., 1990; Bloomberg et al., 1991; Mittelstaedt and Glasauer, 1991; Loomis et al., 1993; Glasauer et al., 1994) have shown that, indeed, the brain can estimate the traveled path from the sole self-generated information, i.e. without external visual nor acoustic cues.

When trying to validate this hypothesis with rodents and with geese required to navigate home after a passive displacement, it was found (Mittelstaedt and Mittelstaedt, 1982; Etienne et al., 1988) that subjects could correctly home after a passive displacement in darkness when they have been transported along angular paths, but not along linear ones. However, passive linear displacement estimation in humans has been studied through verbal estimates (Guedry and Harris, 1963), saccadic eye movements (Israël and Berthoz, 1989), or button-pushing responses (Israël et al., 1993; Mittelstaedt and Glasauer, 1991), and all these studies showed that passive linear motion can be correctly estimated by humans.

We therefore wanted to further examine passive linear distance estimation, investigating not only the distance estimate (the only available parameter with the short, ballistic saccadic or button pushing responses), but also the perception and estimation of the stimulus distance-contingent duration and velocity. The aim of the present experiment was to assess how a human subject can *reproduce*, with self-controlled passive motion, the distance of a passively imposed translation along the X-axis, in darkness (Berthoz et al., 1995). Furthermore, forward passive transport (self-controlled or not) is nowadays a common situation, encountered when driving cars, trains or buses. We therefore wondered about the possible influence of familiarity, and investigated it with two subjects who were passively transported backward, and had to reproduce, also backward, the imposed distance.

METHODS

Experimental Set-up

A mobile automated four-wheeled robot (Robuter™, Robosoft SA, France), equipped with a racing-car seat, was used in the experiment (Fig. 1A). This robot has two motorized wheels and two free wheels. It is velocity-controlled, with 1 m/s maximal linear velocity and 0.9 m/s² maximal linear acceleration. It can be controlled by a remote microcomputer (PC),

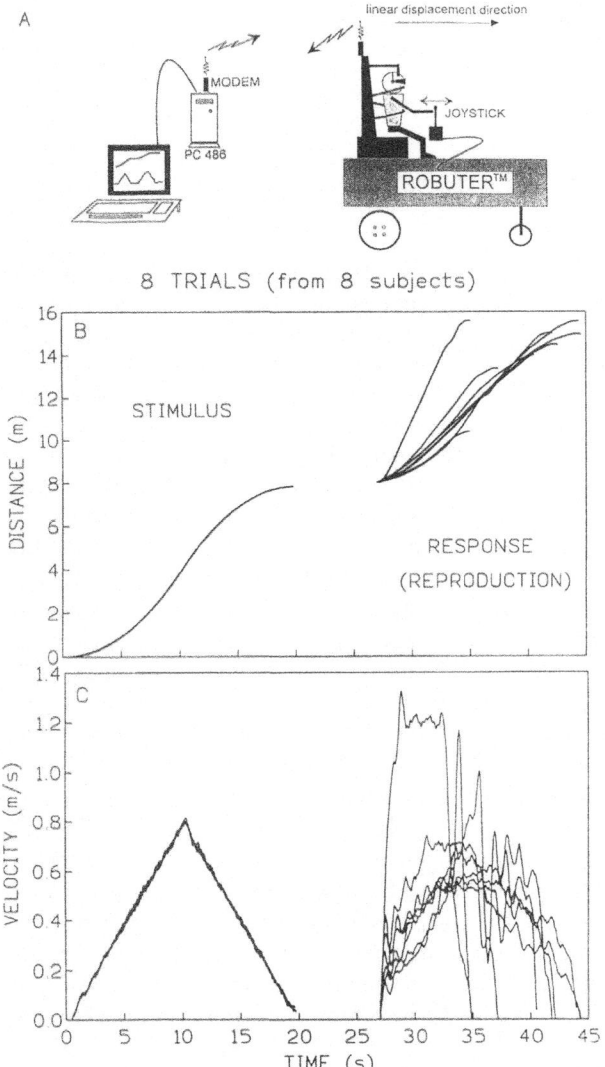

Figure 1. Experimental procedure. A: A mobile automated four-wheeled robot (Robuter™, Robosoft SA, France), equipped with a racing-car seat, was used in the experiment. B: Examples of recordings (odometry) made during the experiment. C: Velocity computed from the position traces of Fig. 1B after low-pass filtering (5 Hz cut-off frequency).

or by a joystick operated by the subject on the robot. Odometry (position on the X and Y axes, angle about the Z-axis, and the time scale) was recorded by the robot during motion, at 50 Hz sampling rate. Input-output signals (commands from the PC, odometry from the robot) are conveyed between the PC and the robot via wireless modems.

The subject was secured with three belts onto the chair of the robot. The head was firmly maintained to the chair between two ear-cushions to prevent yaw rotations, and a small bite-bar which prevented pitch rotations. The subject wore head-phones generating a wide band noise to mask all external acoustic cues, and large black-painted goggles were used to remove external visual information.

The joystick controlled robot velocity, and was set by software configuration so as to deliver only linear movements of the robot, forward or backward along the X-axis. Similarly, all stimuli delivered by the PC were linear displacements along the X-axis (as in natural locomotion). The experiment was performed within a wide corridor, 1.9 m large and 50 m long.

Experimental Procedure

Fourteen healthy volunteers (20-50 yrs) participated in the experiment, which was approved by the local ethical committee.

The subject first learnt to manipulate the joystick by freely traveling (linearly) in the corridor, with vision and audition available. After about 10 min learning, when the subject felt confident with the apparatus, head-phones and black goggles were set.

The subject was passively displaced forward along 2, 4, 6, 8 or 10 m. Velocity profile of the stimulus was triangular, with a constant acceleration of 0.06-1 m/s^2 range, and a peak velocity range of 0.6-1 m/s.

Immediately after the end of the imposed displacement (the stimulus), when the robot had come to a complete stop, the experimenter touched the subject's shoulder. This was the signal for the subject to reproduce, as accurately as possible, the previously imposed distance, using the joystick. Optically-encoded digital odometry was transferred from the robot to the PC (via the modem) after each trial.

The whole test included 13 trials. The order of these trials was systematically changed for the different subjects.

In an additional series of tests, in which two subjects participated (after they had performed the main test), both the imposed displacement and the distance reproduction controlled by the subject were performed backward, in otherwise identical conditions as the main experiment.

RESULTS

The figure 1B shows some examples of recordings (odometry) made during the experiment. Although these trials were identical, they were performed by eight different subjects. It can be seen that the stimulus segments are perfectly superimposed. Therefore, the differences exhibited in the reproduction segments is to be attributed to the subjects manipulating the joystick, and not to the robot. The figure 1C shows the velocity computed from the position traces of Fig. 1B after low-pass filtering (5 Hz cut-off frequency).

Distance Reproduction

All fourteen subjects were able to reproduce accurately the imposed distance. The figure 2A shows, as an example, all the responses of one subject to all trials, and Fig. 2B

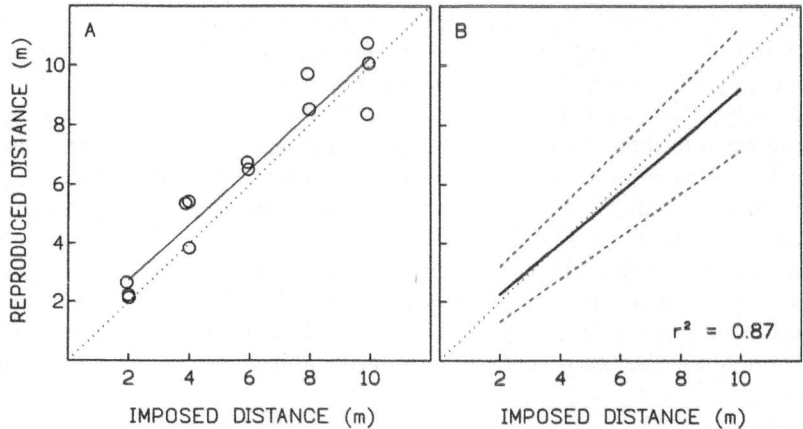

Figure 2. Distance reproduction. A: All the responses of one subject to all trials. B: Mean (heavy line) and mean ± SD (dashed lines) of the slope and intercept of all individual regression lines between stimulus and response distance.

gives the variability of all subjects' responses, with the mean (heavy line) and mean ± standard deviation (SD; dashed lines) of the slope and intercept of all individual regression lines between stimulus (S) and response (R) distance. The average (over the fourteen subjects') linear regression was R (m) = 0.87 x S + 0.51, with a regression coefficient (r^2) of 0.87 ± 0.07. The range of individual regression coefficients r^2 was [0.69-0.98], and the range of slopes of the regression lines was [0.70-1.22].

The reproduction of the shortest distance (2-m) lead to a slight overshoot (the error: R - S, was 0.25 ± 0.54 m, n = 14) and that of the longer distances lead to an undershoot (the error was -0.72 ± 1.28 m for the 10-m stimulus). This was probably a manifestation of the "range effect".

Each stimulus distance was traveled with two or three different peak velocities, and there was no effect of this peak velocity on distance reproduction accuracy.

Duration Reproduction

Although several stimuli of different durations could lead to one identical stimulus distance, when traveled with different peak velocities, there was a significant correlation between stimulus duration and distance (r^2 = 0.79, n = 182). Therefore, the duration of the stimulus could provide some information assisting its reproduction. And indeed, the subjects did reproduce fairly well the duration of the stimulus, although the instruction was only to reproduce the distance. The mean regression line was R (s) = 0.78 x S + 1.61 (r^2 = 0.80 ± 0.09). The figure 3A shows one example of responses duration (of the same subject as in Fig. 2A), and Fig. 3B the mean ± SD of the individual regression lines between stimulus and response duration. The range effect was also exhibited in the duration responses.

Velocity Reproduction

Finally, as subjects could control robot velocity, with the joystick, we examined the relationship between stimulus and response peak velocity. Actually, there was no correlation between stimulus peak velocity and distance. Therefore, peak velocity could not be of any help

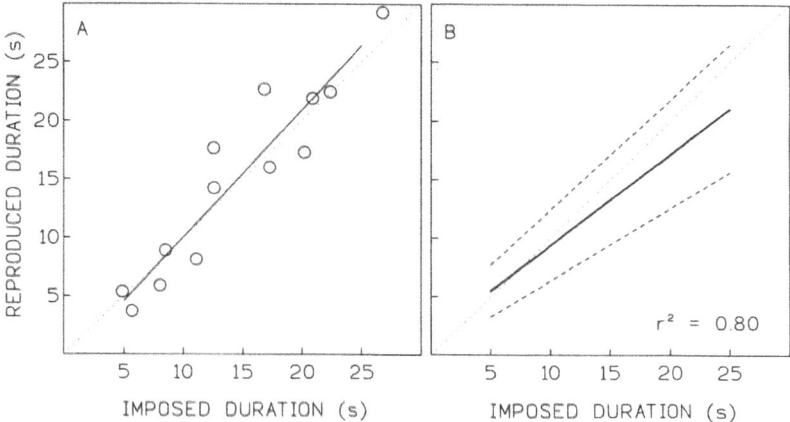

Figure 3. Duration reproduction. A: Responses duration of the same subject as in Fig. 2A. B: Mean ± SD of the individual regression lines between stimulus and response duration.

for the subjects to reproduce the distance. And it seems that the subjects perceived it, since the average regression between stimulus and response maximal velocity was not significant ($p = 0.15$, $n = 14$). Actually, only three subjects exhibited a non-significant regression line, and the others ranged (p) between 0.0268 and 0.0001, with r^2 range of [0.61-0.94] and with the slope value ranging from 0.47 to 1.37. The average regression line of the eleven significant subjects was R (m/s) = 0.89 x S + 0.03 ($r^2 = 0.60 \pm 0.16$). The figure 4A shows the peak velocity of all responses of the same subject as on Figs. 2 and 3, while Fig. 4B shows the mean ± SD of all significant regression lines. It can be seen that the variability among the individual regression lines was much greater than with distance (Fig. 2B) and duration (Fig. 3B). It could be argued that this larger SD was due to the smaller sample size (n of 11 instead of 14), but it was still greater when all 14 subjects were considered.

Whereas the peak velocity was not of interest to the subjects, the shape seemed to be, since most subjects reproduced triangular velocity profiles (Fig. 1C). Visual inspection

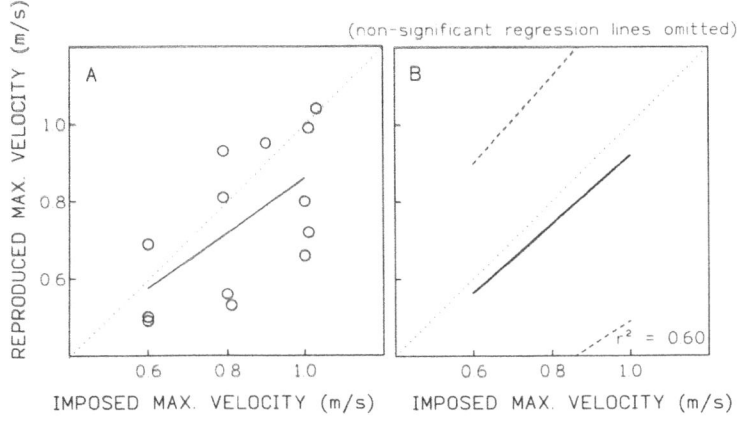

Figure 4. Peak velocity reproduction. A: Peak velocity of all responses of the same subject as on Figs. 2 and 3. B: Mean ± SD of all significant regression lines between stimulus and response peak velocity.

revealed that when subjects did generate a triangular shape of response velocity, there was a good correlation between stimulus and response peak velocity. The absence of such correlation indicated a different velocity profile, in most cases a constant velocity profile.

Backward Reproduction

Two subjects participated also in the second condition, in which the imposed displacements as well as the reproduction were performed backward, still in darkness along the X-axis. The figure 5 shows the mean reproduced distance (5A), duration (5B) and velocity (5C) of these two subjects, with respect to the imposed distance, duration and velocity, respectively. Open circles indicate forward and black circles indicate backward displacement.

Backward distance reproduction was as accurate as forward, even better. Mean slope of the regression line was 0.80 forward and 0.84 backward, thus slightly greater when backward. However, according to the mean regression coefficient r^2 (0.87 forward and 0.83 backward), variability of the responses was slightly greater backward than forward, as can be seen in Fig. 5A. A t-test performed over the distance error (R - S) confirmed that forward error (-0.78 ± 1.2 m) was significantly different ($t = 2.60$, dof = 50, $p = 0.012$) from backward error (0.12 ± 1.3 m). The mean error was smaller in the backward condition, and furthermore it was positive, indicating overshoot of the imposed distance, while it was negative when forward.

About duration reproduction, the mean slope of the regression line was again slightly greater backward (0.75) than forward (0.70), as can be seen in Fig. 5B, and the coefficients r^2 were identical (0.80) in both conditions. A t-test computed on the duration error actually showed that the difference between backward duration error (-0.93 ± 2.88 s) and forward duration error (-2.53 ± 2.90 s) was marginally significant ($t = 1.99$, dof = 50, $p = 0.0512$).

Finally, it can be seen in Fig. 5C that the subjects did not reproduce peak velocity when moving backward, although they did it when forward. For both subjects, there was a significant regression between stimulus and response peak velocity when forward (mean $r^2 = 0.56$, $p = 0.003$), and for both the regression was non-significant in the backward condition (mean $r^2 = 0.17$, $p = 0.2$). While the mean peak velocity errors (R - S = -0.04 m/s forward and -0.02 m/s backward) were not statistically different, the SD of backward velocity error (0.19 m/s) was significantly greater (F = 2.093, $p = 0.035$) than the SD of forward peak velocity error (0.13 m/s).

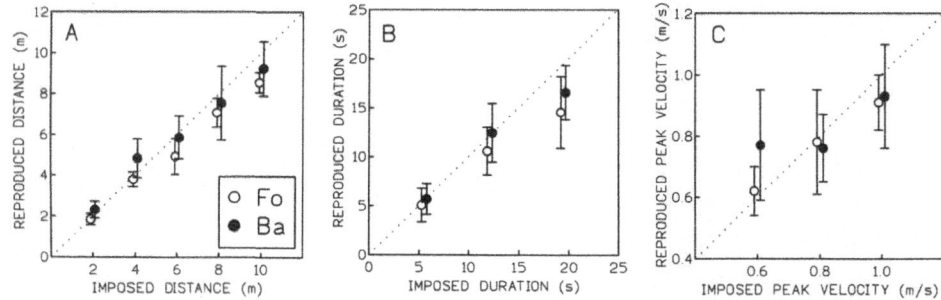

Figure 5. Backward reproduction. Open circles indicate mean values of forward and black circles indicate mean values of backward responses. A: Distance reproduction. B: Duration reproduction. C: Peak velocity reproduction.

DISCUSSION

Subjects were passively displaced linearly forward along the X-axis, and had to reproduce the previously imposed distance through self-controlled passive translation. The main results first show a great accuracy of the reproduced distance. Second, the subjects did also reproduce fairly well the duration of the stimulus. And finally, most subjects reproduced also the peak value and the shape of the imposed velocity profile (triangular). In the backward version of the same test, the subjects reproduced distance and duration with a still greater accuracy than when forward, but did not reproduce the velocity.

This experiment demonstrates that healthy individuals can store and utilize the vestibular and somatosensory information, acquired during passive displacements, to reproduce a given linear trajectory. The ability of human beings to store information about linear displacement and the mechanisms on which it relies is still a matter for discussion. When visual and other allocentric references are eliminated, as in the present experiment, distances cannot be calculated from external cues. In these conditions, information about motion must be acquired dynamically during the motion itself, and needs to be accurate enough to allow a representation of the individual's instantaneous position in the environment. The result showing that the most common strategy employed by our subjects was the reproduction of the dynamics (velocity profile) of the imposed displacement, offers an important interpretative key. We believe that this result strongly favor the hypothesis that the brain stores the dynamic properties of whole-body passive linear motion as provided by specific sensors.

The only experiment, to our knowledge, in which human subjects had to reproduce in darkness a previously traveled distance, i.e. had to perform the same paradigm as the present one but with active walking, is the one reported by Loomis et al. (1993). The subject, blindfolded, was first led by the experimenter to walk along 2 to 10 m, as in the present experiment, and the distance was then reproduced by walking forward without aid. The results are strikingly similar to the present ones: the 2-m distance was overshot by 0.26 m (0.25 m in the present test) and the 10-m distance undershot by 1.02 m (0.72 m here). This is a final point which supports our hypothesis : the subjects were afforded, in the present experiment, to reproduce the whole motion dynamics, and this was as fruitful as if they had actively walked the distance. It could be inferred, from the comparison of these two psychophysics experiments, that the proprioception and efferent copy signals provide mainly information on the dynamics of the current movement.

This similarity of results between reproduction through active walking and active "driving" allows us to suggest that the results of the backward reproduction condition can be extended to backward walking, without the difficulty of backward walking. In other words, self-controlled motion can be considered as a model of active locomotion, which can be better controlled than real walking.

The mean backward distance error was positive (i.e. overshoot) while it was negative when forward, showing that subjects traveled longer distances when moving backward. This could be explained by the uncertainty experienced when moving forward blindfolded (Grüsser, 1983). The subjects all claimed that they felt safe, and they knew that maximal velocity of the robot was close to normal walking velocity (1 m/s). However, moving forward is exposing the face to any obstacle, while moving backward is presenting the back as shield. Therefore, larger distances can be subjectively more safely traveled when backward.

The interesting point is that the subjects did not reproduce velocity, when moving backward. Although they were not aware of they reproducing the velocity profile, when forward, this suggests a change of strategy. The subjects did not move slower when backward, as could have been expected. They rather used an approximately constant peak velocity (and a constant velocity profile), as if they tried to remove one degree of freedom from the motion dynamics.

Another hypothesis is that velocity is not stored, when moving backward. This could be explained by the fact that the vestibular inputs (otolith signals in the present experiment) generated during self-motion reach the medial superior temporal (MST) motion processing areas (Thier and Erickson, 1992), and the visual cells of MST are mostly responsive to "expansion", which is the optical flow experienced when moving forward (Graziano et al., 1994). The velocity profile which was stored when moving forward might be a virtual visual trace. In any case, if MST cells are indeed excited by blindfolded self-motion, then backward motion could hamper the usual visual-vestibular synergy, since the magnitude of otolithic signals is identical in both directions whereas visual MST cells are more tuned to expansion than to contraction.

ACKNOWLEDGMENTS

This work was supported by the EEC Esprit Basic Research programme, project 6615 (MUCOM), and R. Grasso had a grant from the EEC Human Capital and Mobility programme. We are grateful to A. Treffel for the adaptation of the Robuter™, and to S. Glasauer for the development of the computer software.

REFERENCES

Berthoz, A., Israël, I., Georges-François, P., Grasso, R., Tsuzuku, T., 1995, Spatial memory of body linear displacement: What is being stored?, Science (in press).

Bloomberg, J., Melvill, Jones G., and Segal, B.N., 1991, Adaptive modification of vestibularly perceived rotation, *Exp Brain Res* 84:47-56.

Etienne, A.S., Maurer, R., and Saucy, F., 1988, Limitations in the assessment of path dependent information, *Behaviour* 106:81-111.

Glasauer, S., Amorim, M.A., Vitte, E., and Berthoz, A., 1994, Goal-directed linear locomotion in normal and labyrinthine-defective subjects, *Exp Brain Res* 98:323-335.

Graziano, M.S.A., Andersen, R.A., and Snowden, R.J., 1994, Tuning of MST neurons to spiral motions, *J Neurosci* 14:54-67.

Grüsser, O.J., 1983, Multimodal structure of the extrapersonal space. In: Hein A and Jeannerod M (eds.) Spatially oriented behavior, Springer Verlag, Berlin, pp 327-352.

Guedry, , and Harris C.S., 1963, Labyrinth function related to experiments on the parallel swing, Research Report from the NASA.

Israël, I., Chapuis, N., Glasauer, S., Charade, O., and Berthoz, A., 1993, Estimation of passive horizontal linear whole-body displacement in humans, *J Neurophysiol* 70(3):1270-1273.

Israël, I. and Berthoz, A., 1989, Contribution of the otoliths to the calculation of linear displacement, *J Neurophysiol* 62(1):247-263.

Klatzky, R.L., Loomis, J.M., Golledge, R.G., Cicinelli, J.G., Doherty, S., and Pellegrino, J.W., 1990, Acquisition of route and survey knowledge in the absence of vision, *J Mot Behav* 22(1):19-43.

Loomis, J.M., Klatzky, R.L., Golledge, R.G., Cicinelli, J.G., Pellegrino, J.W., and Fry, P.A., 1993, Nonvisual navigation by blind and sighted: assessment of path integration ability, *J Exp Psychol : Gen* 122(1):73-91.

Mittelstaedt, H., and Mittelstaedt, M.L., 1973, Mechanismen der Orientierung ohne richtende Außenreize, *Fortschr Zool* 21:46-58.

Mittelstaedt, H., and Mittelstaedt, M.L., 1982, Homing by path integration. In: Papi F and Wallraff HG, (eds), Avian navigation, Springer Verlag, Berlin, Heidelberg, pp. 290-297.

Mittelstaedt, M.L., and Glasauer, S., 1991, Idiothetic navigation in gerbils and humans, *Zool Jb Physiol* 95:427-435.

Mittelstaedt, M.L., and Mittelstaedt, H., 1980, Homing by path integration in a mammal, *Naturwiss* 67:566-567.

Thier, P., and Erickson, R.G., 1992, Responses of visual-tracking neurons from cortical area MST-I to visual, eye and head motion, *Eur J Neurosci* 4:539-553.

Thomson, J.A., 1983, Is continuous visual monitoring necessary in visually guided locomotion? *J Exp Psychol: Hum Percep Perform* 9(3):427-443.

25

ADJUSTMENT OF THE INTERNAL SENSORIMOTOR MODEL IN THE COURSE OF ADAPTATION TO A SUSTAINED VISUOMOTOR CONFLICT

K. E. Popov,[1] J. P. Roll,[2] B. N. Smetanin,[1] and V. Yu. Shlikov[1]

[1] Institute for Problems of Information Transmission
Russian Academy of Sciences
Ermolova 19
101447 Moscow, Russia
[2] Laboratory de Neurobiologie Humaine
URA CNRS 372
University of Provence
Escadrille Normandie Nemen ave
F-1396 Marseille, France

INTRODUCTION

To control spatially oriented arm movements, human perceptual-motor system must have knowledge of the target location relative to the body as well as the arm position relative to the target. Awareness of the target location is derived from retinal input and complemented by proprioceptive inputs signaling eye-to-head and head-to-trunk positions. Awareness of the moving arm position is derived from proprioceptive input and complemented by visual input signaling the arm position in space to suppress proprioceptive bias. Normally vision and proprioception generate highly congruent sensory inputs that ensure alignment of visual and proprioceptive sensory maps enabling error-free guidance of targeted motions.

Disturbance of natural eye-hand coordination by introducing constant errors into the visual system was shown to be compensated by neural adaptation processes. Recent studies of prism adaptation (Redding and Wallace, 1992a,b) have shown that adaptation to a visuomotor conflict can result from adaptation in either eye-head or arm-head subsystem, depending on the experimental condition. It was hypothesized that the visuomotor system is organized as a chain where adaptation can be distributed between its segments (Rossetti et al., 1993), while parallel organization of the system was suggested by other authors (Jeannerod, 1990).

When the pointing task was used as a test to measure adaptation, performance improved to the control level within a few trials (Welsh and Goldstein, 1972) or a few tens of trials (Weiner et al., 1983), according to different authors. It was reasonable to expect the adjustment of voluntary motor control under the condition of a visuomotor conflict. However, it is less obvious how automatic motor responses will change under similar conditions. To elucidate this issue, we investigated adaptation to visual reversal of proprioceptively induced motor responses crucially dependent on the visual input. Vibration of the upper arm muscles has been shown (Roll et al., 1980) to give rise to either tonic vibration reflex (the TVR) or antagonist vibration response (the AVR) depending on whether subjects were allowed to view their hands or not. We used this phenomenon as a test to study adaptation to visual reversal.

METHODS

Six normal subjects were selected among 12 healthy persons on the basis of their motor responsiveness to vibration. During experiments subjects were seated at a table, their right forearms were placed on a table-mounted horizontal supporting plate with its axis of rotation collinear with that of the elbow. Two compliant springs attached to the support acted to move it back to the neutral position. In addition, the elastic load opposed the forearm motion thus allowing us to evaluate the intensity of the motor response in terms of angular displacement. A DC motor-based vibrator was placed over the muscle belly of either triceps brachii or biceps brachii and secured by elastic bands. The vibration frequency was individually adjusted between 50 and 90 Hz to ensure similar intensity of the TVR and AVR in each particular subject, the vibration amplitude was about 1 mm, the duration was either 4 or 8 sec. The elbow angle was measured by means of a strain gauge-based goniometer. The surface EMG of the two antagonistic muscles was occasionally recorded as well.

Subjects wore goggles with two Dove prisms which produced left-right reversal of vision. The prisms had the angle of vision 25 degrees, thus allowing the subject to view his/her right hand and the distal part of the forearm. Due to the visual reversal, the subject's right hand appeared as the left one.

Involuntary motor responses to the vibration of either the biceps brachii or triceps brachii were tested under normal vision of the hand, under occluded vision, and under the condition of left-right visual reversal. Under the visual reversal, subjects were asked to perform free motions of the right hand and forearm in flexion-extension and pronation-supination accompanied by movements of fingers. In doing so, they were instructed to look at their right hands continuously. Every 1 or 2 min subjects were asked to relax their arms, and motor responses to vibration were tested.

RESULTS

Under normal visual condition, when the subject was allowed to view his/her arm, vibration applied to the biceps brachii or triceps brachii muscle gave rise to a reflex contraction of the muscle under vibration (the TVR), while the same stimulus applied under precluded vision of the arm resulted in the antagonist contraction (the AVR).

The latter response was preceded by an illusory perception of the forearm motion, the motion direction corresponded to elongation of the muscle under vibration. This vision-dependent reversal of the TVR was first described in (Roll et al., 1980).

When subjects wore the inverting goggles, vibration of the biceps brachii muscle resulted in an involuntary elbow extension while vibration of the triceps brachii resulted in an elbow flexion. Thus, left-right reversal of vision by the inverting prisms was similar to precluded vision in that the AVR was observed under this condition.

The angular displacement of the forearm was often larger than that observed under occluded vision.

Free voluntary movements of the forearm, hand and fingers repeatedly performed under visual reversal gave rise to a progressive decrease of the response amplitude (Fig. 1). After a lapse of time, 7 to 12 min in different subjects, the amplitude of the motor response approached zero, then the muscle vibration began to produce the TVR, i. e. an involuntary contraction of the muscle under vibration. The response amplitude increased with time and approached the control level (the angular displacement measured under normal vision) within approximately 5 min. In the experiment illustrated by Fig. 1 the subject ceased moving his hand actively at the time marked by a vertical dotted line. It is seen that this brought the motor response to its original direction very rapidly.

The progressive reversal of the motor response to vibration paralleled with alterations in the subjects' perception of their own arms. Under the visual reversal, the subject's right hand was viewed by him/her as the left one, and vice versa. Early in the experiment, the subject, when looking at his/her hand during its active movements, did not identify the reversed view of his hand with the hand he attempted to move voluntarily. This visuomotor conflict decreased with time. In 5 to 10 min, the subject was able to relate the visible movement of his hand to his internal volitional effort. These perceptual alterations proceeded concurrently with the above changes in the motor response direction.

The duration of voluntary motor activity under the visual reversal affected the persistence of the reversed motor response to vibration. When the subject stopped voluntary movements within 2 to 5 minutes after the motor reversal, as exemplified by Fig. 1, the reversed response persisted for a very short time (about 1 min) and was then replaced by the original motor response (i. e. the AVR). When the subject under visual reversal was allowed

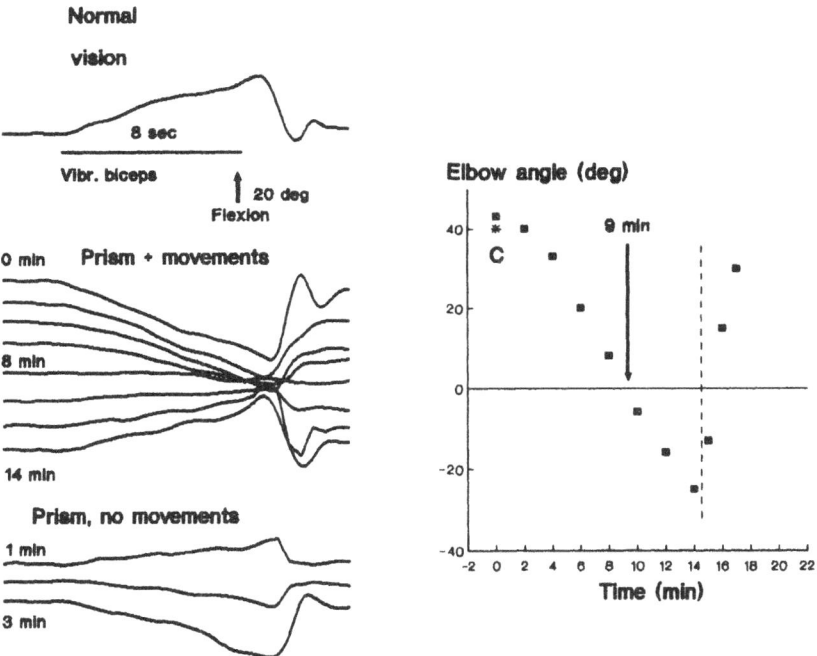

Figure 1. Motor responses to vibration of the biceps brachii muscle tested in the course of adaptation to visual reversal. Left panel shows consecutive responses (elbow goniograms) recorded at 2-min intervals during exposure to reversal prisms. Right panel shows the time-dependence of the response amplitude.

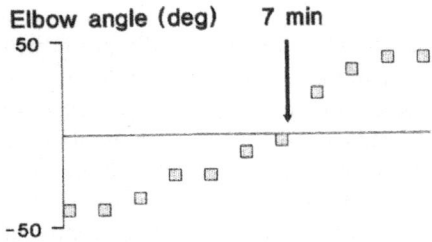

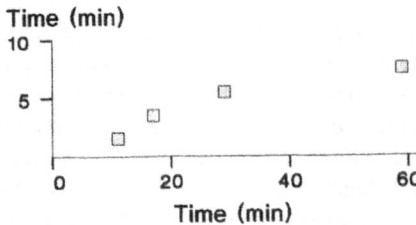

Figure 2. Restoration of the original motor response following adaptation to visual reversal. Upper panel shows the amplitudes of consecutive responses recorded at 1-min intervals following a 60-min adaptation session. Lower panel shows the duration of persistence of the inverted response plotted against the duration of adaptation session.

to perform voluntary movements for a longer time (tens of minutes), the duration of the reversed response persistence increased and reached a plateau of about 7 min, as illustrated by Fig. 2.

DISCUSSION

Automatic motor responses of the upper arm muscles to stimulation of their stretch receptors by vibration were shown to be crucially dependent on the visual input (Roll et al., 198?). In subjects with their eyes closed, the muscle vibration induced illusory motion of the forearm followed by an involuntary contraction of the antagonistic muscle (the AVR). Vision of the hand suppressed illusory motion perception thus resulting in re-addressing the motor response to the muscle under vibration (the TVR). Left-right reversal of vision by the inverting prisms was similar to precluded vision in that the AVR was observed under this condition, i. e. when the subject's right hand was viewed by him/her as the left one. Thus, visual input did not provide a reference for controlling the motor response, when it was decorrelated with other sensory inputs (e. g. proprioceptive) as well as with motor output.

According to current theories of spatially organized motor behaviour (Popov et al., 1986; Gurfinkel and Levick, 1991), correlations between central motor commands and expected multisensory signals are stored in an internal sensorimotor model. It has been shown earlier (Smetanin et al., 1993) that spatially oriented whole-body postural responses to both vestibular and proprioceptive stimuli are governed by an error signal which represents the difference between the model's prediction and actual multisensory input. The error signal, when present for a long time because of a sensorimotor conflict, can give rise to adjustments of the internal model. The above findings provide an example of the adjustment of the internal sensorimotor model when challenged by a sustained visuomotor conflict. These observations strongly suggest that this sensorimotor conflict can be resolved by virtue of establishment of a new correspondence between motor commands and expected sensory inflow, and that active sensorimotor behaviour under conflicting conditions is a prerequisite for an extremely fast re-arrangement of the internal model.

ACKNOWLEDGMENTS

This was work supported by the Russian Foundation of Fundamental Research under grant # 93-04-20520.

REFERENCES

Gurfinkel, V.S., and Levick, Yu.S., 1991, Perceptual and automatic aspects of the postural body scheme. In Brain and Space, (ed. J. Paillard), Oxford Univ. Press: 145-162.

Jeannerod, M., 1990, The representation of the goal of an action and its role in the control of goal-directed movements. In Computational Neuroscience (ed. E. L. Schwartz), MIT Press, Cambridge, Mass.: 352-368.

Popov, K.E., 1986, Smetanin B. N., Gurfinkel V. S., Kudinova M.P., and Shlikov V. Yu., Spatial perception and vestibulomotor response in man, *Neirophiziologiya (Kiev)* 16:779-787.

Redding, G.M., and Wallace, B., 1992a, Effects of pointing rate and availability of visual feedback on visual and proprioceptive components of prism adaptation, *J. Mot. Behav.* 24:226-237.

Redding, G.M., and Wallace, B., 1992b, Cognitive load and prism adaptation, *J. Mot. Behav.* 24:238-246.

Roll, J.P., Gilhodes, J.C., and Tardy-Gervet, M.F., 1980, Effects perceptifs et moteurs des vibrations musculaires chez l'homme normal. Mise en evidence d'une reponse des muscles antagonistes, *Arch. Ital. Biol.* 118:51-71.

Rossetti, Y., Koga, K., and Mano, T., 1993, Prismatic displacement of vision induces transient changes in the timing of eye-hand coordination, *Percept. Psychophys.* 54:355-364.

Smetanin, B.N., Popov, K.E., and Schlikov, V.Yu., 1993, Postural reactions evoked by vibratory stimulation of neck muscles in man, *Neirofiziologiya/Neurophysiology (Kiev)* 1:101-108.

Weiner, M.J., Hallet, M., and Funkenstein, H.H., 1983, Adaptation to lateral displacement of vision in patients with lesions of the central nervous system, *Neurol.* 33:766-772.

Welsh, R.B., and Goldstein, G.D., 1972, Prism adaptation and brain damage, *Neuropsychol.* 10:387-394.

26

FINGERTIP TOUCH AS AN ORIENTATION REFERENCE FOR HUMAN POSTURAL CONTROL

J. J. Jeka and J. R. Lackner

Ashton Graybiel Spatial Orientation Laboratory and
National Center for Complex Systems
Brandeis University
Waltham, Massachusetts 02254

INTRODUCTION

Somatosensory[*] information from cutaneous and proprioceptive receptors serves a dual role in the control of upright posture. Unlike the vestibular receptors which are located in the head and therefore must infer information about the head relative to the trunk and other body components, somatosensory receptors are distributed throughout the skin surface (e.g., mechanoreceptors) and musculature (e.g., spindles), providing crucial information about the relative position of different body components. At the same time, somatosensory information may also provide functional information about contact of a body part to an external object or surface, such as texture and compliance as well as the forces applied to the surface, allowing for accurate and flexible control of handled objects and tools.

The dual nature of somatosensory function is apparent when we perform everyday activities that require precise control of body posture such as making contact through fine adjustment of hands or fingers with a stationary object or surface. To grasp and smoothly pick up an object, one's body position must be precisely controlled to maintain hand contact with the object while simultaneously exerting the adequate amount of force to grasp and lift the object without dropping it. Under such conditions, we rely heavily upon somatosensory information to simultaneously orient our own body position and manipulate the object. Until the "precision grip" studies of Johansson and colleagues (for a review see Johansson, 1991), such functional behavior had not been investigated.

Intuitively, we conceive of contact with a stable surface, such as leaning against a wall with the hand and arm, as providing postural stabilization through passive physical reaction forces that balance those imposed by movement of the body. A crucial concern,

---

[*] "Somatosensory" is meant here as it is often referred to in the posture literature (Horak and Macpherson, 1995) and the way "haptic" is used in the literature concerning object manipulation (Klatzky et al., 1987) as a combination of cutaneous and proprioceptive cues.

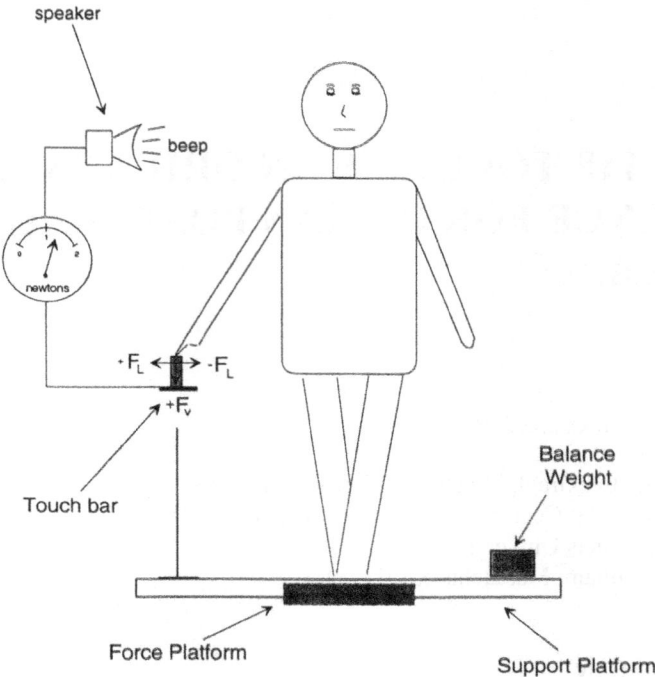

Figure 1. Subject depicted in tandem Romberg posture on the force platform in a touch contact condition with her right index fingertip on the touch bar. For the sake of illustration, the subject is shown exceeding a typical threshold force of 1 N and the alarm is sounding. In actual experiments, this occurred in less than 5% of all touch contact trials. In the force contact conditions, the auditory alarm was turned off and the subject could apply as much force as desired. In the no contact conditions, the subject's arms hung passively by her side.

however, is to separate physical support of the body provided by the forces applied though the hand from the sensory information about the external surface and one's own body position provided by cutaneous and proprioceptive receptors embedded in the fingertips, hand and arm. Recently, we have developed a paradigm to do just this; it allows us to separate purely physical support from the use of somatosensory cues to enhance postural control while touching an external object. Fig. 1 depicts our test situation. A subject is in the tandem Romberg stance (heel-to-toe) on a force platform touching a device used to measure the forces applied by the right index fingertip. The tandem stance is used to increase postural instability in the medial-lateral direction. The touch apparatus consists of a rigid horizontal metal bar attached to a rigid metal stand. It is positioned laterally to the subject with the bar parallel to the sagittal plane of the subject. The subject places his/her right index finger on the middle of the bar while strain gauges mounted on the metal bar transduce the forces applied by the fingertip. A feedback circuit measures the magnitude of applied force (vertical and lateral directions) and triggers an auditory alarm if the applied force exceeds a predefined threshold. Subjects touch the bar with their right index fingertip and are required to keep the applied force below the threshold (i.e., keep the alarm from sounding). This allows control of the applied force below any desired level. Before the experimental trials begin, subjects are allowed to push on the bar to ascertain the force level that triggers the alarm. In over 95% of the experimental trials to date, subjects were able to complete the trial without exceeding threshold force levels. We have performed a series of investigations with this paradigm, the main results of which we summarize below.

1. CONTACT OF THE INDEX FINGERTIP WITH A STATIONARY BAR REDUCES HUMAN POSTURAL SWAY AT CONTACT FORCES FAR BELOW THOSE NECESSARY FOR PHYSICAL SUPPORT OF THE BODY

We have found that touch contact of a fingertip to a stable surface reduces postural sway[*] in subjects standing unilaterally (Holden et al., 1994) and bilaterally (Jeka and Lackner, 1995). Subjects were tested under six conditions: two visual conditions, (V)ision, eyes open, and (D)ark, eyes closed; and three fingertip contact conditions: no contact, during which the subject's arms hung passively by their side; (T)ouch contact, in which the subject could apply up to 1 N ($\approx$ 100 grams) of force on the touch apparatus before an auditory tone signaled the threshold of applied force; and (F)orce contact, during which subjects could apply as much force as desired. The six experimental conditions will be identified as follows: V = vision - no contact, VT = vision - touch contact, VF = vision - force contact, D = dark - no contact, DT = dark - touch contact and DF = dark - force contact.

Fig. 2a shows the CPx displacement of one subject in a dark-no contact (D) trial overlaid upon a dark-touch (DT) trial, illustrating the reduction in center of pressure displacement due to the addition of touch contact. Fig. 2b displays the mean CPx displacement in each condition collapsed across five subjects: medial-lateral (CPx) center of pressure mean displacement was highest in the dark-no touch (D) condition and significantly lower in all other conditions (p < .01). The important finding is that touch and force contact lowered mean CPX displacement equivalently, with or without vision present, despite mean force levels which were over 10 times greater with force contact ($\approx$ 400 grams) than touch contact ($\approx$ 40 grams). In a model designed to evaluate the reduction in body sway due to passive mechanical forces at the fingertip (Holden et al., 1994), contact forces of 40 grams would produce a maximum reduction of sway of < 5%. In fact, touch contact reduced sway by 50-60% for all subjects.

2. THE TEMPORAL RELATIONSHIP BETWEEN POSTURAL SWAY AND FINGERTIP CONTACT FORCE CHANGES WITH DIFFERENT LEVELS OF CONTACT FORCE

Fig. 3a-b shows in a typical force contact (Fig. 3a) and touch contact (Fig. 3b) trial, the time series and respective correlations between CPx displacement and lateral fingertip contact force (FL), along with the respective time lags at which maximum correlations occurred. Correlations between CPx and FL were highest with force contact ($\approx$ 0.9), with very small time lags between the two signals (< 50 ms). This means that fingertip contact forces in the force contact condition were in-phase with body sway; subjects were partially leaning on the contact surface with their finger for support. Correlations between center of pressure (CPx) and lateral contact force (FL) were lower with touch contact ($\approx$ 0.8) and changes in fingertip force led

---

[*] These first studies have estimated postural sway through center of foot pressure movements on a force platform, which tend to be larger and of higher frequency than center of mass movements (cf., Winter et al., 1990). Center of pressure is a linear measure while body sway is an angular measure and therefore not completely analogous. However, we measured the relationship between center of pressure and center of mass movements in our experimental paradigm and found their magnitude to be equivalent in each condition and their average correlation to be larger than 0.8 (Jeka and Lackner, 1995). Thus, at the low amplitude of body sway observed in these experiments, center of pressure displacement can be considered to be approximately equivalent to angular body sway.

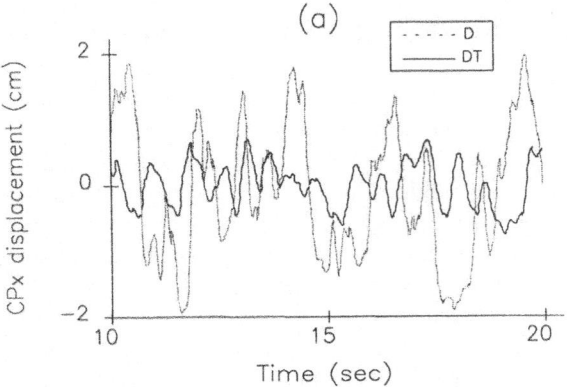

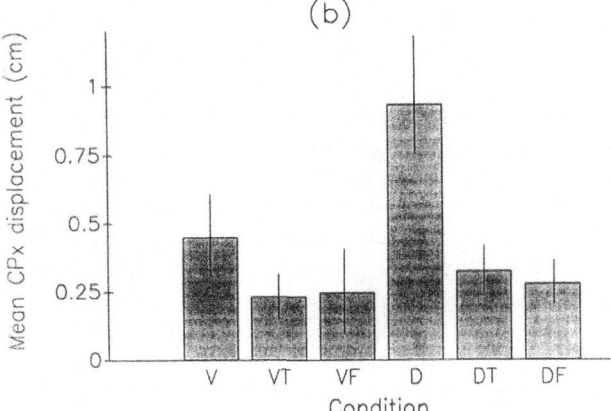

Figure 2. (a) Overlaid time series of CPx displacement from dark - no contact (dotted line) and dark - touch contact (solid line) conditions illustrating the reduction in CPx displacement due to fingertip touch contact. The middle ten seconds are shown for visual clarity. (b) Mean CPx displacement collapsed across subjects for each experimental condition. CPx displacement was highest in the no contact - dark (D) condition and lowest with any form of fingertip contact. Error bars correspond to the standard error of the mean.

changes in body sway by ≈300 ms. This suggests that as subjects swayed towards the touch bar with only very light touch, contact forces initially increased, but as sway continued towards the bar, contact forces began to decrease so as not to trigger the alarm threshold. This means that subjects must use musculature remote from the fingertip to arrest sway towards the touch bar, because touch contact forces alone are inadequate to damp sway of the body. Therefore, the stabilization provided by touch contact is due to a sensory-motor relationship: forces generated by the musculature remote from the fingertip (legs, trunk, etc.) are guided by sensory information provided by cutaneous receptors in the fingertip (Johansson, 1991) and proprioceptive information about arm position (e.g., Matthews, 1988).

Evidence in support of this interpretation can be found in "precision grip" studies (cf. Johansson, 1991) in which maximal dynamic cutaneous afferent activity is found at approximately 35-50 grams of load force (Westling and Johansson, 1987). Interestingly, this is the same range of contact force subjects spontaneously adopted in our touch contact conditions, even though up to 100 grams of force was allowed. This means that subjects were adjusting touch force to levels where neurophysiological sensitivity was greatest to provide the highest resolution of contact force vectors. They were making use of "precision contact" of a single finger with a surface, analogous to the "precision grip" of Johansson and his colleagues, in which two fingers contact the external object.

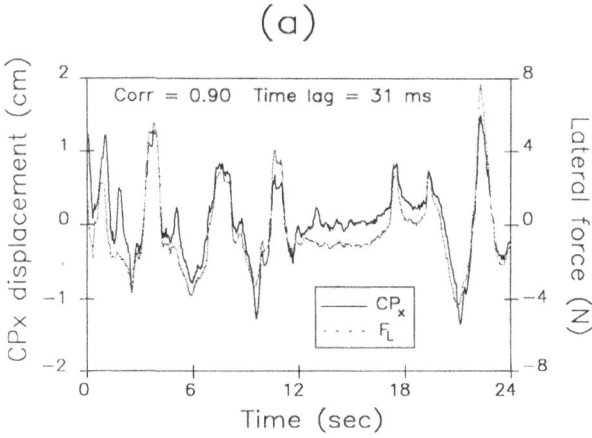

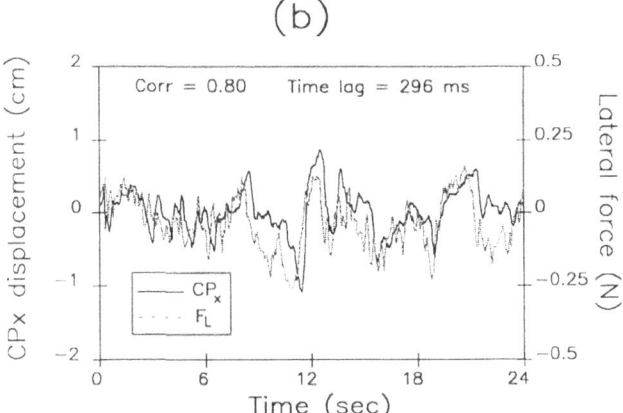

Figure 3. Overlaid time series of CPx displacement (solid line) and lateral fingertip force (dotted line) in (a) vision-force contact (VF) and (b) vision-touch contact (VT) conditions. Individual correlations and time delays for each trial are shown.

3. LEG MUSCLE EMG ACTIVITY WAS LOWER WITH FORCE CONTACT THAN WITH TOUCH CONTACT

In a subsequent study (Jeka and Lackner, 1995), we measured EMG activity in the peroneal muscles, which are particularly important in stabilizing lateral body sway in the tandem Romberg stance, to determine whether leg muscle activity changed with different finger contact force levels. We predicted that muscle activation should be higher with touch contact than force contact, because touch contact was providing spatial cues about body orientation, rather than the physical support associated with force contact of the fingertip. We found that EMG amplitude was 50% higher with touch contact than with force contact, indicating that postural leg musculature played a larger role in reducing postural sway when subjects applied very small contact forces than when subjects were partially leaning on the bar for support.

Timing relationships between CPx displacement and EMG activity in each leg indicate that EMG activity led CPx displacement by ≈150 ms in each condition. This means that with touch contact, the changes in lateral contact force at the fingertip began 150 ms

(a) Touch contact

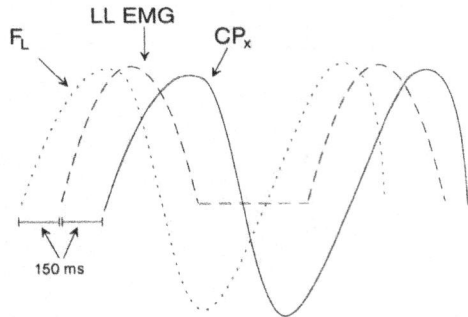

(b) Force contact

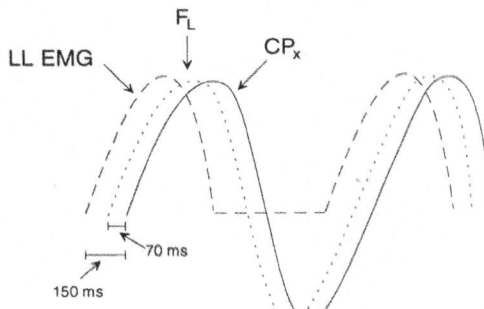

Figure 4. Schematic representation of the temporal relationship between CPx displacement, left leg EMG activity and lateral contact force (FL) in the (a) touch contact and (b) force contact conditions. On average, FL preceded CPX by 300 ms with touch contact and 70 ms with force contact. LL EMG preceded rightward CPX displacement by 150 ms with both touch and force contact. Right leg EMG activity is not shown, but was active with leftward CPX displacement (bottom half of the cycle) and also preceded CPX displacement by 150 ms.

ahead of correlated changes in EMG activity, enough time for a stabilizing long-loop reflex to be initiated (Diener and Dichgans, 1986) or for conscious anticipatory innervations to be employed. By contrast, in the force contact conditions, fingertip contact forces were approximately in-phase with CPx displacement and lagged behind leg muscle EMG activity, indicating that the contact forces were not precuing a particular muscle activity pattern but physically counteracting body sway. Fig. 4 illustrates schematically the change in timing relationships between CPx displacement, lateral contact force (FL) and left leg EMG activity from the touch contact to the force contact condition.

4. CHANGING THE FRICTIONAL PROPERTIES OF THE CONTACT SURFACE INFLUENCES POSTURAL CONTROL WITH FORCE CONTACT BUT NOT TOUCH CONTACT

To probe the differences between touch contact and force contact further, we evaluated the influence of different frictional properties of the finger contact surface on postural equilibrium by applying a lubricant to the contact surface (Jeka and Lackner, 1995). A slippery contact surface renders shear forces on the finger mechanically less effective to counteract body sway. This means that with physically supportive levels of fingertip force

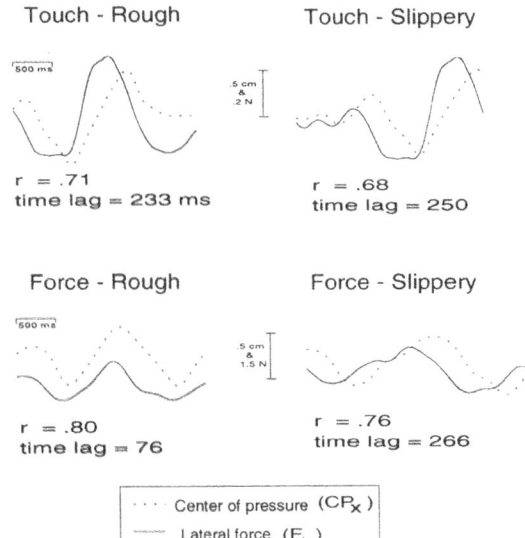

Figure 5. Overlaid time series segments (3 s) of CPX displacement and lateral fingertip contact force (FL). The CPX- FL correlation was highest and had the shortest time lag with fingertip force contact on a rough surface. With force contact on a slippery surface, the CPX - FL correlation and time lag was equivalent to the touch contact conditions.

on a slippery surface, subjects might: 1) sway more than with contact of a rough surface; or 2) switch to use a coordinative strategy more indicative of "touch contact", in which fingertip contact force changes lead body sway by about 200-300 ms. We did not expect contact surface characteristics to influence postural stability with light touch contact of the finger, because sensory information about the fingertip is available through cutaneous and proprioceptive inputs. Consequently, with light touch contact of the fingertip, we predicted that the sway reduction and the timing relationships between fingertip forces and body sway should not be affected by contact surface properties.

The results showed that changing the touch bar surface from rough to slippery had no effect on postural sway amplitude with either touch contact or force contact. However, Fig. 5 shows that as we predicted time delays between center of pressure (CPX) and lateral contact force (FL) with force contact of the fingertip changed from 76 ms on a rough surface to ≈266 ms on a slippery surface. Thus, the time delays for CPX displacement relative to fingertip force for force contact on a slippery surface were equivalent to those of the touch contact conditions, suggesting that subjects perceived the lack of horizontal physical support from the slippery surface and switched to a "light touch strategy" to enhance postural stability.

SUMMARY

Contact with the environment through touch of the hand can profoundly influence our body movements and sense of body orientation (cf. Lackner, 1981; 1992). Such observations suggested to us that touch contact could provide spatial information about body orientation that subjects might use to enhance their postural stability. We have found, in fact, that touch of the fingertip to a stationary surface at force levels far below those adequate to provide physical support can enhance the perception of body orientation and stabilize postural control. The crucial pieces of evidence which distinguish postural control with light touch contact from physically supportive contact forces are:

- Time delays between body sway and fingertip contact forces are much longer with touch contact (300 ms) than with force contact (<100 ms), suggesting that light touch forces are precuing postural musculature to stabilize body sway.
- Postural muscle activity is much higher with touch contact than force contact, indicating that fingertip contact forces in the light touch conditions were inadequate to counteract center of mass movements, accordingly, leg muscle activity was necessary to attenuate lateral sway.
- Changing the surface properties of the contact surface from a rough to a slippery texture influenced postural control with force contact but not touch contact, indicating that postural stabilization with touch contact is guided by sensory information provided by receptors in the fingertip and arm and proprioceptive information about arm configuration, rather than by physical support of the body.

These results suggest potential rehabilitation methods for patients with various sensory-motor deficits and disorders. For example, we have found that once contact with an external surface is available, postural control in patients with bilateral loss of vestibular function is equivalent to that of healthy individuals, indicating that touch cues can substitute for vestibular cues in the control of body orientation (Jeka et al., 1995a). Touch contact may also provide spatial cues for the orienting abilities of blind individuals. We have evidence that contact cues derived through common ambulatory aids such as the long cane provide stabilizing orientation cues for upright posture as well as information about obstacles and surfaces in the surrounding environment (Jeka et al., 1995b). Our findings also may have direct relevance to patients with gait disorders and imbalance attributable to neurological deficits (e.g. patients with hemiparesis, Parkinson's disease and some elderly individuals). Such populations are often capable of generating the appropriate level muscular forces to maintain stable balance and ambulation, but are unable to do so appropriately when unsupported. It is important to note that using contact of the hand with only small applied forces while standing requires innervation of leg musculature to stabilize posture. From this perspective, light touch contact is not a balance support, but a true balance aid that could potentially lead to improved long-term recovery of function by allowing appropriate recruitment of postural musculature involved in the maintenance of voluntary upright stance.

AUTHOR NOTES

We thank Joel Ventura and Art Larson III for technical assistance with the force platform and touch apparatus and Paul DiZio for discussions of the experimental paradigms and results. J.J. Jeka was supported by NIH postdoctoral fellowship 1 F32 NS09025-02 and J.R. Lackner by NASA grant NAG9-515 and a grant from the Naval Training Systems Center and the Air Force Office of Scientific Research.

REFERENCES

Diener, H.C. and Dichgans, J., 1986, Long loop reflexes and posture. In W. Bles and T. Brandt (Eds.), Disorders of Posture and Gait. pp. 41-51 Elsevier, London.

Holden, M., Ventura, J., and Lackner, J.R., 1994, Stabilization of posture by precision contact of the index finger, *Journal of Vestibular Research* 4:285-301.

Horak, F.B., and Macpherson, J.M., 1995, Postural orientation and equilibrium, In: Sehaprd, J. and Rohwell, L (Eds.), Handbook of physiology, University Press, New York, Oxford.

Jeka, J.J., Dizio, P., Horak, F.B., Krebs, D. and Lackner, J.R., 1995a, Haptic cues stabilize postural sway in individuals with bilateral vestibular loss (submitted).

Jeka, J.J., Easton, R.D., Bentzen, B.L., and Lackner, J.R., 1995b, Haptic cues for postural control in sighted and blind individuals. Perception and Psychophysics (in press).

Jeka, J.J., and Lackner, J.R. 1995, The role of haptic cues from rough and slippery surfaces in human postural control, Experimental Brain Research 103:267-276.

Jeka, J.J., and Lackner, J.R., 1994, Fingertip contact influences human postural control, *Experimental Brain Research* 100:495-502.

Johansson, R.S., 1991, How is grasping modified by somatosensory input? In: D.R. Humphrey and H.-J. Freund (Eds.), Motor control: Concepts and issues, John Wiley and Sons Ltd., pp. 331-355.

Klatzky, R.L., Lederman, S., and Reed, C., 1987, There's more to touch than meets the eye: The salience of object attributes for haptics with and without vision. *Journal of Experimental Psychology: General* 116(4):356-369.

Lackner, J.R., 1981, Some contributions of touch, pressure, and kinesthesis to human spatial orientation and oculomotor control, *Acta Astronautica* 8:825-830.

Lackner, J.R., 1992, Multimodal and motor influences on orientation: implications for adapting to weightless and virtual environments, *Journal of Vestibular Research* 2:307-322.

Matthews, P.B.C., 1988, Proprioceptors and their contribution to somatosensory mapping: complex messages require complex processing, *Can J Physiol Pharmacol* 66:430-438.

Westling, G., and Johansson, R.S., 1987, Responses in glabrous skin mechanoreceptors during precision grip in human. *Experimental Brain Research* 66:128-140.

Winter, D.A., Patla, A.E., and Frank, J.S., 1990, Assessment of balance control in humans, *Medical Progress through Technology*, 16:31-51.

27

EFFECTS OF NECK MUSCLE VIBRATION AND CALORIC VESTIBULAR STIMULATION ON THE PERCEPTION OF SUBJECTIVE 'STRAIGHT AHEAD' IN MAN

M. Fetter and H. -O. Karnath

Department of Neurology
Hoppe-Seyler Str. 3
72076 Tübingen
Germany

INTRODUCTION

For accurate motor behavior, like grasping or fixating a target, the correct perception of the target's spatial location relative to the body is essential. That is, the spatial location of a target has to be transformed into an egocentric, *body-centered coordinate system*. In recent years, strong evidence has been found that for this purpose the brain uses abstract, neural representations of space interposed between sensory input and motor output (Andersen et al., 1993). These representations seem to be organized in non-retinal, *egocentric coordinates*. Several authors (Ventre et al., 1984; Biguer et al., 1988; Karnath et al., 1991; 1993) have shown that the perception of '*straight ahead' body orientation* appears to be very closely connected with the neural generation of the reference frames that underlie the subject's mental *representation of space*. The processes behind this generation of a neural representation of egocentric spatial information is relatively complex. Several coordinate transformations are necessary. For example, when reaching or grasping for a stationary object, the positions of the eyes and head may vary from moment to moment although the relevant spatial location of the target with respect to the body may not change. If the visual target location was coded in retinal coordinates only, then each time the eyes moved, the coded location would change as well. For a quick and accurate response, the retinotopic coordinates of the target must be transformed into a coordinate system based on a non-retinal, body-centered frame of reference. To locate the direction of gaze in space and to relate this information to the orientation of the body, the input from the retina has to be combined with eye-position signals as well as head-position information. Therefore, it can be expected that the perception of 'straight ahead' is influenced by different external information sources.

One source is visual information. With *optokinetic stimulation* normal subjects displace their subjective 'straight ahead' toward the drum motion (Brecher et al., 1972). A second source is vestibular information. Passive rotatory acceleration of the body about its

Multisensory Control of Posture, Edited by T. Mergner and F. Hlavačka
Plenum Press, New York, 1995

vertical axis is long known to displace the apparent 'straight ahead' in a direction opposite to that of body acceleration (Fischer and Kornmüller, 1931; Morant, 1959) . Together with the subjective displacement of the 'straight ahead' position goes the impression that a stationary light appears to move in the direction of the acceleration, which has been termed '*oculogyral illusion*' by Graybiel and Hupp (1946). Similar effects of vestibular influence on visual 'straight ahead' have been described for caloric stimulation of one labyrinth by Hamann and coworkers (1992). With *caloric stimulation* the deviation was always toward the relatively colder ear, corresponding to the direction of the slow phase of the nystagmus. A third source is *proprioceptive information*. The influence of proprioceptive signals from neck muscles in computing egocentric coordinates of visual space has been extensively investigated (Lackner and Levine, 1979; Biguer et al., 1988). The latter authors investigated normal subjects sitting with head and body oriented straight ahead toward a central stationary visual target. During vibration of the left posterior neck muscles subjects reported an apparent motion and displacement of the stationary visual target toward the right. When requested to point to the target, subjects showed a consistent error in pointing, and this error was in the same direction as the illusory displacement. All these findings indicate that visual, vestibular and neck proprioceptive input contribute to the neural generation of the reference frames that underlie the subject's mental representation of space in egocentric coordinates. Accordingly, one should expect interactive effects of the different input channels. Evidence for *visual-vestibular interaction* on the *spatial orientation* in man has been reviewed by Dichgans and Brandt (1978). When vision is excluded, the vestibular system, which registers position and motion of the head in space, and proprioceptive information from the neck region act together to relate trunk to space. This concept of *vestibular-proprioceptive interaction* was originally introduced by von Holst and Mittelstaedt (1950) and was further elaborated by Roberts (1973). Since then several lines of evidence from experimental work in animals have supported this idea, even though there is still an ongoing debate on how the neck (that is muscle spindles and cervical joint receptors) contributes to the maintenance of body orientation and balance. The aim of the present study was to investigate the interaction of neck muscle proprioception and vestibular input on the subjective 'straight ahead' orientation in human subjects.

METHODS

We investigated 17 normal subjects with an age range of 25 to 69 years with a median of 31 years, seven subjects were female. The subjects were seated upright in a light-bulb-shaped cabin with the head in the center of the upper spherical part of the bulb with a diameter of 195 cm in complete darkness with the exception of a dim laser spot, which was reflected onto the inner surface of the cabin by a mirror galvanometer system. To measure subjective straight ahead, the subject was asked to move a laser point toward where he thinks straight ahead is. For that purpose, the subject could move the laser point by pressing one of four small directional buttons mounted crosswise on a small box. When a button was pressed, the laser point moved smoothly in the indicated direction with a velocity of 1°/s.

For *neck muscle vibration* an experimental vibrator was used. Its frequency was fixed at 100 Hz with an amplitude of 0.4 mm. The tip of the vibrator, a flat disk 2.3 cm in diameter, was placed on the subject's posterior neck on the right or the left. Its position was individually adjusted to achieve an illusion of horizontal displacement of the stationary laser point. The testing procedure started as soon as the subject had a clear illusion of target movement. The stationary spot was then extinguished and vibration continued. The spot immediately reappeared pseudorandomly at one of four eccentric positions (in the for corners of a quadrant with horizontal and vertical deviations of 10° from straight ahead) and was then required to be adjusted to subjective straight ahead.

Vestibular stimulation was applied by 30 cc ice water irrigation of the left external auditory canal. During stimulation the subject sat upright in the chair with the head tilted approximately 60° backward. In all subjects a vigorous nystagmus was induced with the slow phase to the left side. Testing procedures were started 2 min after irrigation again in upright position, at that time most of the caloric nystagmus had decayed.

The test conditions were: (I) no stimulation ('baseline condition'); (II) vibration of the left posterior neck muscles; (III) vibration of the right posterior neck muscles; (IV) ice water calorics of the left external auditory canal; (V) ice water calorics of the left external auditory canal plus vibration of the left posterior neck muscles; (VI) ice water calorics of the left external auditory canal plus vibration of the right posterior neck muscles.

RESULTS

The subjective 'straight ahead' positions were very closely scattered around the objective straight ahead body orientation. In test condition I, without stimulation, the 17 subjects steered the laser to an average position of 0.75° to the right of straight ahead with a SD of 2.5° in the horizontal plane and to 0.48° down from straight ahead (at eye level) with a SD of 3.5° in the vertical plane. For analysis of the stimulated conditions the angular deviation of each subject's judgment from that in baseline condition I was calculated.

Figure 1 shows the mean angular displacement of subjective 'straight ahead' judgments in the horizontal plane for the different stimulation conditions. Included in this figure are the results of nine subjects who reported an illusion of motion with both left-sided and right-sided vibration. The principal direction of apparent motion of the target was mainly to the right with vibration on the left and mainly to the left with vibration on the right, although some vertical motion was sometimes reported in addition. Accordingly, all nine subjects showed a horizontal displacement of their subjective straight ahead body orientation to the left with left-sided vibration (condition II) and to the right with vibration on the right side (condition III), that is opposite to the perceived movement of the stationary target. A corresponding effect was seen after vestibular stimulation (condition IV). Icewater calorics on the left side induce slow phases to the left, that leads to the impression that the stationary target is moving to the right and a corresponding shift of subjective straight ahead to the left. In condition V, when neck muscle vibration and vestibular stimulation were simultaneously applied on the left, the leftward displacement of subjective straight ahead body orientation was larger than the effect of either stimulation alone. The opposite effect was achieved when left-sided vestibular stimulation was combined with right-sided neck muscle vibration (condition VI). This combination seemed to neutralize the effects that both types of stimulation had when exclusively applied.

Four subjects had no illusion of target motion either with vibration of the left or the right neck muscles. Another four subjects only had an illusion of motion to one side. In contrast, all 17 subjects experienced an illusion of movement and displacement of the stationary laser point after caloric stimulation. With caloric stimulation, again, all subjects experienced the principal direction of motion in the horizontal plane (that is to the right), additionally some minor vertical motion was sometimes reported. In all subjects illusion of movement of the stationary target was highly correlated with their subjective displacement of body orientation when their neck muscles were vibrated.

In addition to the principal horizontal direction of motion, the subjects sometimes also reported a vertical motion component of the stationary laser point with vibration and/or

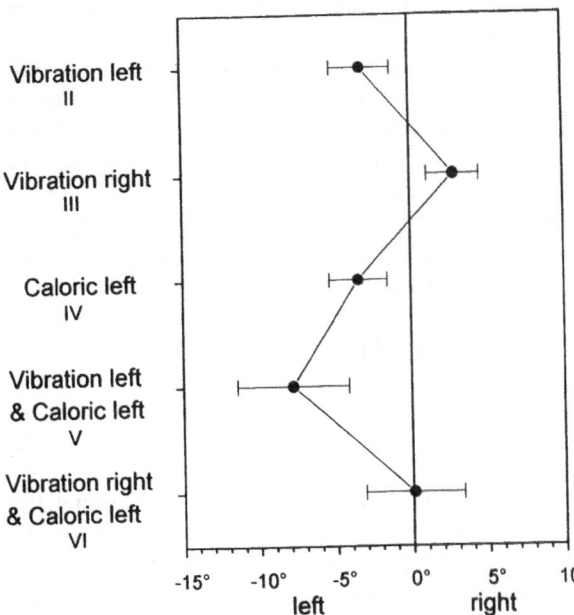

Figure 1. Mean values (± 1 standard deviation) of subjective 'straight ahead' judgments in the horizontal plane for test conditions II to VI of nine subjects who reported an illusion of motion with both left-sided and right-sided vibration (angular deviation of each subject's judgment from that in 'baseline' condition I).

vestibular stimulation. In contrast to the clear horizontal displacements of subjective 'straight ahead' with proprioceptive and vestibular stimulation, in the vertical plane only a mild predominance of downward displacements was detected during caloric stimulation.

From the data shown it appears that the combined effect of neck muscle vibration and vestibular stimulation on the horizontal deviation of subjective body orientation is additive. To test the hypothesis of a linear relationship between the two types of stimulation, the observed angular deviations of the 'straight ahead' orientation in conditions V and VI of all 17 subjects were correlated with the sum of the respective judgments when tested alone. This calculation resulted in a rather linear correlation with a Pearson product-moment coefficient of 0.85.

CONCLUSIONS

Our data clearly support the notion of a profound impact on subjective body orientation in humans by proprioceptive input from the neck region, even though the effect is quite variable and less effective than is vestibular stimulation. The astonishing accuracy with which normal human subjects can estimate their straight ahead body orientation under normal conditions argues for a stable, body-centered reference frame for the evaluation of body orientation in space and further shows that under normal conditions the sensory systems tested act together in a very precise manner, supplying us with a close to optimal estimate of body orientation.

Interestingly, the effect of caloric stimulation on the perception of subjective body orientation could be measured even after most of the vestibular nystagmus had decayed, indicating that the central neural processing that leads to a deviation of straight ahead perception probably has a longer time constant than the induced caloric nystagmus. In agreement with our results, a linear interaction during combined vestibular, neck and visuo-oculomotor stimulation has also been described by Mergner and coworkers (1992) for the perception of visual object motion in space.

In the monkey, posterior parietal cortex is the most likely area of the brain involved in such transformation and integration processes as suggested by Andersen and coworkers (1993). Interestingly, there are patients with right sided parietal lobe lesions who have a profound difference between objective and subjective straight ahead resulting in the clinical syndrome of hemineglect. In these patients the major component leading to the syndrome seems to be a disturbed neural computation of the egocentric, body-centered frame of reference. Also in these patients subjective straight ahead orientation can be shifted by the same vestibular or proprioceptive stimuli used in this study. This could provide an interesting new approach for the rehabilitation of neglect patients. On the basis of the presented results and the findings of Karnath and coworkers (1993) and, for example, Cappa et al. (1987) one could speculate that a periodic, systematic therapy over several weeks with stimulation of the contralesional posterior neck muscles by vibration and/or the activation of the vestibular system by, for example, caloric stimulation, may lead not only to a transient but also a permanent reduction of neglect symptoms via a correction of the displaced body-centered frame of reference in these patients.

ACKNOWLEDGMENTS

This work was supported by the Deutsche Forschungsgemeinschaft SFB 307-A2.

REFERENCES

Andersen, R.A., Snyder, L.H., Li, C.-S., Stricanne, B., 1993, Coordinate transformation in the representation of spatial information, *Curr. Opin. Neurobiol.* 3:171-176.

Biguer, B., Donaldson, I.M.L., Hein, A., Jeannerod, M., 1988, Neck muscle vibration modifies the representation of visual motion and direction in man, *Brain* 111:1405-1424.

Brecher, G.A., Brecher, M.H., Kommerell, G., Sauter, F.A., Sellerbeck, J., 1972, Relation of optical and labyrinthean orientation, *Opt Acta* 19:467-471.

Cappa, S., Sterzi, R., Vallar, G., Bisiach, E., 1987, Remission of hemineglect and anosognosia during vestibular stimulation, *Neuropsychologia* 25:775-782.

Dichgans, J., Brandt, T., 1978, Visual-vestibular interaction: effects on self-motion perception and postural control. In: Held, R., Leibowitz, H.W., Teuber, H.-L. (eds.) Handbook of Sensory Physiology, Vol. VIII. Perception. Springer, Berlin, Heidelberg, New York, pp. 755-804.

Fischer, M.H., Kornmüller, A.E., 1931, Egozentrische Lokalisation. 2. Mitteilung (optische Richtungslosigkeit beim vestibulären Nystagmus), *J Psychol Neurol* 41:383-420.

Graybiel, A., Hupp, D.I., 1946, The oculo-gyral illusion, a form of apparent motion which may be observed following stimulation of the semicircular canals, *J Aviat Med* 17:3-27.

Hamann, K.F., Strauss, K., Kellner, M., Weiss, U., 1992, Dependence of visual straight ahead on vestibular influences. In: Krejcová, H., Jerábek, J. (eds.) Proceedings of the XVIIth Bárány Society Meeting, pp 65-66.

Holst, E. von, Mittelstaedt, H., 1950, Das Reafferenzprinzip (Wechselwirkungen zwischen Zentralnervensystem und Peripherie), *Naturwissenschaften* 37:464-475.

Karnath, H.-O., Schenkel, P., Fischer, B., 1991, Trunk orientation as the determining factor of the 'contralateral' deficit in the neglect syndrome and as the physical anchor of the internal representation of body orientation in space, *Brain* 114:1997-2014.

Karnath, H.-O., Christ, K., Hartje, W., 1993, Decrease of contralateral neglect by neck muscle vibration and spatial orientation of trunk midline, *Brain* 116:383-396.

Lackner, J.R., Levine, M.S., 1979, Changes in apparent body orientation and sensory localization induced by vibration of postural muscles: vibratory myesthetic illusions, *Aviat Space Environ Med* 50:346-354.

Mergner, T., Rottler, G., Kimmig, H., Becker, W., 1992, Role of vestibular and neck inputs for the perception of object motion in space, *Exp Brain Res* 89:655-668.

Morant, R.B., 1959, The visual perception of median plane as influenced by labyrinthian stimulation, *J Psychol* 47:25-35.

Roberts, T.D.M., 1973, Reflex balance, *Nature* 244:156-158.
Ventre, J., Flandrin, J.M., Jeannerod, M., 1984, In search for the egocentric reference. A neurophysiological
    hypothesis, *Neuropsychologia* 22:797-806.

28

BODY LEANING INDUCED BY GALVANIC VESTIBULAR AND VIBRATORY LEG MUSCLE STIMULATION

F. Hlavacka and M. Krizkova

Institute of Normal and Pathological Physiology
Slovak Academy of Sciences
Bratislava
Slovak Republic

INTRODUCTION

The aim of human stance stability is to maintain the body's center of gravity over its base of support with minimal postural muscle activation. Multisensory information, mainly from vestibular, visual and somatosensory afferents, is integrated in a continual process of adjusting the body's center of gravity in relation to an internal space reference frame (Gurfinkel et al., 1988). Thus, the body is continuously and slowly leaning away from and towards the vertical position.

Vibratory stimulation of leg muscles in standing subjects induces compensatory postural reactions which cause the body to lean in the direction of the vibrated muscles (Eklund, 1972). These vibration-induced postural responses probably result from a change of somesthetic body position information (Popov et al., 1986; Roll et al., 1989). Direction-ally-specific postural reactions have been elicited by unilateral vibration of different lower leg muscles in standing subjects (Saling and Hlavacka, 1988). Similarly, application of electrical current between the human mastoid processes during stance also evokes body lean, mainly in the lateral direction towards the side of the anode (Njiokiktjien and Folkerts 1971).

The aim of this experiment was to analyze how these two compensatory postural reactions elicited from different afferent inputs (vestibular and proprioceptive) interact. We wanted to experimentally test the hypothesis that a "vector summation" could account for the postural effects of combined vestibular and proprioceptive stimulation (Hlavacka et al., 1992).

METHODS

Body sway was recorded from healthy subjects with their heads oriented straight ahead and eyes closed during vibration of tibialis anterior (TA) muscle, paired with galvanic vestibular stimulation of either polarity. Subjects stood on a force platform with their heels together and feet at a 30 degree angle to one another, and their arms at their sides. Each trial was 50 s with each of 8 conditions repeated 4 times. Ten subjects participated (6 men and 4 women, mean age 30.9 years, range 20-48 years). Informed consent was obtained from all subjects.

Vibratory stimulation with duration 5 s, frequency 90 Hz and amplitude 1 mm was sequentially applied to the belly of the TA muscle. A small dc-motor with unbalanced load (5 grams) was used as a vibrator. The vibrator was fastened to the muscle of the right lower leg with a rubber strap.

The vestibular system was stimulated using bipolar, binaural electrical current through electrodes placed on both mastoid processes. For vestibular "right" stimulation, which induced body lean to the right, the positive electrode (anode) was placed on the right mastoid and the cathode on the left mastoid. For vestibular "left" stimulation, polarities were reversed. Kidney-shaped, silver/silver chloride electrodes (10 cm^2) were enveloped in gauze and moistened with physiological saline solution. A constant current stimulator produced cosine bell-shaped galvanic current with peak value 1 mA and duration 3.3 s. A cosine bell waveform was chosen because of its similarity with actual physiological changes of vestibular information during body inclination. Four possible stimuli were initiated in random trials which were 50 s in duration. Each vestibular stimulus was started 1.75 s prior to muscle vibration onset in order to evoke the peak of sway response to vestibular and vibration stimulation simultaneously. In pilot studies, the timing of the galvanic stimuli was varied, either 1.5 s or 0.75 s before or 0 s, 0.75 s or 1.5 s after the onset of the vibration.

The center of foot pressure (COP) of subjects in upright stance was measured with a force platform equipped with automatic weight correction. Both anterior-posterior and lateral directions were digitized at 41 Hz (2048 samples over 50 s for one signal). Data were stored and evaluated using a PC/AT 286 computer.

The COP in both directions was analyzed over 10 s intervals which included 2 s before the vibration onset and 3 s after each 5 s stimulation interval. The four postural responses during each experimental condition for each subject were averaged. Each subject's postural responses were adjusted such that the mean COP position for 2 s before stimulation was set to zero. The final postural response evoked by galvanic stimulation, vibratory stimulation, or combined galvanic-vibratory stimulation was averaged for all 10 subjects (10 records for sagittal component and 10 records for the lateral component of body tilt). Final, averaged results are presented as mean values ±SD (standard deviation) from the data of all ten subjects.

RESULTS

Figures 1 and 2 show that vibration of the TA muscle induced changes in the position of the COP (dashed lines) which started with a small shift contralateral to the side of the vibration, and then increased in magnitude toward the ipsilateral side and reached a plateau position. The vibration of the right TA muscle induced body lean forward and to the right. The averaged magnitudes of the COP position shift measured at 4 s following the onset of TA vibration were 2.05±0.83 cm to the right (Fig. 1, dashed lines), and 2.64±1.37 cm forward (Fig. 2, dashed lines).

Vestibular stimulation produced lateral shifts in the position of the COP, with the direction depending on the polarity of the galvanic current, followed by a return to the initial position (Fig. 1, thin line). Postural leans were to the right when the anode was placed at the right ear (GS right) and to the left when the anode was at the left ear (GS left). The averaged maximal changes of COP position induced by galvanic stimulation were 2.18±1.60 cm to the right (Fig. 1 above, thin line) and 2.04±1.05 cm to the left (Fig. 1 below, thin line). The COP position shift forward was clearly smaller than the lateral shift. The lateral COP position shifts produced by either proprioceptive or vestibular stimulation alone had similar magnitudes.

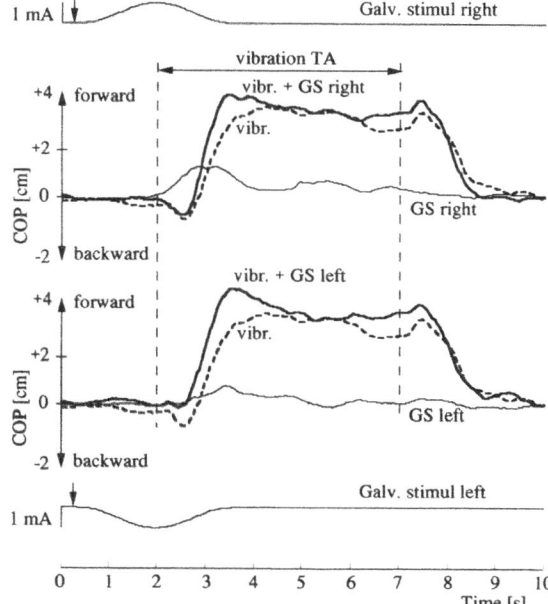

Figure 1. Left-right shifts of COP as a function of time for vestibular stimulation only (GS right or GS left - thin line), for muscle vibration only (vibr. dashed line, TA - m. tibialis ant.) and for paired stimuli (vibr. + GS - heavy line). Arrows on the galvanic stimulus curves indicate the onset of vestibular stimulation. Vertical dashed lines indicate the onset and termination of muscle vibration. Curves are formed by averaging the results from 10 subjects.

With combined galvanic-vibration stimulation, the larger, significant effect of the vestibular stimulation was in the lateral direction (Fig. 1, interval 2-4s). A smaller, non-significant influence of vestibular stimulation was observed in forward-backward COP position shifts (Fig. 2). Since the anterior-posterior sway induced by vestibular stimulation was minimal, the interaction of proprioceptive and vestibular input could not be seen in this

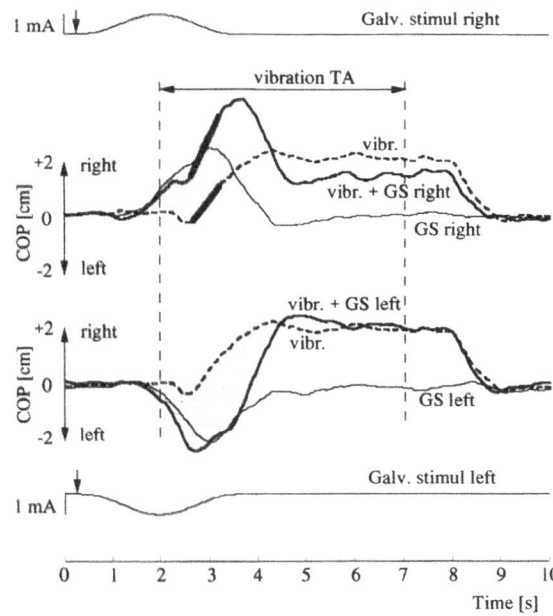

Figure 2. Forward - backward shifts of COP as a function of time for vestibular, vibratory and paired stimulation. Other description as in Fig. 1.

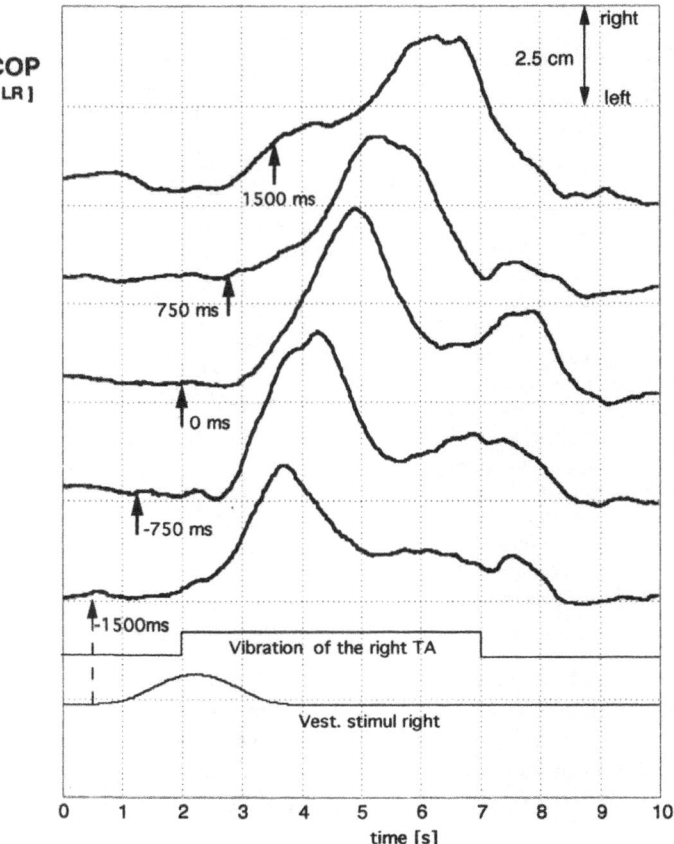

Figure 3. Position of the center of pressure (COP) in the lateral direction as a function of time for paired stimulation (vibration + GS right). Arrows on the curves indicate the onset of the vestibular stimulation.

direction. That is, the combined effect of both stimuli produced a COP response which was close to the COP response to the vibration stimulus alone. Comparison of the slopes of COP curves with only proprioceptive vibration to the paired vestibular and proprioceptive stimulation showed that the COP slope increased during combined stimulation (Fig. 1). A summation of the effects of vestibular stimulation alone and proprioceptive stimulation alone resulted in the increased slope of the COP curves, particularly in cases where the direction of the postural responses was similar (Fig. 1 above).

The results of the varied timing of the paired stimuli (vibration TA + GS right) are shown in Fig. 3 for lateral and in Fig. 4 for anterior-posterior body leaning. It is evident that in the lateral direction the time of the peak response due to the combined galvanic and vibration stimulus is related to the onset time of the galvanic stimulus (Fig. 3). This was not observed in the anterior-posterior direction (Fig. 4).

DISCUSSION

Results of our experiment show that two modalities of afferent inputs (vestibular and proprioceptive) combine to produce an additive effect on the postural orientation of standing

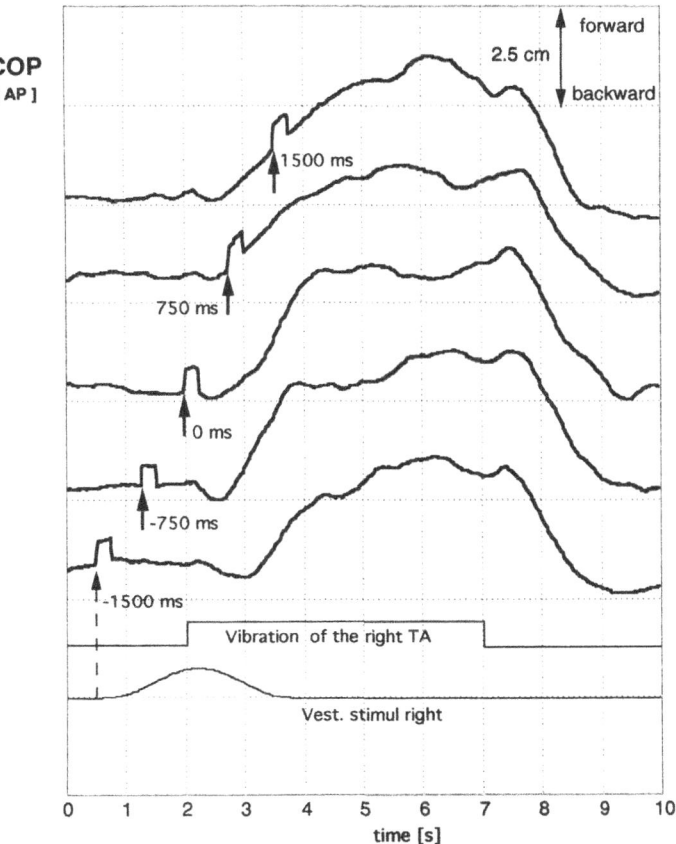

Figure 4. Position of the center of pressure (COP) in anterior-posterior direction as a function of time for paired stimulation (vibration + GS right). Arrows on the curves indicate the onset of the vestibular stimulation.

subjects. Our results show that the direction and magnitude of postural responses to leg muscle vibration can be modulated by simultaneous changes in vestibular input activated by galvanic current. These results are consistent with a "vector summation" hypothesis in which proprioceptive and vestibular input each provide an internal representation vector of vertical orientation which are continually summed together.

Leg muscle vibration has been used to perturb postural orientation by many investigators (Eklund, 1972; Gurfinkel et al., 1988; Popov et al., 1986; Roll et al., 1989; Saling and Hlavacka, 1988). Body inclination induced by leg muscle vibration during stance cannot be explained by the spinal reflex - tonic vibration reflex (Eklund, 1972). Vibration-induced muscle afferent signals arriving at suprasegmental equilibrium centers are more likely responsible for the unintentional displacement of the center of gravity (Hagbarth, 1973). The influence of central mechanisms in the interpretation of body vertical based on perceived lengthening of the Achilles tendon by vibration has been called "proprioceptive vertical" (Gurfinkel et al., 1988).

Leaning induced in response to vestibular stimulation by bipolar, binaural current may also be representative of changes in the internal estimate of body vertical. Galvanic stimulation with the negative electrode placed near the left ear and the positive electrode near the right ear induces an increased discharge rate in the left vestibular-nerve afferents and a decreased discharge rate in the right vestibular-nerve afferents (Goldberg et al., 1982). These changes in activity of both vestibular-nerves are similar to vestibular afferent changes during body lean to the left. Because neck and body proprioception afferent inputs are not changed during the artificial vestibular stimulation, the asymmetry in the vestibular nerve afferent activities may be perceived by the CNS as a mismatch between the new internal representation of the direction of gravity (vestibular reference of vertical tilt to the right) and the actual proprioception information about the body orientation. In order to match the natural proprioceptively-derived body orientation with the artificially altered vestibular representation of vertical to the right, subjects compensate by leaning to the right. Thus, a new vertical reference, altered by galvanic stimulation, is adopted and the body's position is properly aligned with the new internal perception of vertical. This explanation is consistent with the fact that freely standing subjects perceive themselves to be aligned with gravity as they lean with galvanic stimulation and with the hypothesis that vestibular stimulation produces a change in the estimate of the direction of gravity (Inglis et al., 1995).

Previous studies on the effects of externally supporting a subject during vestibular stimulation support our view. If some part of a standing subject's body is connected to a fixed external base, subjects perceive body tilt in the opposite direction they would actually tilt with vestibular stimulation during free stance (Fitzpatrick et al., 1994; Popov et al., 1986). Thus, perception of body lean induced by vestibular stimulation depends on interpretation of an altered vestibular vertical reference based on a combination of proprioceptive as well as vestibular information.

The postural responses to the paired vestibular and proprioceptive inputs in our study show the clear additive effect of the two modalities in the lateral direction where the vestibular influence was greater. The difference between COP response to the vibration input alone and to vibration with vestibular stimulation was statistically significant. Others have shown additive effects when several proprioceptive inputs were activated by simultaneous vibration of two or three muscles in different body segments (Roll et al., 1989b). They observed that the effects of combined vibration of "synergistic" muscle groups were additive, but the summation was not linear since the combined effects were clearly smaller than the sum of the individual effects.

Our findings show evidence supporting the hypothesis of a vector (both direction and magnitude) summation of the postural response to both vestibular and proprioceptive inputs. It is likely that vector summation of sensory inputs is valid only for a very limited range of input intensities and for displacements around the equilibrium point of stance where the body moves as an approximate inverted pendulum. The vector summation more accurately describes the direction of orientation vectors than the magnitude of the vectors. Pilot studies, in which time relations between vibration and galvanic stimuli were varied, showed a larger vestibular influence than can be explained by linear addition (Fig. 3).

From the results, we conclude that the adjustment of the vertical body position is a continual process using an integrative body reference of vertical derived, in our case, from vestibular and somatosensory vertical references. This long-latency, adaptive control of postural orientation is centrally initiated with the aim to keep the body aligned with gravity and to minimize postural muscle activation during erect stance. The body position is continually adjusted to maintain alignment with an internal reference of body vertical whenever information from one or more sensory channels is altered.

A simplified and hypothetical explanation of our findings is presented in Fig. 5 for lateral plane orientation. Consider that the CNS creates the internal reference of body vertical (thick line with arrow) as an average of the sensory references to vertical from the visual, vestibular and somatosensory systems. With eyes closed, only the vestibular and somatosensory reference are involved (Fig. 5A, dashed and dotted lines with arrow). Fig. 5B illustrates the condition in which the galvanic vestibular stimulation tilted the vestibular reference of vertical to the right and the proprioceptive reference was not changed. The resulting internal reference of body vertical lies between this altered vestibular reference and the unaltered proprioceptive reference. The compensatory body tilt to the right matches the internal reference. Fig. 5C illustrates a similar situation in which proprioceptive input is changed by muscle vibration such that the proprioceptive vertical reference is tilted to the right and the vestibular reference is unchanged. Combining activation of both vestibular and propriocep-tive systems results in increased postural tilt magnitude as it matches the larger tilt of the internal vertical reference resulting from summation of the vestibular and proprioceptive references (Fig. 5D).

This schematic interpretation suggests that the observed final body position repre-sents the position of the internal reference to vertical. For example, if eyes are open in situation Fig. 5E, a strong visual reference is also involved, and the shift of the final body reference is smaller. This is consistent with results showing smaller postural response to muscle vibration [1] or to galvanic stimulation (Njiokiktjien and Folkerts, 1971) with eyes open compared to eyes closed. Similarly, if subjects stand on an unstable support which decreases proprioceptive reference information, the effect of vestibular stimulation is in-creased such that final body position is closer to the vestibular vertical reference (Fitzpatrick et al., 1994). The results of the present study showed that both vestibular and proprioceptive signals play important roles in the estimation of the internal representation of body vertical. By continual adaptive control, the directionally-dependent interaction of vestibular and proprioceptive inputs enables humans to stand erect and to maintain a body position aligned with gravity.

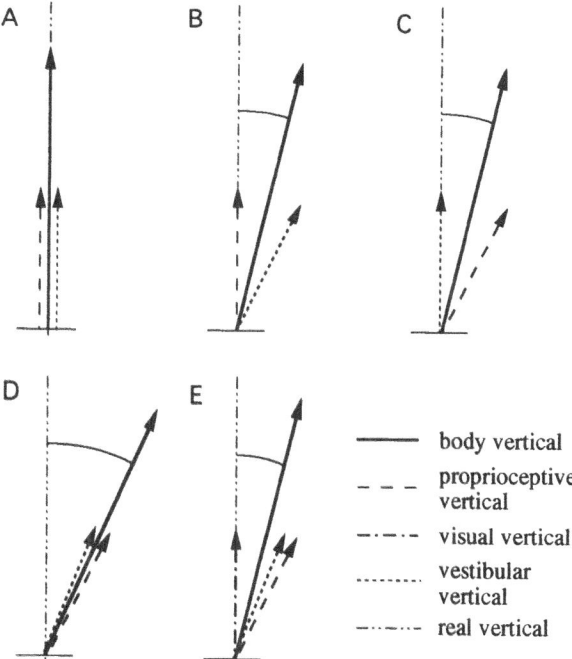

Figure 5. Schematic representation of "vector summation" hypothesis in the lateral plane (see explanation in text).

ABSTRACT

In order to understand proprioceptive and vestibular contributions to human stance posture, the effect of electrical vestibular stimulation on body lean induced by leg muscle vibration was investigated. The magnitude and direction of postural responses were registered as changes in center of foot pressure (COP) with a force platform. Vibratory stimulation of tibialis anterior muscle in the right leg was used to alter proprioceptive input. Vestibular stimulation consisted of galvanic cosine bell-shaped current of both polarities in a binaural, bipolar electrode position.

Vibration of the tibialis muscle resulted in body lean forward-right. Vestibular stimulation produced postural responses in the lateral direction to the right (anode at right ear) or to the left (anode at left ear) and then back to the initial position. The vestibular influence on vibratory-induced body lean was observed only in the lateral direction. Upon combined vibratory and galvanic stimulation, the evoked body lean showed a trajectory that could essentially be described as a vector summation of the two individual effects. However, preliminary experiments indicate that with a different timing of the two stimuli the interaction may become non-linear in that the vestibular responses become enlarged by ongoing proprioceptive response. The results of the present study showed that both vestibular and proprioceptive signals play important roles in the estimation of internal representation of body vertical.

REFERENCES

Eklund, G., 1972, General features of vibration-induced effects on balance, Uppsala *J. Med. Sci.* 77:112-124.

Fitzpatrick, R., Burke, D., and Gandevia, S., 1994, Task-dependent reflex responses and movement illusions evoked by galvanic vestibular stimulation in standing humans, *J. Physiol.* 478:363-372.

Goldberg, J.M., Fernandez, C., and Smith, C.E., 1982, Responses of vestibular-nerve afferents in the squirrel monkey to externally applied galvanic currents, *Brain Res.* 252:156-160.

Gurfinkel, V.S., Levik, Yu.S., Popov, K.E., Smetanin, B.N., and Shlikov, V.Yu., 1988, Body scheme in the control of postural activity. In: Stance and Motion: Facts and Concepts. V.S. Gurfinkel, M.E. Ioffe, J. Massion, and J.P. Roll (eds.), Plenum Press, New York and London, pp.185-193.

Hagbarth, K.-E., 1973, The effect of muscle vibration in normal man and in patients with motor disorders. In: New Developments in Electromyography and Clinical Neurophysiology, J.E. Desmedt (ed.), Vol. 3, Karger, Basel, pp. 428-443.

Hlavacka, F., Mergner, T., and Schweigart, G., 1992, Interaction of vestibular and proprioceptive inputs for human self-motion perception, *Neuroscience Letters* 138:161-164.

Inglis, J.T., Shupert, C.L., Hlavacka, F., and Horak, F.B., 1995, The effect of galvanic vestibular stinulation on human postural responses during support surface translations. *J. Neurophysiol.* 73:896-901.

Njiokiktjien, Ch., and Folkerts, J.F., 1971, Displacement of the bodys center of gravity at galvanic stimulation of the labyrinth, *Confinia Neurol.* 33:46-54.

Popov, K.E., Smetanin, B.N., Gurfinkel, V.S., Kudinova, M.P., and Shlykov, V. Yu., 1986, Spatial perception and vestibulomotor responses in man, *Neurophysiol.* (translated from Neirofiziologiya) 18:548-554.

Roll, J.P., Vedel, J.P., and Ribot, E., 1989, Alteration of proprioceptive messages induced by tendon vibration in man: a microneurographic study, *Exp. Brain Res.* 76:213-222.

Roll, J.P., Vedel, J.P., and Roll R., 1988b, Eye, head and skeletal muscle spindle feedback in the elaboration of body references. In: Progress in Brain Research, Vol. 80, J.H.J. Allum and M. Hulliger (Eds.), Elsevier Science Publishers B.V. (Biomedical Division), pp.113- 123.

Saling, M., and Hlavacka, F., 1988, Postural responses evoked by unilateral vibration of lower limb muscles in standing subjects. In: Mechanoreceptors: Development, Structure and Function. P. Hnik, T. Soukop, R. Vejsada, and J. Zelena (Eds.), Plenum Press, New York, pp.407-409.

29

VESTIBULAR-SOMATOSENSORY INTERACTIONS FOR HUMAN POSTURE

F. B. Horak,[1] F. Hlavačka,[2] and C. L. Shupert[1]

[1] R.S. Dow Neurological Sciences Institute of Legacy Good Samaritan
   Hospital
1120 NW 20th Ave.
Portland, Oregon
[2] Slovak Academy of Sciences
Institute of Normal and Pathological Physiology
Sienkiewiczova 1
813 17 Bratislava, Slovakia

INTRODUCTION

The postural control system is organized around two main behavioral goals: equilibrium and orientation. The task of controlling equilibrium requires balancing all the forces acting on the body to control the position of the body's center of mass. The task of maintaining orientation requires interpretation of sensory information from many sources to align body parts with reference to gravitoinertial forces and other environmental features. Automatic postural responses that restore equilibrium in response to external perturbations in stance return the body to a particular orientation; that is, a particular body position which is selected based on particular orientation references (Massion, 1992).

The roles of somatosensory and vestibular information in the tasks of maintaining equilibrium and orientation are not completely understood. A simple summation of vestibular and somatosensory reflexes is unlikely to provide the vast behavioral flexibility required for postural coordination, the demands of which can change depending on the environmental or behavioral context. It is more likely that postural control is based on integration of sensory information from multiple systems to produce a centrally coordinated response. For example, the correct interpretation of sensory information concerning body movement from distal segments of a multilink body (like vestibular information from the head) requires knowledge of the state of the intermediate segments (like the trunk and the neck). This somatosensory and vestibular information must be correctly integrated by the CNS if information from the vestibular system is to be used to encode and control motion of the body.

This kind of sensory integration is shown in experiments involving stimulation of the vestibular system using galvanic currents delivered to the inner ear. Galvanic current introduces a vestibular signal that results in body sway, the direction of which depends upon the orientation of the head and trunk (Nashner and Wolfson, 1974; Hlavacka and Njiokiktjien, 1985). Also, the particular muscles activated by galvanic vestibular stimulation depend on a subject's initial orientation and external support condition (Storper and Honrubia, 1992; Fitzpatrick et al, 1994). Thus, the CNS transforms sensory information from a vestibular coordinate system (head in space) to a somatosensory coordinate system (body sway over its base of support) during responses to galvanic stimulation (Lund and Broberg, 1983).

The present study examines the interaction between vestibular and somatosensory signals for maintaining equilibrium and orientation for postural control by introducing galvanic vestibular stimulation prior to and during automatic equilibrium responses to platform translations. Natural vestibular and somatosensory stimulation associated with passive, forward postural sway and an active, backward postural response to a backward surface translation were combined with galvanic vestibular stimulation that induced either a forward or backward sway. If the vestibulospinal system participates in the formulation and initiation of automatic postural responses to restore equilibrium, the earliest automatic postural responses should be altered by introducing a vestibular signal prior to and during the perturbation. If the vestibular system provides an orientation reference or alters vestibular reafference during postural movements, galvanic stimulation during an external perturbation should change the later postural responses that move the body to its orientation goal.

METHODS

Seven healthy adults stood with eyes closed and heads turned to the right on two computer-controlled force platforms. Subjects were exposed to very low levels (.1-.5 mA) of bipolar galvanic current applied to the mastoid bones using 9 cm^2 electrodes, either while standing still or while undergoing backward translations of the force plates. The current intensity chosen for each subject was the lowest at which the subject reported a slight feeling of disorientation and demonstrated a slight body sway. The polarity of the current resulted in perceived sway either congruent with (anode on the forward ear) or opposed to (anode on the backward ear) the forward sway generated by actual body displacements in response to backward platform translations. Subjects were exposed to randomized trials (5 per condition) of platform translation alone at three velocities (slow (1.4 cm/s), medium (4.2 cm/s), and fast (14 cm/s), galvanic stimulation alone (8 s duration), and combinations of platform translations and galvanic stimulation. The relative time of application of the galvanic current was also varied and occurred at 2500, 1500, 500 or 0 ms prior to platform translation onset.

Postural responses were quantified by recording the sum of right and left foot center of pressure (CoP), calculating the body's center of mass from kinematics (CoM), and determining the onset latency and integrated area (0-75 ms) of surface electromyographic activity in gastrocnemius, soleus, tibialis, quadriceps, hamstrings, paraspinalis and rectus abdominus muscles. To clearly display the effect of vestibular stimulation on postural responses without the component of the response due to the translation alone, the "net" effect of vestibular stimulation on the CoM and CoP was calculated by subtracting responses to posterior vestibular stimulation (negative values) from the corresponding responses to anterior vestibular stimulation (positive values) and dividing the result by 2. Analyses were performed on individual trials and then averaged to produce mean results which were used for statistical comparisons using MANOVA and post-hoc analyses.

RESULTS

Vestibular stimulation applied prior to and during a platform translation caused subjects to alter their response to the translation such that they leaned either forward (anode anterior ear) or backward (anode posterior ear) from their initial postural orientation (Figure 1A). The change in final CoP and CoM position due to vestibular stimulation was larger after a surface displacement (1.5 +/-.4 cm) than during quiet stance (.8 +/- .5 cm). The final CoP and CoM positions for galvanic stimulation paired all translation velocities were significantly different from vestibular stimulation alone (all p's <.007 for CoP and all p's <.05 for CoM). The final CoP and CoM positions induced by vestibular stimulation, however, were not different for different velocities of translation.

The effect of vestibular stimulation was larger on the later components of the responses to platform translation than on the earliest components. Figure 1B shows that distinct differences in the postural responses due to vestibular stimulation first occurred at 1 s after initiation of the CoP response. There was no difference in the initial rate of change of CoP response in trials with and without vestibular stimulation. However, vestibular stimulation resulted in a large, phasic change in CoP response at 1-2 s following onset of translation as well as the smaller, tonic change in final orientation position.

The effects of pairing galvanic vestibular stimulation with platform translations were larger than a simple, algebraic summation of the effects of platform translation and galvanic stimulation alone.

Figures 2 A and B show the net effect of vestibular stimulation on the CoP and CoM responses to platform translation. This net effect was calculated by subtracting out the component of the response due to the platform alone.

The magnitude of the effect of vestibular stimulation increased with increasing velocity of platform translation. The differences in the peak net CoP for the different

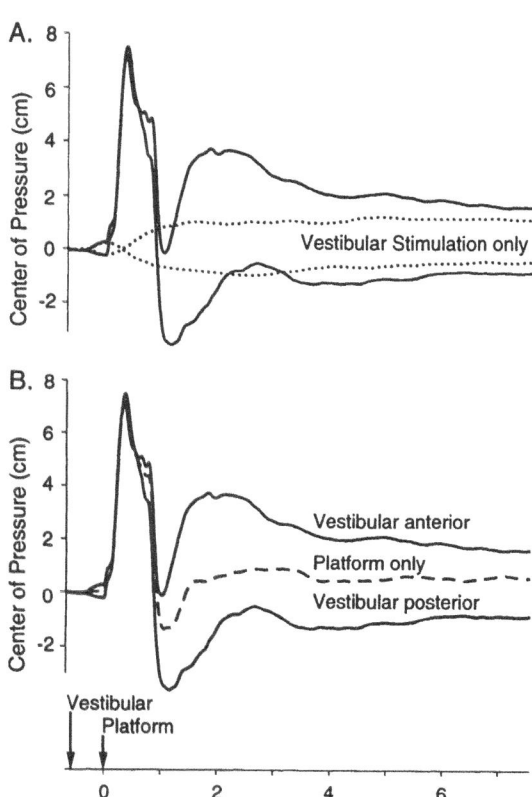

Figure 1. A. Position of the CoP as a function of time when anterior and posterior galvanic vestibular stimulation are combined with a surface translation (solid lines) and for anterior and posterior vestibular stimulation alone during quiet stance (dotted lines). B. Position of the CoP response to platform translation alone (dashed line) compared with responses to translation during anterior and posterior galvanic vestibular stimulation (solid lines). The figure shows average group responses to fast velocity (14 cm/s) platform translations. In this and the following figure, group means are displayed and positive numbers indicate anterior movement.

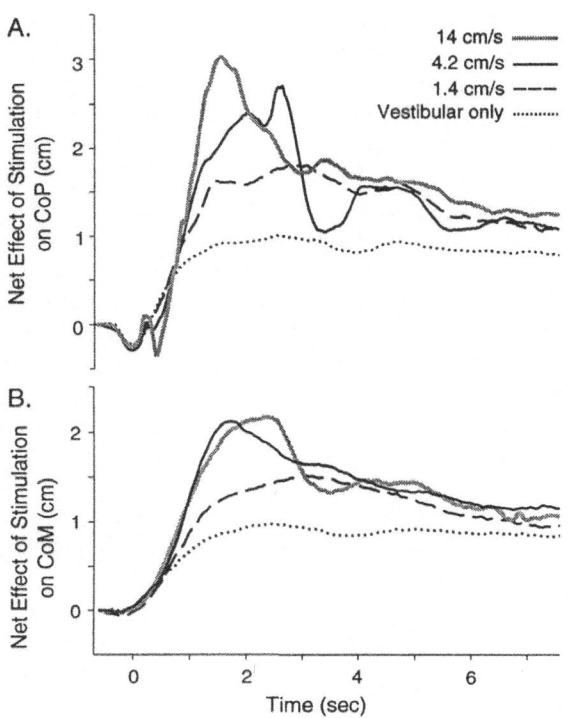

Figure 2. The net effect of vestibular stimulation on the CoP response (A) and CoM response (B) to 14, 4.2, and 1.4 cm/s translations compared with vestibular stimulation only during quiet stance (dotted). The effect of vestibular stimulation depended on translation velocity and occurred after the initial postural response seen in Fig. 1. The net effect of vestibular stimulation is formed by subtracting the response to posterior stimulation from the response to anterior stimulation, which eliminates the effect of platform translation alone.

velocities were statistically significant (all p's<.05). The net CoM change due to vestibular stimulation was also larger for larger velocities (all p's<.05 with one exception, the difference between 14 and 4.2 cm/s, p>.8).

Leg and trunk EMGs were examined to investigate the effects of vestibular stimulation on the earliest postural responses and to determine how the vestibulospinal signal implements a change in postural orientation. Vestibular stimulation alone during quiet stance at currents less than 1 mA produced a very small change in tonic background activity (<10 μV), which is largest in gastrocnemius, less in hamstrings, less in paraspinals and absent in soleus, tibialis, quadriceps and rectus abdominus. Averaged EMG responses to a step of galvanic vestibular stimulation shows a biphasic response with a small, initial phasic response followed by a larger, tonic response in the opposite direction, lasting the duration of the stimulation (8 s). For example, stimulation with the anode at the forward ear resulted in an initial inhibition of tonic gastrocnemius followed by a tonic excitation while the subject leaned forward against gravity.

Vestibular stimulation did not have a significant effect on the initial latency or integral of the first 100 ms burst of gastrocnemius, hamstrings, and paraspinal EMGs activated in response to backward platform translations. After the first burst, however, vestibular stimulation resulted in either an increase (vestibular anterior) or decrease (vestibular posterior) in tonic gastrocnemius activity (Figure 3).

In order to determine whether the lack of an effect of vestibular stimulation on the earliest postural response was due to the relative time of vestibular stimulation prior to the translation, the onset of vestibular stimulation was varied. The largest effects of vestibular stimulation on the postural response occurred when the 8 s galvanic step was initiated 500 ms prior to platform translation. The effects were less significant when stimulation was

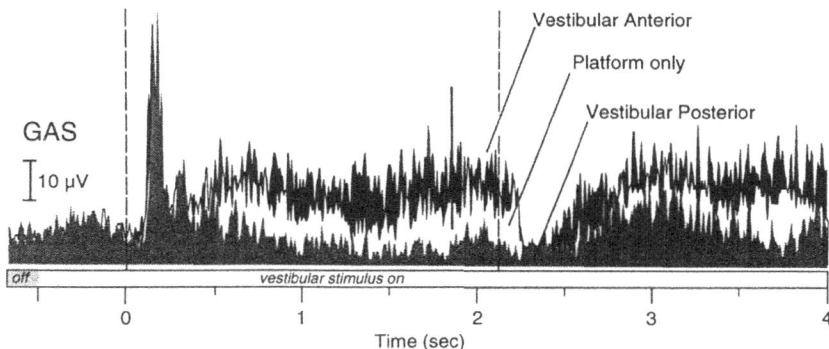

Figure 3. Group averaged gastrocnemius EMG comparing responses to platform translation (4.2 cm/s) during anterior or posterior vestibular stimulation with response to platform translation only. A step of tonic galvanic vestibular stimulation did not affect the earliest EMG burst but did alter later tonic activity. Onset and offset of backward platform translation is indicated with a line.

initiated at 2500 or 1500 ms prior to platform translation and least when the vestibular stimulation was initiated at the same time as the platform translation.

DISCUSSION

These results support a very important role for the vestibular system in maintaining body orientation but not in the formulation or triggering of automatic postural responses to dysequilibrium in this task. A tonic step of galvanic vestibular stimulation prior to and during an equilibrium response during stance did not affect the triggering, temporal organization, or amplitude of the earliest automatic postural responses to dysequilibrium. In contrast, tonic steps of galvanic vestibular stimulation altered the targeted final body position, suggesting a change in an internal vertical orientation reference (Inglis et al., 1995).

The effect of galvanic stimulation on body orientation was largest during the fastest platform translations; thus, the postural control system appears to be more sensitive to a vestibular error signal during rapid postural movements than during quiet stance. The change in final equilibrium position was more than the simple algebraic sum of the effect of platform translation and galvanic stimulation separately. These results are consistent with previous findings showing that the effects of galvanic vestibular stimulation are greater during voluntary movement than during quiet stance (Popov et al., 1986). The results are also consistent with the hypothesis that the postural control system relies more on somatosensory information for low frequencies of movement and on vestibular information for higher frequencies of movement (Peterka and Benolken, 1992; Horak and Shupert, 1994).

When subjects experienced an illusory change in the direction of vertical induced by galvanic stimulation prior to a postural perturbation, they altered the magnitude of the later components of their postural responses in order to achieve a new final body orientation which was tilted away from upright. The peak effect of vestibular stimulation on the CoP at 1.5-2.5 s after the onset of platform translation reflects a modification of the postural response to bring the body into this new orientation. The effort to maintain the new position is reflected in the tonic increase or decrease in EMG activity after the initial postural response. For example, in order to maintain a forward leaning CoM position after the surface translation, subjects increased later plantarflexion torque which was reflected in a forward CoP position and larger, tonic gastrocnemius EMG.

These findings imply that the vestibular system may be critically involved in providing the sensory reafference during postural movements which permits the CNS to determine whether the initial equilibrium response will be effective in producing the desired body orientation (von Holst and Mittelstaedt, 1973). Thus, the galvanic vestibular stimulation may have altered either the natural vestibular afference during postural movements, the targeted vertical orientation reference, or both, resulting in altered postural responses.

The observation that vestibular stimulation was most effective when initiated 500 ms prior to platform translation suggests that the tonic vestibular error signal must be centrally integrated over time to result in a new internal representation of vertical and, thereby, exert an influence on posture. Initiating the vestibular stimulation later, coincident with platform movement, produced much smaller effects. Initiating vestibular stimulation earlier did not result in larger effects, nor did it result in changes in the initial equilibrium response, implying that 500 ms is adequate for this integration to occur and that habituation or adaptation might occur when stimulation is prolonged. The results of this study are most consistent with the hypothesis that the earliest components of the postural response may be shaped by somatosensory inputs, whereas vestibular inputs may require longer integration times and may shape the later components of the postural response.

ACKNOWLEDGMENTS

Our thanks to Dr. J.T. Inglis who initiated the pilot studies upon which this work is based. We also thank Jill Knop for her technical assistance. Drs. Horak and Shupert were supported by NIH grants R01-DC01849 and P60-DC02072.

REFERENCES

Fitzpatrick, R., Burke, D., and Gandevia, S., 1994, Task-dependent reflex responses and movement illusions evoked by galvanic vestibular stimulation in standing humans, *J. Physiol.* 478:363-372.

Hlavacka, F., and Njiokiktjien, C., 1985, Postural responses evoked by sinusoidal galvanic stimulation of the labyrinth, *Acta Otolaryngol. (Stockh.)* 99:107-112.

Horak, F.B., and Shupert, C.L., 1994, Role of the Vestibular System in Postural Control. In: Vestibular Rehabilitation, S. J. Herdman, (ed.), F.A. Davis Company, Philadelphia.

Inglis, J.T., Shupert, C.L., Hlavacka, F., and Horak, F.B., 1995, The effect of galvanic vestibular stimulation on human postural responses during support surface translations, *J. Neurophysiol.* 73:896-901.

Lund, S. and Broberg, C., 1983, Effects of different head position on postural sway in man induced by a reproducible vestibular error signal, *Acta Physiol. Scan.* 117:307-309.

Massion, J., 1992, Movement, posture and equilibrium: Interaction and coordination, *Prog. Neurobiol.* 38:35-56.

Nashner, L.M. and Wolfson, P., 1974, Influence of head position and proprioceptive cues on short latency postural reflexes evoked by galvanic stimulation of the human labyrinth, *Brain Research* 67:255-268.

Peterka, R.J., and Benolken, M.S., 1992, Role of somatosensory and vestibular information in attenuating visually-induced human postural sway. In: Posture and Gait: Control Mechanisms Vol. 1, edited by M. Woollacott and F. Horak. Eugene, OR: University of Oregon Press, pp. 272-275.

Popov, K.E., Smetanin, B.N., Gurfinkel, V.S., Kudinova, M.P., and Shlikov, V.Yu., 1986, Spatial perception and vestibulomotor responses in man, *Neurophysiol.* (translated from Neirofiziologiya) 18:548-554.

Storper, I., and Honrubia, V., 1992, Is human galvanically induced triceps surae electromyogram a vestibulospinal reflex response, *Otolaryngol. Head Neck Surg.* 107:527-536.

von Holst, E., and Mittelstaedt, H., 1973, The reafference principle: interaction between the central nervous system and the periphery. In: The Collected Papers of Erich von Holst, translated by Robert Martin. Coral Gables, FL: University of Miami Press, pp. 139-173.

30

EYE-HEAD COORDINATION BEFORE AND AFTER CANAL PLUGGING IN MONKEY

C. Siebold,[1] L. Ling,[1] J. Phillips,[1] S. Newlands,[1] T. Mergner,[2] and
A. F. Fuchs[1]

[1] Regional Primate Research Center, and Department of Physiology and
    Biophysics and Otolaryngology
University of Washington
Seattle, Washington 98195
[2] Neurological Clinic
University of Freiburg
79104 Freiburg, Germany

INTRODUCTION

To direct the line of sight toward an interesting target in space, a combination of eye and head movement is used under natural conditions. The accurate coordination of such a "head-free" gaze movement requires information about the ongoing head movement. It is not yet understood if this information is derived from a signal originating in the vestibular apparatus. To address this problem, we compared normal gaze shifts in the head-free primate with gaze shifts after the vestibular signal was removed by plugging both horizontal semicircular canals.

METHODS

One rhesus monkey was surgically prepared for behavioral experiments under deep anesthesia and sterile conditions. The magnetic search coil technique was used for eye movement recording. A head-stabilizing post was embedded in dental acrylic and secured to the skull with stainless steel screws. The position of the head post was directly over the atlanto-axial joint. During the experiment, the monkey sat in a primate chair with its head connected to a holder that allowed head movements constrained to the horizontal plane. In front of the monkey was an array of 15 LEDs (in 5° steps from 0 to ±30° and at ±40°) arranged on a horizontal line. The monkey was trained to direct the line of sight to each illuminated LED 'target' in turn, first with the head restrained ('head-fixed') and later with the head free to move ('head-free'). Gaze, or eye position in space, was derived from the demodulation of the induced voltage in the search coil, while a potentiometer signal attached to the head holder gave head position in space ('head'). Eye position in the head ('eye') was derived from on-line subtraction of those two signals. After recording head-free gaze shifts in the

Multisensory Control of Posture, Edited by T. Mergner and F. Hlavačka
Plenum Press, New York, 1995

normal monkey ('pre-plug') over a period of several weeks, both horizontal semicircular canals where plugged in a second surgery ('post-plug'). To verify the effectiveness of the surgery, the gain of the vestibulo-ocular reflex (VOR) was measured in the dark. One day after plugging there was no measurable VOR gain. The animal was kept in complete darkness for one night before the first experiment was performed.

RESULTS

Examples of three gaze shifts of comparable size before, one day after (Day 1), and 50 days after canal plugging are shown in Fig. 1. In the pre-plug condition the gaze movement started with a rapid eye movement followed slightly later by a slower head movement (Fig. 1A). At the time when the eye movement ended, gaze continued in the direction of the target, being carried solely by the head. During the period between 'eye end' and 'gaze end' the eye started to counter-rotate in the orbit to compensate for the head movement; however, the compensation, as measured by the velocity "gain" of eye velocity/head velocity was < -1.0. After gaze reached the target, the head movement still continued to advance. Until the end of the head movement, the velocity gain was approximately -1, so that gaze was perfectly stable in space. This perfect stability was the result of the well-known VOR. Further

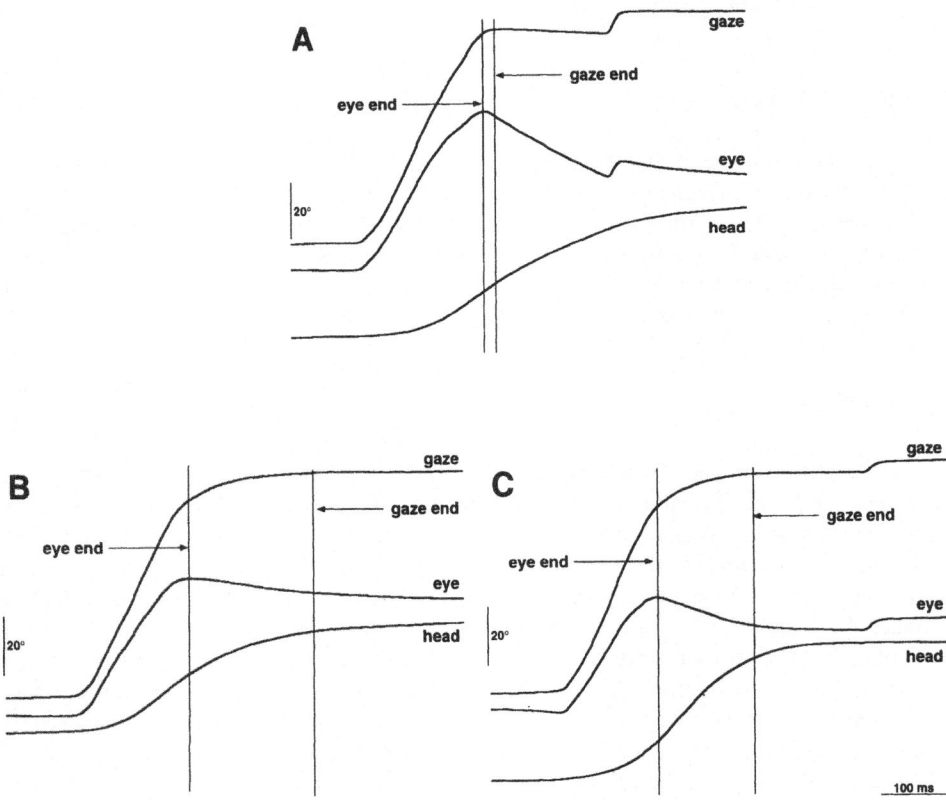

Figure 1. Gaze, eye, and head movement traces of three gaze shifts with comparable amplitude. (A) Pre-plug condition, (B) first day after canal plugging, and (C) 50 days after canal plugging. Vertical lines indicate the end of the eye or gaze movement.

information on the characteristics of normal head-free gaze shifts can be found in Phillips et al. (in press).

One night after canal plugging (Fig. 1B, 'Day 1'), the initial part of the gaze movement was similar to the pre-plug condition. It started with a rapid eye movement, which, however, was much smaller than normal and, therefore, contributed considerably less to overall gaze amplitude. In the interval between 'eye end' and 'gaze end', there was a counterrotation of the eye, which again was far less than normal. The ratio between the eye velocity and head velocity 60 ms after the eye end yielded values around 0.35. However, gaze did reach the target with normal accuracy. Therefore, the time between 'eye end' and 'gaze end' increased considerably compared to the pre-plug condition (to about 230 ms for 60° gaze shifts). The existence of any compensatory eye movement at all was surprising since passive vestibular stimulation in the dark with the head fixed yielded no measurable VOR. The presence of these compensatory eye movements suggests that either centrally generated signals or other sensory inputs are available to allow the eye to compensate partially for the ongoing head movement. In another experiment (not shown), we interrupted the head mechanically during that part of the gaze movement where the eye had already started to counterrotate. Although the head trajectory stopped abruptly, the eye continued to advance without any noticeable disturbance. This finding can be explained only if pre-programmed information about the head movement is available to drive the eye. Similar conclusions have been reached by others (Dichgans et al., 1973, 1974).

Several weeks after canal plugging (Fig. 1C, 'Day 50'), some changes in eye-head coordination emerged. In the initial part of the gaze movement, gaze again was carried mostly by a rapid eye movement. Although this eye movement was similar in size to that of 'Day 1', it reached a higher peak velocity (on average 550 deg/s compared to 400 deg/s for a 60° gaze shift). By the time the rapid eye movement was over, the head again had reached a considerable velocity. At this point, the eye again started to counterrotate in the orbit to compensate partially for the ongoing head movement. The gain of the compensation 60 ms after eye end averaged about 0.5. The result of (i) the higher gain of the eye counterrotation and (ii) the faster head movement in the 'Day 50' condition reduces the time between eye end and gaze end considerably (by an average of about 120 ms for 60° gaze shifts). Since the gain of the VOR measured on the same day in the dark was around 0.2, the gain increase of the horizontal compensatory eye movements with time is thought to reflect a contribution of the vertical canals, which are activated with our animal in the upright test position. The considerable eye counterrotation gain of 0.5, therefore, may reflect the summation of two effects (a) a pre-programmed signal, which was already present at 'Day 1' and (b) a head movement signal originating in the vertical canals. Finally, the head movement is much faster (higher peak velocity) and of shorter duration than on 'Day 1'.

CONCLUSIONS

Taken together, plugging the horizontal semicircular canals has considerable influence on the coordination of eye and head movements during gaze shifts. Eye amplitude is reduced and the time between eye end and gaze end is increased. Also, gaze duration is considerably increased. Over 7 weeks, peak eye and head velocity increase, while the time between 'eye end' and 'gaze end' and gaze duration decrease. Many of the changes in eye-head coordination with time may be attributed to strategic behavior. Such changes include decreasing the eye amplitude to reduce the possibility for overshooting the target and decreasing the time between 'eye end' and 'gaze end' to reduce the time where clear vision is disturbed by uncompensated changes in the head trajectory. However, the emerging counterrotation of the eye clearly reflects a central mechanism. Pre-programmed information is utilized to drive the eye immediately after plugging and over time is augmented by information from the intact vertical vestibular system. A more complete description of the results produced by canal plugging and head braking will be found in a future publication.

REFERENCES

Dichgans, J., Bizzi, E., Morasso, P., and Tagliasco, V., 1973, Mechanisms underlying recovery of eye-head coordination following bilateral labyrinthectomy in monkeys, *Exp. Brain Res.* 18:548-562.
Dichgans, J., Bizzi, E., Morasso, P., and Tagliasco, V., 1974, The role of vestibular and neck afferents during eye-head coordination in the monkey, *Brain Res.* 71:225-232.
Philips, J.O., Ling, L., Fuchs, A.F., Siebold, C., and Plorde, J.J., 1995, Rapid horizontal gaze movement in the monkey, J. Neurophysiol. (in press).

31

SUBJECTIVELY PERCEIVED EGO-MOTION AND ITS RELATION TO CENTRIFUGE INDUCED MOTION SICKNESS

A. H. Wertheim and B. S. Mesland

TNO Human Factors Research Institute
PO Box 23
3769 DE, Soesterberg
The Netherlands

INTRODUCTION

Perceived ego-velocity (PEV) derives from an interactive process between various kinds of information. Obviously, one important source of PEV is optic flow, because a coherent flow pattern across the retinae may generate a sensation of self-motion or "vection" (unless the flow pattern is present only very briefly or moves too fast - see e.g. Berthoz et al., 1975; Dichgans and Brandt, 1978; Berthoz and Droulez, 1982). Henceforth we will call this the "visual component" of PEV.

Non-visual contributions to PEV derive from vestibular afferents (in the case of linear self-motion, from the otoliths), proprioception and possibly even from cognition (see Guedry, 1974 for a review). We will reserve the general term "extraretinal component" for these combined non-visual contributions to PEV. In darkness (i.e. in the absence of visual information) this extra-retinal information is the only component within PEV. It was the purpose of the research presented here to measure the gain of PEV when it is generated purely on the basis of extra-retinal information.

To measure this extra-retinal PEV we used a method derived from a theory proposed by Wertheim (Wertheim, 1994; Wertheim and Bles, 1984). The basic idea underlying this method is, that when a moving observer perceives a stimulus as stationary in space, the visual system 'assumes' that the moving image of the stimulus across the retinae is an artefact of ego-motion. We thus would be able to use the velocity of the image of a stimulus that is seen as stationary, as an operational measure of PEV. However, matters are slightly complicated by the fact that image velocity always depends on two other factors as well: stimulus distance from the perceiver and the visual angle between the stimulus and the perceivers' line of sight. Hence, to decide whether image velocity is indeed only caused by selfmotion (i.e. whether it equals PEV) or not, the perceptual system must first annul the latter two factors - by scaling back retinal image velocity in proportion to distance and visual angle. We assume that the perceptual system can make fairly accurate estimates of visual angle and stimulus distance

Multisensory Control of Posture, Edited by T. Mergner and F. Hlavačka
Plenum Press, New York, 1995

(see e.g., van de Grind et al., 1992) and that it does indeed rescale image velocity more or less properly. Mathematically, image velocity, when thus corrected for visual angle and distance, equals the difference between stimulus velocity in space and the velocity of (the eyes of) the observer in space. Henceforth to denote this difference we will refer to the term "corrected retinal image velocity". It thus follows, that when a stimulus is perceived as stationary in space, the perceptual system of the observer assumes that corrected retinal image velocity stems only from, i.e. is equal to, ego velocity. Consequently, the corrected retinal image velocity of a stimulus which is seen as stationary in space, can be used as an operational measure of PEV.

EXPERIMENT I

Method

In the experiment reported here, we measured PEV of a subject moving sinusoidally on a linear track sled in a totally darkened environment. On each side next to the sled's track we positioned a video monitor, such that the monitors faced each other (their screens parallel to the sled's tracks, the sled passing in between them - see Fig 1). On these monitors we presented a moving stimulus pattern (checkerboard pattern) which could be made visible for 200 ms each time the subject moved on the sled past the monitor. The subjects did not look sideways at the monitors, but kept their eyes focused on a small fixation point straight ahead. Thus the checkerboard patterns were always perceived peripherally. The reason for presenting the stimulus patterns so briefly was that care should be taken that the stimulus patterns themselves could not affect PEV, i.e. we should prevent them from causing a visual component in PEV. We assumed that 200 ms was brief enough, as such brief instances of optic flow are unlikely to have a vection inducing potential.

The two patterns always moved synchronously in the same or opposite direction as the sled. Their velocity on the monitors was varied systematically until the subject reported that they were perceived as stationary in space. At that point (the Point of Subjective Stationarity, PSS) corrected retinal image velocity was calculated as the difference between stimulus and sled velocity. This reflects PEV.

Sled velocity, which was the input to the perceptual system that generates PEV, was sinusoidal. Hence PEV, which is the output of this system, was likely to be sinusoidal as

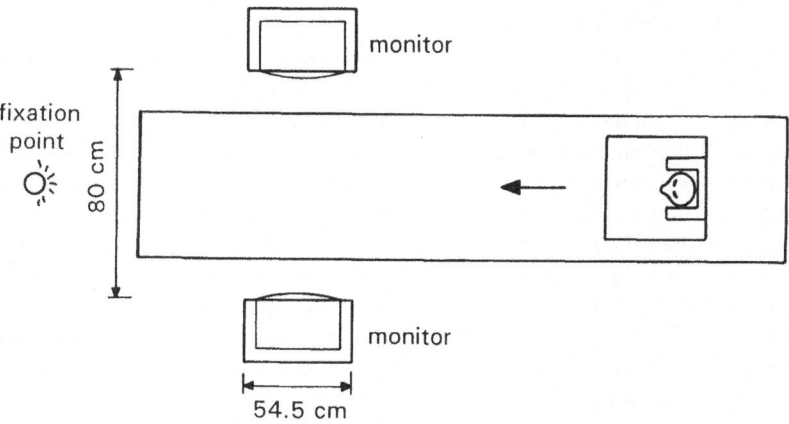

Figure 1. Schematic representation of the experimental set-up.

well. To test this hypothesis, and to investigate the phase and gain relations between PEV and actual ego-velocity (i.e. sled velocity), we measured PEV (the retinal signal at the PSS) at six different phases of the sled-velocity sinus, that is, we replicated the experiment with the monitors positioned at three different positions along the sled's track and with the sled moving either in forward or in backward direction.

Apparatus

Subjects were placed on the seat of a sled, which could be moved backward and forward along the linear track with a maximal peak-to-peak displacement of 3.20 m (for a technical description of the sled, which was originally designed and used by ESA for vestibular research in space, see Soons et al., 1981). The subject's head was kept fixed to the sled with a vacuum cushion draped around the head and neck and attached to the seat. The subjects were instructed to keep their eyes focused on a small fixation LED that was positioned straight in front of the eyes, at 2 m in front of the foremost end point of the sled's track. Thus, when moving forward or backward on the sled, no eye movements relative to the head were made (apart from some negligibly small vergence eye movements). EOG was measured continuously, registered on line by computer, and monitored on line with an oscilloscope to verify whether the gaze remained stable indeed.

The sled moved sinusoidally along the linear track with a frequency of 0.145 Hz and a peak-peak displacement amplitude of 2.40 m. Thus peak sled velocity was 109.5 cm/sec.

Two video monitors (40 x 40 cm) were positioned at eye level, facing each other from opposite sides of the sled's track (see Fig. 1). As mentioned above, the stimulus pattern on the screens was always perceived peripherally. It consisted of a moving checkerboard pattern of 5 * 5 cm black and white squares. Its presentation was triggered by a photoreceptor attached to the moving sled itself (which passed over an adjustable infrared light source attached to the sled's track). The velocity of the stimulus pattern across the video monitor was variable and controlled by computer.

Procedure

The experiment consisted of 6 conditions, each one consisting of a PSS measurement at a different phase of the sled-velocity sinus. The conditions were chosen as follows:

Condition A: The stimulus was presented when the sled had accelerated to a forward velocity of 54.8 cm/sec (phase: 30 deg of the sled-velocity sine).

Condition B: The stimulus was presented when the sled had reached its maximum forward velocity (phase: 90 deg).

Condition C: The stimulus was presented when the sled had decelerated again to a forward velocity of 54.8 cm/sec (phase: 150 deg).

Condition D: The stimulus was presented when the sled had accelerated backwards to reach a backward velocity of 54.8 cm/sec (phase: 210 deg).

Condition E: The stimulus was presented when the sled had reached its maximum backward velocity (phase: 270 deg).

Condition F: The stimulus was presented when the sled had decelerated again in backward direction to reach a backward velocity of 54.8 cm/sec (phase: 330 deg).

Immediately after presentation of the stimulus pattern, subjects reported whether or not they had peripherally perceived motion of the stimulus pattern across the screen and, if they had, in what direction (forward or backward). They reported with the help of a small hand held box which contained three response buttons: one for perceived forward motion, one for perceived backward motion and one for no motion perceived. They were instructed to use the 'no motion' button also when in doubt about the direction of movement of the stimulus. Care was taken to remind subjects that we wanted to know if they saw the patterns move in space, that is, on the screens (i.e. not relative to themselves).

For each PSS measurement two thresholds were measured: one for perceiving motion of the checkerboard pattern on the screens in the same direction as the sled (with-threshold), and one for perceiving the pattern as moving in the direction opposite to sled (against-threshold). In each of the six conditions, stimulus velocity at the PSS was calculated as the midpoint between these two thresholds. The corrected retinal image velocity at the PSS (which equals PEV) was then calculated as the difference between stimulus velocity at the PSS and sled-velocity at the moment of stimulus presentation.

The two opposite thresholds were measured separately, using software (Zeppenfeld, 1991) that consisted of an adapted version of what is known as the PEST procedure (see Taylor and Creelman, 1967) with stimulus velocities that could vary within a range of 70 cm/sec (with-motion) and -70 cm/sec (against-motion).

Within each condition the PSS measurement was replicated three times, which implies that a total of six threshold measurements were taken: three with- and three against-threshold measurements. The six thresholds were obtained in three blocks, each containing one with- and one against threshold (in random order). Within a condition the PSS was defined as mean stimulus velocity across these three blocks. The six conditions were presented to each subject in a different (random) order. All blocks were separated by resting intervals.

In this study 4 male and 2 female paid volunteer students, aged between 22 and 24 years, from Utrecht University, participated as subjects.

Results

With one subject some of the PSS measurements of conditions B and E could not be taken because of problems with the perceived direction of stimulus motion. During threshold measurements in these conditions the stimulus velocities were either perceived all as with-motion or all as against-motion.

For the remaining 5 subjects the absolute values of PEV in conditions A,C,D and F differed significantly (ANOVA: $p < 0.01$) from those in conditions B and E, but not from each other. Conditions B and E did not differ significantly from each other either. A paired samples T-test revealed that PEV in conditions B and E differed significantly from input velocity ($p < 0.01$), whereas in conditions A,C,D and F it did not.

Figure 2 shows the best fitting sinus drawn through the 6 group mean values of PEV obtained at the various phases of the sled-velocity sine. The mathematical equation which describes this best sine shows that at this particular sled frequency of 0.145 Hz., PEV gain was approximately 0.8 and PEV has a phase lead of approximately 4 degrees relative to the sled-velocity sine. Also, PEV was slightly larger during forward motion than during backward motion (1.6 cm/sec.). The amplitude of the PEV sine was 83.3 cm/sec.

Discussion

The findings are interesting for various reasons. First, as expected, the six mean PEV values closely fit a sinusoidal curve. The fact that the total PEV sinus could be measured with the help of this method, suggests that the method is suitable indeed to obtain the transfer-function of the perceptual system that generates PEV.

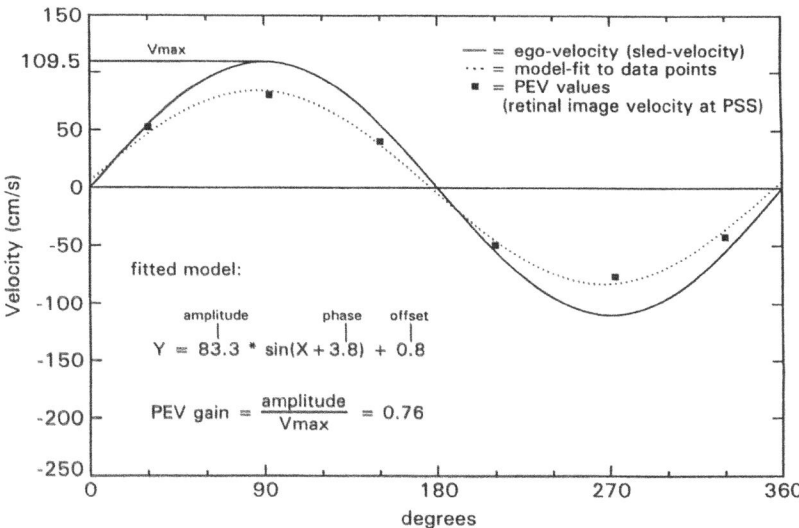

Figure 2. Physical ego-velocity (drawn line) and model fit (dotted line) through the experimentally derived mean group PEV values (solid line). The relation of PEV to physical ego-velocity is given by the inset formula.

Second, PEV seems to be slightly higher during forward sled motion as compared to backward sled motion (see the slight offset value of 0.8 in Fig 2). This was also observed in some earlier studies at our laboratory (Wertheim, 1993; Wertheim and Mesland, 1993).

Third, we found a phase lead of approximately 4 deg. According to a mathematical model of the otoliths (Grant and Best, 1987, as mentioned in Marcus, 1992) the otolith response should indeed have a small phase lead relative to ego-velocity. Given this particular sled frequency, that model predicts a phase lead of approximately 6 deg. The observed phase shift is of the same order of magnitude as this prediction, which hints at the possibility that we might actually have measured the otolith-response to linear accelerations here. However, until we can control for possible proprioceptive and cognitive contributions to PEV (e.g. from the vibrations of the sled, auditory cues, etc.), this claim cannot be made with certainty.

Fourth, the fact that we observed only such a small phase shift, suggests that in future experiments in which we use the same sled frequency, we can establish the gain of PEV quite reliably by measuring PEV only in one condition: at maximum sled velocity.

EXPERIMENT II

Introduction

Centrifugation of human subjects for a period of at least 1 hour, with a constant Gx load of 3g may result in postural instability and severe motion sickness at the end of the centrifuge run, lasting for several hours (Bles et al., 1989; Bles and de Graaf, 1993). These effects may be a result of otolith adaptation resulting in a distorted visual-vestibular interaction. Why some subjects are susceptible to Sickness Induced by long duration Centrifugation (SIC) while others are not, is not clear. Since both in the centrifuge and on the sled the otoliths are stimulated in the Gx direction, there may be a relationship between extra-retinal PEV as measured on the sled and susceptibility to SIC. To explore this possibility the following investigation was carried out.

Apparatus and Procedure

This study was carried out within the context of a much wider research enterprise in which vestibular effects of constant long duration centrifugation were evaluated. Our PEV measurements formed part of this evaluation. They were always taken before and in some cases also after centrifugation.

In the centrifuge, the subjects were placed in supine position inside its free swinging gondola (the centrifuge is located at the Netherlands Aerospace Medical Centre in Soester-berg). Centrifuge runs lasted for one hour and the Gx-load was 3g.

The amount of sickness experienced by the subject afterwards was quantified with a standard motion sickness scale, known as the Misery Scale (MISC), which was filled out by the subjects themselves. It consists of a simple rating scale varying from 0 (no problems) to 10 (vomiting). The MISC was filled out at several points in time after centrifugation.

Extra-retinal PEV was measured following the exact method and sled frequency as described in experiment I, with three exceptions: first, given the very small phase shift observed in the first experiment, we decided in the present experiment to measure PEV only at maximum forward sled velocity. Second, we did not always use a maximum sled velocity of 109.5 cm/s. In a number of cases PEV was measured with a maximum forward velocity of 40 cm/s. PEV gain was calculated as PEV (defined, as in the earlier experiment, as the difference between stimulus velocity at PSS and sled velocity) divided by sled velocity. Third, in this experiment we used a different stimulus pattern, which consisted of a vertical grating of four alternating dark and luminous stripes with blurred edges.

A total of 18 subjects (4 female, 14 male) participated in this study. Several of these subjects participated more than once as subjects both in the centrifuge and on the sled.

Results

In replication of an earlier study (Wertheim 1993) in those cases where we obtained PEV measurements both before and after centrifugation the data showed no systematic effect of centrifugation (i.e. no evidence of otolith adaptation). Therefore, the measurements obtained after centrifugation will be disregarded, and from here on, when mentioning PEV gain values we refer either to those measured immediately before centrifugation, or to those measured independently (days or even weeks before or after the centrifugation experiment).

MISC scores were obtained at several points in time after centrifugation. However, they all correlated highly. Therefore we will here report only on the MISC scores obtained one hour after centrifugation.

In terms of the MISC it appeared that subjects were either very sick or scored (close to) zero. There were almost no intermediate scores. Therefore the subjects could easily be separated into two groups: one group of 5 subjects with MISC scores higher than 5 (SIC) all of whom had vomited at one point in time, and another group with 12 subjects with MISC scores lower than 5 (NO SIC), none of whom had vomited.

Both with respect to their MISC scores and with respect to their PEV scores, those subjects who participated more than once, showed to be highly consistent, the differences between repeated measurements being quite small. Hence, to facilitate statistical treatment, for each of these subjects we used one mean PEV and one mean MISC score per subject.

ANOVA showed that the NO SIC group had a significantly ($p < .01$) higher PEV gain (mean: .81, s.d: .28) than the SIC group (mean: .44, s.d.: .17) (see Fig. 3). Although this is a significant difference at the group level, there was one exceptional NO SIC subject, whose

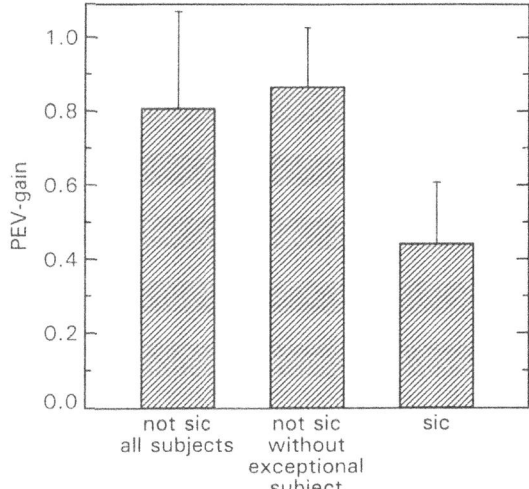

Figure 3. Group mean PEV gain for the SIC and the NO SIC subjects (with and without the one exceptional subject). Line segments represent plus or minus one SD.

MISC scores were low (0 and 1), but whose PEV gain was also very low (close to zero). Without this subject the groups differed even more significantly (p < .001) with respect to their PEV values, and the individual mean MISC and PEV scores correlated highly (r = -.79, see fig. 4).

Discussion

With the exception of one case, there appeared to be a strong relationship between the amount of sickness experienced after long duration centrifugation with a Gx load of 3g and individual extra-retinal PEV. This finding is a bit counter-intuitive, since we would expect otolith adaptation effects to be stronger in people who have a more sensitive

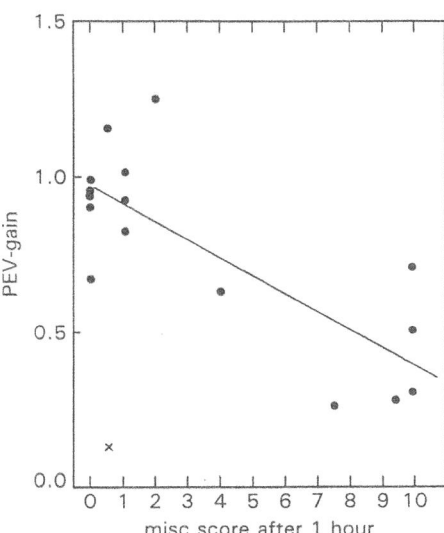

Figure 4. PEV gain as a function of MISC score. Cross represent the data from the one exceptional subject. Mean MISC and PEV gain values for each subject.

extra-retinal (otolith) system (i.e. whose extra-retinal PEV gain is high). In stead, the relationship seems to be exactly the other way around.

One possible hypothesis could be that people with a low PEV gain are, in normal circumstances (i.e. in a normally illuminated environment), strongly dependent on visual information. In other words, their PEV may normally contain quite a large visual component. If so, our results seem to suggest that visual information plays an important role in the propagation of SIC. After centrifugation most subjects report that the visual world appears perceptually to "lag" behind head movements. Could it be that subjects who highly depend on visual information suffer more from such perceptual distortions than others, and for that reason are more susceptible to SIC?

Given the fact that we did not observe effects of centrifugation on PEV gain, this seems to imply in more general terms that long duration centrifugation may affect visual-otolith interactions rather than only otolith responsiveness (see Wertheim, 1993 for a similar hypothesis).

It should be mentioned here, that recently we were in the position to test two more subjects prior to centrifugation. The results of these two subjects were similar to those of the one exceptional subject mentioned above. We have been hesitant to include these subjects in the present data analysis for the following reason. In these experiments we reduced the brightness of the grating to such an extent that the rims of the screens were not any more discernible. The reasoning for doing this was to ensure cues from relative motion between the stripes and the rims of the monitor screens were completely absent. In a separate study, performed especially to test the effect of the absence of such cues, PEV gain appeared to be dramatically reduced in 5 out of 7 subjects. When a pattern of only one stripe was used (which never came close to the borders of the monitor screens because of the brief presentation time), PEV gain was even reduced to zero for all subjects. Although we do not yet fully understand the reasons for this phenomenon, one hypothesis might be that PEV gain was actually still the same (i.e. much higher), but by reducing all visual cues from perception (apart from the stimulus stripes themselves) we may have reduced the ability of the brain to estimate stimulus distance. In such a case, the perceptual system may use a default value of "infinity". If so, corrected retinal image velocity becomes zero. In fact, this is what might have happened to the exceptional subject mentioned earlier. At the present stage, however, these ideas are still rather speculative, and their main value is that they provide a further challenge for future research.

SUMMARY

During linear acceleration of human subjects (otolith stimulation) the subjects' Perceived Ego Velocity (PEV) was investigated with a new psychophysical method, proposed earlier by Wertheim (1990, 1994). The method consists of measuring the perceptual threshold for detecting object motion (in the dark) during ego-motion. It was found that during 0.15 Hz sinusoidal ego-velocity, PEV was also sinusoidal, but had a gain of .8 and a phase-lead of 4 deg. The results from these experiments appeared to be related to individual susceptibility to motion sickness, as induced by long duration centrifugation, SIC (Bles et al., 1989).

REFERENCES

Berthoz, A., and Droulez, J., 1982, Linear self motion perception. In: Tutorials on Motion Perception. A.H. Wertheim, W.A. Wagenaar and H.W. Leibowitz (Ed's). Plenum, New York.

Berthoz, A., Pavard, B., and Young, L.R., 1975, Perception of linear horizontal self motion induced by peripheral vision (linear vection). Basic characteristics and visual vestibular interaction, *Experimental Brain Research* 23:471-489.

Bles, W., Bos, J.E., Furrer, R., Graaf, B. de, Hosman, R.J.A.W., Kortschot, H.W., Krol, J.R., Kuipers, A., Marcus, J.T., Messerschmid, E., Ockels, W.J., Oosterveld, W.J., Smit, J., Wertheim, A.H., and Wientjes, C.J.E., 1989, Space Adaptation Syndrome induced by a long duration +3Gx centrifuge run. Institute for Perception Technical report, IZF-TNO-1989-25, Soesterberg, The Netherlands.

Bles, W., and Graaf, B. de., 1993, Postural consequences of long duration centrifugation, *Journal of Vestibular Research* 3:87-95.

Dichgans, J., and Brandt, T., 1978, Visual-vestibular interaction: Effects on self-motion perception and postural control. In R. Held, H.W. Leibowitz and H.L. Teuber (Ed's), Handbook of Sensory Physiology Vol. VIII: Perception, Berlin: Springer Verlag.

Grant, W., and Best, W., 1987, Otolith-organ mechanics: Lumped parameter model and dynamic response, *Aviation Space and Environmental Medicine* 58:970-976.

Grind van de, W.A., Koenderink, J.J., and Doorn van, A.J., 1992, Viewing-distance invariance of movement detection, *Experimental Brain Research* 91:135-150.

Marcus, J.T., 1992, Vestibulo-ocular responses in man to gravito-inertial forces, Ph.D. thesis. TNO Institute for Perception, Soesterberg, The Netherlands.

Soons, A.F.L., Burden, D.F., Garvin, M.J., and Wyn-Roberts, D., 1981, The Space Sled - A European facility for life-science experiments in Spacelab, *ESA Journal* 5:99-108.

Taylor, M.M., and Creelman, C.D., 1967, PEST: Efficient estimates on probability functions, *Journal of the Acoustical Society of America* 4(4):782-787.

Wertheim, A.H., 1990, Visual, Vestibular and Oculomotor interactions in the perception of object-motion during egomotion. In R. Warren and A.H. Wertheim (Ed's), Perception and control of self-motion (pp. 171-216) Hillsdale, NJ: Lawrence Erlbaum.

Wertheim, A.H., 1993, Pilot studies on object motion perception during linear self-motion after long duration centrifugation of human subjects. Report IZF 1993 B-3, TNO Institute for Perrception Soesterberg, The Netherlands.

Wertheim, A.H., 1994, Motion perception during self-motion: The direct versus Inferential controversy revisited, *Behavioral and Brain Sciences* :293-355.

Wertheim, A.H., and Bles, W., 1984, A reevaluation of cancellation theory: Visual, vestibular and oculomotor contributions to perceived object motion. Report IZF 1984-8, TNO Institute for Perception, Soesterberg, The Netherlands.

Wertheim, A.H., and Mesland, B.S., 1993, Movement Perception during Sinusoidal Linear Egomotion. Report IZF 1993 A-3, TNO Institute for Perception, Soesterberg, The Netherlands.

Zeppenfeld, P., 1991, Determination of thresholds for the detection of differences between visual and vestibular stimulation. Thesis, Technical University of Delft, The Netherlands.

32

ANGULAR VELOCITY ESTIMATION UNDER VARYING LINEAR ACCELERATION

M. -L. Mittelstaedt

Max-Planck-Institut für Verhaltensphysiologie
Seewiesen
Germany

INTRODUCTION

This study investigates the effect of angular and translatory acceleration on the estimation of angular displacement during and after constant passive rotation. Subjects were tested in standing or lying attitude. Because these experiments were part of a study on human navigation (Mittelstaedt and Glasauer, 1991) experimental conditions remained largely within the range of everyday movements.

A. CONSTANT ROTATION IN A VERTICAL ATTITUDE

Method

Subjects (Ss) were positioned on a rotatable disk, at the center as well as at a radial distance of r=015, 0.8 or 1.6 m. Outside the center they stood with their median plane oriented tangentially, and gripped a radial bar with both hands. They were rotated forwards or backwards respectively. At the center they stood free. Ss were blindfolded and earphoned with pink noise. In Exp.A subjects were accelerated with 1.74 radsec^{-2} (100 degsec^{-2}) to a constant rotation of variable angular velocity (ω) of 0.35, 0.52, 0.70 and 0.87 radsec^{-1} (20-50 degsec^{-1}= 3.33-8.33 rpm). The acceleration phase was always smaller than 0.8 seconds.

The centripetal acceleration ranged from zero to 0.124 G. The latter causes a (roll) tilt of the resulting vector of 7.08 deg.

Ss were payed or employed volunteers of both sexes aged 18 to 53. Their task was to indicate successively whenever they felt turned through 180 degrees. There was no further instruction on how to manage this task, because their reports showed that all Ss relied on the perception of their angular displacement relative to the imagined location of structures within the well known room.

The problems of this method have often been discussed (Benson et al., 1966; Guedry, 1974,) and must be kept in mind. But it is the difference in performance under different experimental conditions that matters here, and this can be assessed however the Ss managed to cope with their task if it is verified that the S used the same strategy in all tasks.

Results

At first the indications were fairly veridical but lagged progressively, as though subjective velocity declined exponentially to zero. For plots of ordinal number (n) of indication over time (t) of indication a curve fit was made with

$$n = A_0(1 - e^{-t/\tau})$$

where A_0 is the asymptote of the fit and τ the time constant. This exponential fit was fairly good in all tests; an example is shown in Fig. 1. The value of the time constant τ differed individually (e.g. at r=0 from τ =22 to 45 sec), but was fairly constant in the same S in a given test paradigm.

At a given angular velocity the idiosyncratic time constant τ increases with radial distance (r) (Fig 1). Except at r=0, τ depends also on disk velocity. Under all experimental conditions τ increases with $r\omega^2$, that is, it depends nearly linearly on centripetal acceleration (a) in the range from zero to 1.3 msec^{-2} (Fig. 2).

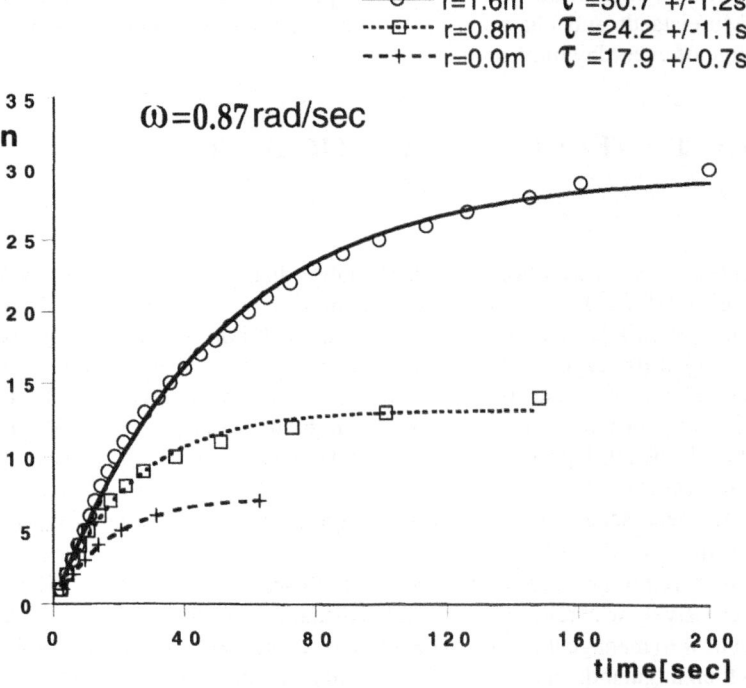

Figure 1. Estimation of subjective angular orientation by a subject S who was rotated, after an acceleration of 100 degsec^{-2}, to a constant velocity of 0.87 radsec^{-1} at three different radii. The S indicated successively whenever she/he felt rotated through 180 deg. The ordinal number (n) of such indications is plotted against time (t) of indication. Points are fitted with $n = A_0(1 - e^{-t/\tau})$. Note that the time constant τ, the number of indications and the time of the last indication increase with the radius.

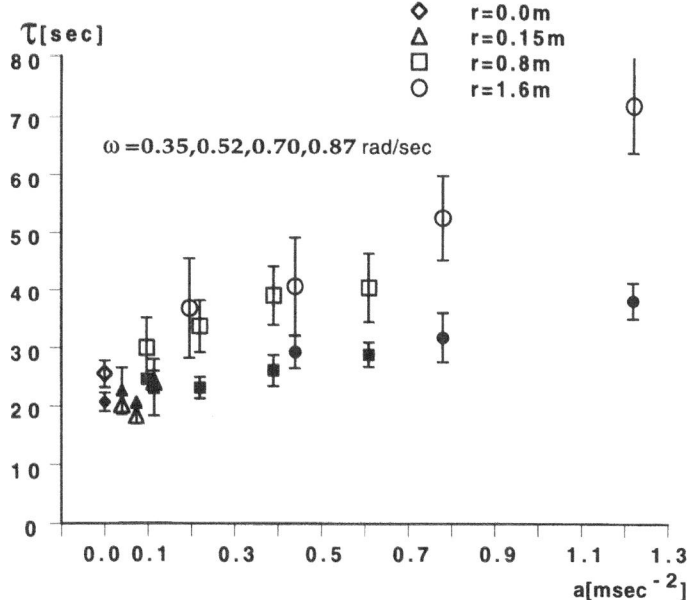

Figure 2. For 2 subjects (S1: open symbols, 113 runs, S2: filled symbols, 53 runs) the time constant τ (in sec), evaluated as in Fig. 1, is plotted against centripetal acceleration (a, with a = rω^2), at all experimental radii (r) and angular velocities (ω). Brackets: Standard error of the mean. Note that τ is nearly linearly dependent on centripetal acceleration (rω^2). This holds for all 10 Ss.

In some Ss the magnitude of τ during forward rotation was larger than τ at backward rotation (possibly due to cognitive problems in angular estimation). The dependency on centripetal acceleration, however, was the same.

B. AFTER-ROTATION IN A VERTICAL ATTITUDE

Method

After constant rotation of at least 2 minutes backwards, the Ss were decelerated with 50 or 100 degsec^{-2} to a full stop within maximally 1.2 sec. Consequently they felt rotated forward ("forward after-rotation"), and were thus able to indicate 180-deg-turns as above.

Results

In all Ss, the time constant τ of the forward after-rotation was somewhat smaller (τ between 5 and 22 sec) than τ at r=0 on the rotating disk. After low disk velocities it could not even be determined in some Ss, because the number of indications was so small. This occurred predominantly at r=1.6 m. However, whenever τ could be determined it was independent of radius and angular velocity of the preceding rotation (p>0.5). This independence is to be expected if it is in fact the additional orthogonal force that causes the prolonged time constant in the former paradigm.

C. CONSTANT ROTATION IN A HORIZONTAL ATTITUDE

C1. Varied Position of the Rotation Axis

Method. Ss were positioned with the spinal (Z-) axis radially, lying on their side, legs extended, face in turning direction. The binaural axis (BA) was placed in 3 positions: 1. BA at r = 1.5 m, 2. BA collinear with the rotation axis (r=0) and 3. BA at r=0.4-0.5 m. Angular velocity was always ω = 0.78 rad sec⁻¹. Ss had to indicate imagined 180 deg-turns as above.

Results. Other than in the above experiments, the time constant τ at a given angular velocity is lowest at position 3 (r=0.4 - 0.5m) rather than at r=0 (Fig. 3) and seems to increase with the unsigned (possibly squared) distance of the RA from r=0.45, that is, from a point near the last ribs. This raises the question whether the effect of centripetal acceleration on angular velocity estimation in this paradigm may be (also) mediated by somatic graviceptors in the trunk: That the centroid of the mass(es) governing these graviceptors are located near the last ribs is indicated by experiments with neuromectomized, paraplegic and nephrecto-mized Ss (Mittelstaedt, 1992).

C2. Legs Flexed and Extended

This special paradigm was chosen, because the above mentioned experiments on somatic graviceptors (Mittelstaedt, 1992) have shown that there is an amplifying effect of leg extension on bias and gain of somatic graviceptors.

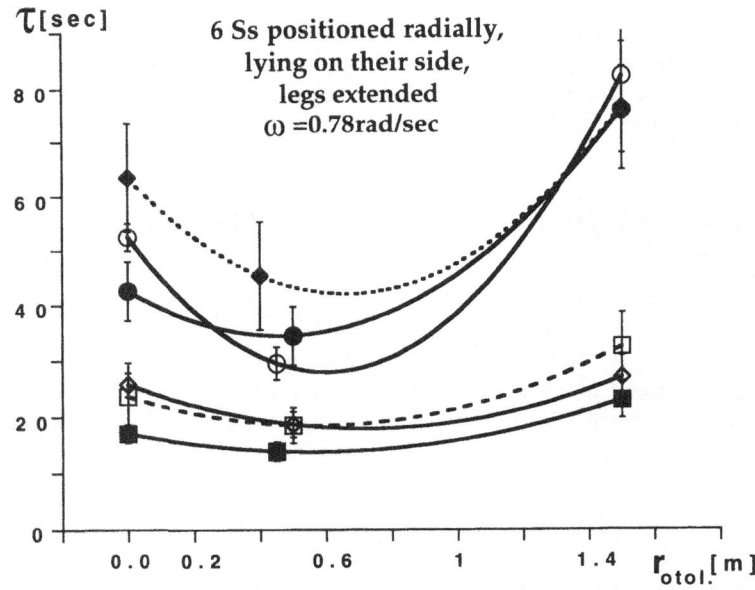

Figure 3. Subjects (6 Ss) were positioned with the spinal (Z-) axis radially, lying on their side, legs extended, face in turning direction. The binaural axis (BA) was placed at 3 different radii(r$_{otol}$): 1. BA at r$_{otol}$ = 1.5 m, 2. BA collinear with the rotation axis (r$_{otol}$ = 0) and 3. BA at r$_{otol}$ = 0.4-0.5 m. Angular velocity was always ω = 0.78 rad/sec. Ss had to indicate imagined 180 deg-turns. The time constant τ (in sec), evaluated as in Fig. 1, is plotted against r$_{otol}$. Points were fitted with a second-order-polynom. Brackets: Standard error of the mean. Note that the time constant τ is lowest at position 3 (r$_{otol}$ = 0.4 - 0.5 m), that is, at the location of the centroid of the mass(es) governing the somatic graviceptors rather than at r$_{otol}$ = 0, the location of the otoliths.

Method. The BA was placed at r = 0 m, such that only the somatic graviceptors are exposed to centripetal acceleration. The Ss were lying with legs flexed or extended, respectively. Angular velocity was always ω = 0.78 rad/sec. Ss had to indicate imagined 180 deg-turns as above.

Result. In all Ss the time constant τ, ranging from 17 to 64 sec, was larger (20 to 40%) with legs extended than with legs flexed.

This change is identical in sign with and comparable in amount to the effect of leg extension on the subjective horizontal position in the above mentioned experiments, and therefore supports the hypothesis that somatic graviceptors are involved.

DISCUSSION

The size of the time constant during constant rotation, after a given constant rotatory acceleration, depends on the centrifugal force caused by centripetal acceleration (s. Exp. A). The conclusion that this force in turn causes the increase of τ is corroborated by the results of Exp. B, where, without additional centrifugal force, the time constant is independent of the angular velocity and the radius of the preceding rotation.

It must be considered, however, that Ss are exposed to yet another additional translatory force during the experiments, namely, tangential acceleration. In Exp. A or Exp. B it depends only on the radius, because the applied steplike torque varies merely in duration, but not in magnitude. Yet τ in Exp. B turned out to be independent of the radius. Consequently, tangential acceleration cannot have caused the additional effect.

It may be concluded, then, that it is the additional translatory acceleration in the direction of the y-axis that causes the prolonged sensation of turning with increasing radius. Because the stimulus to the semicircular canals was identical in the respective conditions of Exp. A and B, the difference must be due to an effect of sense organs measuring translatory acceleration, namely, otoliths and/or somatic graviceptors, on the central nervous processing of the canal afference.

Experiments C1 and C2 with Ss in a horizontal attitude have led to the conclusion that somatic graviceptors are involved. A candidate may be the kidney (Mittelstaedt, 1992, and this volume p. 147). The dependence of τ on centripetal acceleration, present in normal Ss, was in fact missing in the first S of a planned series with bilaterally nephrectomized Ss. Taken together all results point to a prevalent role of somatic graviceptors (at least in horizontal attitude) in this context.

Obviously, in order to gain angular information from the canals, temporal integration of the canal afference is necessary. The present result thus means that this integration becomes the more veridical the larger the product of radius and angular acceleration. The dependence may make sense if indeed this product stays at a high and more or less constant level whenever navigation is called for in real life.

The result is another piece of evidence for the interaction of the effect of linear and rotatory acceleration (Benson, 1974; Bronstein and Gresty, 1991; Guedry, 1992). In a paradigm re-analyzed and discussed by Guedry (Guedry, 1992), which was similar to Exp. A but kept radius and acceleration profile constant, Ss felt that they were rotated at a large radius during the phase of rotatory acceleration, whereas at a small (or even zero) radius during the deceleration phase after a period of constant rotation. No quantitative data are available, however, and no reports are made during the period of constant rotation, whereas in the present study no data could be gained from the very short periods of acceleration. Therefore it remains an open question whether the results of the two studies are compatible.

There are extant theories (for reviews see Merfeld et al, 1993) of the otolith-canal-interaction. The results presented here (and also those of Guedry, 1992) are not predicted by the extant theories, including those developed in our own institute (Mittelstaedt et al, 1989, Glasauer, 1992). The reason may be that the output in our experiments is processed on a higher psychophysical level than, for instance, eye movements, or that an influence of somatic graviceptors has so far not been taken into account.

ACKNOWLEDGMENTS

Thanks are due to R. Ströbele and K. Fischer for the construction of the rotating disk and to W. Jensen for the development of the instruments and the software to control it. Thanks also due to the subjects for their cooperation.

REFERENCES

Benson, A.J., Goorney, A.B., and Reason, J.T., 1966, The effect of instructions upon post-rotational sensations and nystagmus, *Acta oto-laryng* 62:442-452.

Benson, A.J., 1974, Modification of the response to angular accelerations by linear accelerations, In: Handbook of Sensory Physiology. Kornhuber H.H. ed. 6, Springer Verlag, New York, Heidelberg, Berlin, pp. 128-320.

Bronstein, A.M., and Gresty M.A., 1991, Compensatory eye movements in the presence of conflicting canal and otolith signals, *Exp. Brain Res.* 85:697-700.

Glasauer, S., 1992, Interaction of semicircular canals and otoliths in the processing structure of the subjective zenith, *Annals of the New York Academy of Sciences* 656:847-849.

Guedry, F.E., 1974, Psychophysics of vestibular sensation, In: Kornhuber, H.H, (edt.) Handbook of Sensory Physiology 6, Springer Verlag, New York, Heidelberg, Berlin pp. 1-154.

Guedry, F.E., 1992, Perception of motion and position relative to the earth, *Annals of the New York Academy of Sciences* 656:315-328.

Guedry, F.E., Rupert, A.H., McGrath, B.J., and Oman, C.M., 1992, The dynamics of spatial orientation during complex and changing linear and angular acceleration, *Journal of Vestibular Research* 2:259-283.

Merfeld, D. M., Young, L. R., Oman, C. M., and Shelhamer, M. J. 1993, A multidimensional model of the effect of gravity on the spatial orientation of the monkey, *Journal of Vestibular Research* 3:141-161.

Mittelstaedt, H., 1992, Somatic versus vestibular gravity reception in Man, *Annals of the New York Academy of Sciences*, 656:124-139.

Mittelstaedt, H., 1995, Determinants of the visual and the postural vertical, this volume p 147.

Mittelstaedt, H, Glasauer, S., Gralla, G., and Mittelstaedt, M.-L., 1989, How to explain a constant subjective vertical at constant high speed rotation about an earth horizontal axis, *Acta Otolaryngol.* 468:295-299.

Mittelstaedt, M.-L., and Glasauer, S., 1991, Idiothetic navigation in gerbils and humans, *Zool. Jb. Physiol.*, 95:427-435.

33

EVIDENCE FOR SOMATOSENSORY COMPONENTS IN MOVEMENT-EVOKED BRAIN POTENTIALS

K. Bötzel, C. Ecker, and S. Schulze

Department of Neurology
Ludwig-Maximilians-Universität
81366 Munich
Germany

INTRODUCTION

Electric potentials can be recorded non-invasively from the human scalp with surface electrodes. With digital signal processing the ongoing electroencephalogram is converted into the evoked potentials. These waveforms represent the electric brain potentials in defined time epochs around a stimulus. Thus, the electric brain potentials corresponding to the perception of a stimulus or related to a reaction of the subject can be evaluated. The issue of this paper is the reafferent information elicited by a simple limb movement. Do evoked potentials give evidence for sensory feedback about limb movements? The potential associated with the preparatory phase before the start of a movement (Bereitschaftspotential) has been the subject of numerous investigations (Kornhuber and Deecke, 1965; Bötzel et al., 1993). The Bereitschaftspotential is followed by the "motor potential", a negative peak at the onset of the movement with the maximum over the hand motor cortex of the hemisphere contralateral to the movement (Kornhuber and Deecke, 1965). At frontal electrodes a potential was described with a peak at 90 ms after the begin of the movement and named "N+50" by Shibasaki (1980) and "frontal peak of the motor potential" by Tarkka and Hallet (1991). These potentials are suspected to signify reafferent somatosensory information. A recent source analysis of the Bereitschaftspotential (Bötzel et al., 1993) revealed a potential at 90 ms after movement onset which was tentatively called reafferent potential. In this paper we report experiments which were designed to elucidate the nature of this potential, its origin, and its behaviour after lesions of the somatosensory pathways.

METHODS

Evoked potentials were recorded from healthy students during a recording session which lasted up to 4 hours. Each subject performed all experiments in one session. Thus, the results of the different experiments could be compared easily because the electrodes were at the same locations in all experimental conditions. Potentials were recorded under three conditions: I. active finger extensions of the right middle finger every three to four seconds, II. passive movements of the same finger at the same velocity and amplitude (the finger was moved by the experimenter pulling on a string). III. somatosensory evoked potentials after electric stimulation of the median nerve at the right wrist. Since the origin of these electric evoked potentials is well-known (Allison et al., 1989) they could serve as a landmark for the potentials of the other two experimental conditions. We recorded the EEG of the subjects from 32 scalp electrodes and additionally the signal of an accelerometer mounted on the middle finger. The same procedure was applied to a patient with loss of the position sense of the left hand which was caused by a demyelinating lesion of the dorsal columns in the cervical spinal cord. To localize the cerebral sources of the recorded potentials, we performed a source analysis with a computer program (BESA) which repetitively changes the position and orientation of assumed intracranial dipoles (current sources) until the resulting surface potentials match with the recorded data (Scherg, 1990). These results cannot however, be interpreted as the only dipole constellation that can generate the recorded data. This is due to the fact that there is no unique solution to the problem of how to calculate the sources of a given surface potential distribution (the inverse problem) (Wood, 1982). So, it becomes clear that this procedure is influenced by certain assumptions of the investigator (electrode placement, number of dipoles). These assumptions have to be guided by anatomic and physiologic knowledge which may also stem from PET investigations and from fMRI-studies.

RESULTS

Only the potentials evoked by the active movement showed the well-known slowly increasing negativity before the actual movement onset: the Bereitschaftspotential (Fig. 1). Surprisingly, the potentials of the active and the passive task were very similar in the epoch around 90 ms after the start of the movement. At frontal electrodes, they showed a negative deflection and at parietal sites a positivity at the same time. This potential configuration had been reported to be caused by a dipole in the central sulcus with maximum activity at 90 ms which was assumed to reflect reafferent somatosensory processes (Bötzel et al., 1993). Only at the electrodes overlying the contralateral motor cortex the potentials differed considerably, probably due to the potential corresponding to the motor command (Fig. 1). In a further step, the somatosensory potentials after median nerve stimulation at the wrist were recorded in the same subjects. It is known, that the N20 component of these potentials originates in Brodmann's area 3b of the primary somatosensory cortex (Allison, 1989). Thus a comparison of the surface potentials (Fig. 2) and the dipoles caused by active movement, passive movement and electric stimulation should enable one to localize precisely the origin of the proposed reafferent dipoles. The corresponding dipoles are depicted in Fig. 3. Although the polarity of the potentials is different, it becomes clear that all dipoles are located in the same brain region. This is a strong argument for the assertion that the potential components at 90 ms are related to somatosensory processes.

In a further recording we were able to demonstrate, that the potentials evoked by electric stimulation and the 90 ms components are abolished by the same anatomical lesion. A woman aged 45 noticed difficulties with the use of her left hand. On clinical examination she could move the wrist and the fingers well but had no sense of position of these when visual control of finger movements was prevented. This impairment was due to a demyelinating lesion in the upper cervical cord which impeded the dorsal column function. When we recorded the potentials after active and passive stimulation of the left hand, no potentials at 90 ms were recognized. Correspondingly, the potentials after electric stimulation of the

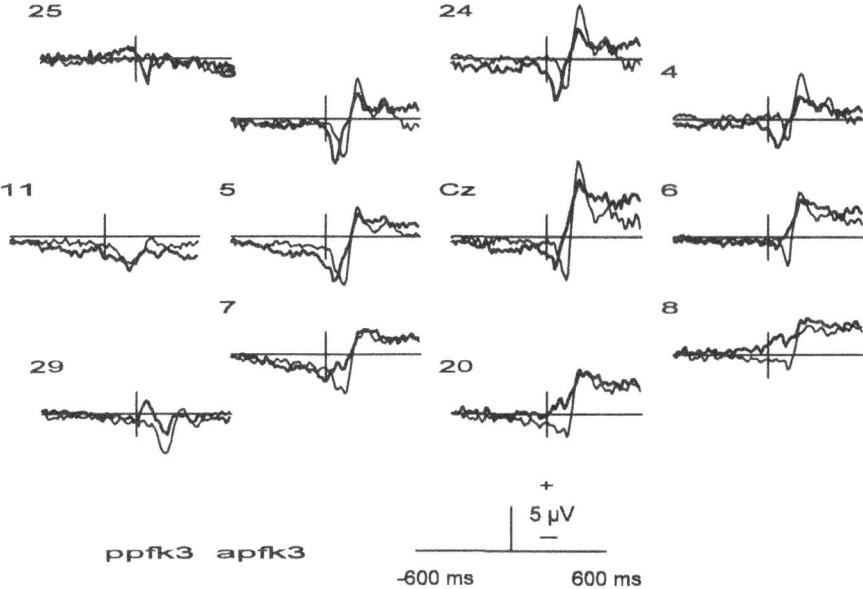

ppfk3 apfk3

+
5 µV
—

-600 ms   600 ms

Figure 1. Potentials of a representative subject recorded in the active (thick line) and in the passive paradigm (thin line). Topographic display of potentials at frontal (24), central (Cz) and electrode overlying the motor cortex (5). A slowly increasing negative potential (the Bereitschaftspotential) is seen with the maximum at electrode Cz before the active movement begins. Potentials after movement onset (vertical bar) are similar in the two conditions.

left median nerve were completely absent (Fig. 4). This was taken as an argument for the common anatomical pathways of these potentials.

DISCUSSION

Somatosensory information about limb position is conveyed via the muscle spindle afferents and the dorsal column pathway to the somatosensory cortex. Joint receptors and tendon organs do not seem to play a major role here. Limb position can well be estimated in the absence of joint receptors, as can be demonstrated in patients with a knee joint

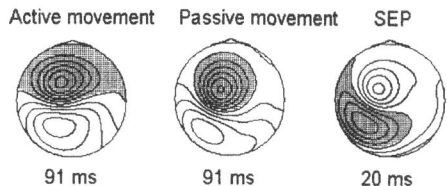

Active movement   Passive movement   SEP

91 ms   91 ms   20 ms

Figure 2. Maps of the surface potentials obtained within the three experimental conditions. Data of the active condition were high-pass filtered.

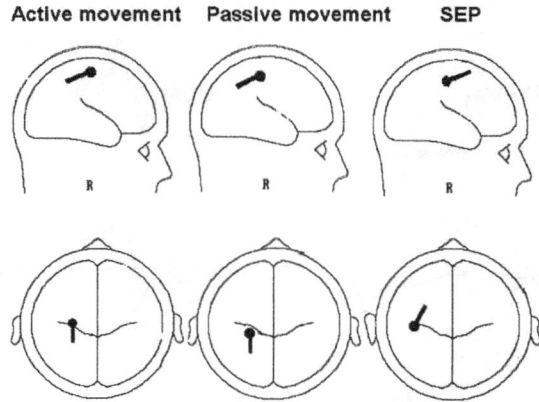

Figure 3. The dipoles which can explain the surface maps (Fig. 2) have similar location in the three conditions.

prosthesis. Tendon organs only signal the force exerted on a tendon. It is well established that after stimulation of mixed peripheral nerves the same fast conducting pathway via the dorsal columns of the spinal cord is used, as during natural stimulation. The identification of the potential field at 90 ms was achieved by separating the different components of movement evoked potentials with a dipole analysis program. Dipole analysis programs which use the spatiotemporal approach, as the one we used, are well apt to separate the contributions of different sources to the observed potentials. By the mere inspection of the recorded waveforms, dipolar potential distributions are not always easily recognized. This is even more evident when large potentials as the Bereitschaftspotentials overlap with other potentials. In the case of the putative reafferent potential described here, a clear dipolar surface distribution was seen in the surface maps. This is also true for the N20 component of the potential after electrical stimulation. Potentials of this kind are reliably analysed with a dipole analysis program. Furthermore, the origin of the N20 component of the potential after electrical stimulation is well established. Thus, the dipole which models this surface

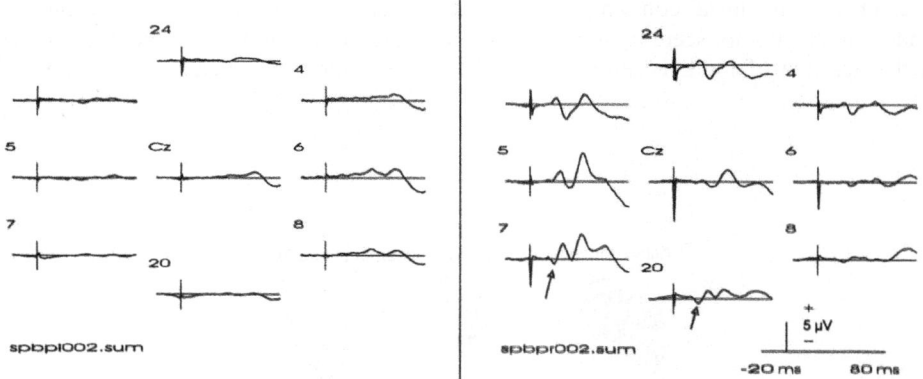

Figure 4. Somatosensory potentials of a patient with dorsal column lesion. Panals on the right side show potentials after electric stimulation of the right median nerve (intact hand). The parietal N20 peak is marked by arrows. Panels on the left show the absence of potentials after stimulation of the left median nerve.

component can be taken as a landmark for the potentials observed in the two other tasks. The results are compatible with sources in the central sulcus for the 90 ms components of the active as well as the passive potentials. Since the potentials occurred during active as well as passive movements, the origin can hardly be a motor command. Therefore a somatosensory origin must be assumed. This is confirmed by the recordings from a patient with absent position sense, in whom both the electrical stimulation and the mechanical stimulation caused no visible surface potentials.

The description of the movement related potentials demonstrate that dipole analysis is a valuable instrument for the evoked potential researcher. Although any analysis of these potentials should start at the description of the measured data, the following dipole analysis helps to interpret them, which generates another view of the data and supports or rejects hypotheses about physiological processes. However, one must bear in mind that subjective factors enter the results of the analysis. These are assumptions on the number of generators or rejections of dipole constellations which seem unlikely on the basis of what one assumes to know. Thus the interchange with other methods which provide anatomical data becomes a necessary precondition for the analysis of intracranial dipole sources.

REFERENCES

Allison, T., Mc Carthy, G., Wood, C. C., Darcey, T. M., Spencer, D. D., and Williamson, P. D., 1989, Human cortical potentials evoked by stimulation of the median nerve. 1. Cytoarchitectonic areas generating short latency activity, *J. Neurophysiol* 62:694-710.

Bötzel, K., Plendl, H., Paulus, W., and Scherg, M., 1993, Bereitschaftspotential: Is there a contribution of the supplementary motor area? *Electroenceph Clin Neurophysiol* 89:187-196.

Kornhuber, H.H., Deecke, L., 1965, Hirnpotentialänderungen bei Willkürbewegungen und passiven Bewegungen des Menschen: Bereitschaftspotential und reafferente Potentiale, *Pflügers Arch ges Physiol* 284:1-17.

Scherg, M.,1990, Fundamentals of dipole source potential analysis. In: Grandori F, Hoke M, Romani GL (eds): Auditory evoked magnetic and electric potentials. Adv. Audiol., Vol 6. Karger, Basel, pp 40-69.

Wood, C. C., 1982, Application of dipole localization methods to source identification of human evoked potentials, *Ann NY Acad Sci* 388:139-155.

34

CONTROL OF HUMAN HEAD POSTURE

Michael A. Gresty and Adolfo M. Bronstein

Medical Research Council Human Movement and Balance Unit
National Hospital
Queen Square, London WC1N3BG
Great Britain

INTRODUCTION

The head as a monobloc has 6 degrees of freedom which can be usefully designated with navigational terms 'yaw, pitch, roll' for angular motions and 'bob' (up-down), 'heave' (side to side) and 'surge' (face backwards-forwards) for linear motions (Findley and Gresty, 1988). Characterisation of head motion can therefore be a relatively simple affair with the aid of modern transducers such as the 6xdimensional 3SPACE FASTRAK (Polhemus, Kaiser Aerospace and Electronics) which is used for detecting body motion in 'virtual reality' applications. However, head motion can be the end product of complex combinations of trunk, limb and neck movement which are more difficult to define. In particular the most pertinent problem is how the head is controlled by the multiplicity of neck muscles acting about the multiple jointed geometry of the cervical spine which comprise the suspension system of the head.

In practice, most studies of human head motion have simplified the problem considering the head to be a mass, suspended in inverted pendulum fashion, on the body by means of a multidirectional rotational joint and visco-elastic restraints (muscles and connective tissue) and have exploited, as a first approximation, linear systems analysis, using a second order differential equation to describe passive head dynamics (Barnes and Rance, 1975, Viviani and Berthoz, 1975) viz; for equilibrium

$$M\ddot{H} + v\dot{H} + kH + F = 0$$

where M = moment of inertia of the head, H= angular displacement of head to trunk, v = viscosity, k = elasticity, F = additional muscle forces acting, . and .. = the differential operators.

The relationships between the physical entities involved in this model have important implications for the question of how, or 'how well', the head is controlled, particularly with respect to its stability in response to a perturbation of the body. The tendency for the head to oscillate (resonant frequency) if given a jolt is proportional to the damping ratio $(v/2*M)/(\sqrt{[k/M]})$. The head will oscillate significantly if this ratio is <1.0 with larger and more sustained oscillations occurring proportionately the less the damping ratio. The actual frequency of oscillation is determined by $\sqrt{(k/M)}$. More realistic models of head dynamics involving higher order differential equations and non-linearities have been developed but, as yet, have had limited practicable application in experimental biophysics (Zangenmeister et al., 1992). Perhaps the most important complicating considerations for modeling are that the head neck system is multijointed with the joints and connections (muscle tendons) themselves having significant mass so that 'resonance' may be encountered at a variety of frequencies (Viviani and Berthoz, 1975) and, secondly, that rotations about the joints are not perfectly symmetrical for all directions (Barnes and Rance, 1975) making the system susceptible to cross axis transmission (see below).

Active neuro-muscular processes also determine the head's response to perturbation as well as the characteristics of 'voluntary' movement. There is some evidence which will be detailed below that the mechanisms for head stabilization in humans are the vestibular-collic reflex which assists stabilization of the head in space and the cervico-collic reflex (acting like, and possibly of the same mechanism as a stretch reflex) which realigns the head with the trunk in response to displacement. Evidence will be presented that there are other automatic neuromuscular mechanisms which contribute to postural control of the head. A further feature which the reader should bear in mind is that the passive characteristics of the head neck system are themselves variable; in particular being affected by muscle contraction. By way of example, co-contracting neck muscle by 'tensing' the neck will affect the elastic and viscous characteristics to give higher damping and a higher potential resonant frequency.

Much discussion of human head movement control has focused on the interplay between the passive characteristics of the head-neck system and the more automatic, 'reflex' modes of active control. The types of investigations undertaken have depended largely upon the availability of the formidable nature of the equipment needed to move large segments of the human body in a controlled fashion. The variety of experimental scenarios deployed are discussed below, classified according the plane or direction in which motion stimuli have been delivered.

HEAD STABILIZATION IN RESPONSE TO SUSTAINED PREDICTABLE AND RANDOM OSCILLATION OF THE TRUNK

Yaw

Rotational stimuli in Yaw are arguably the easiest to achieve because of the relatively low moment of inertia of the body seated upright and rotating about an axis aligned with the spine. The type of head movement obtained for yaw stimuli depends upon the instruction to the subject on the dimension - keep head still in space ranging to keep head fixed to trunk - and the frequency content of the stimulus.

The linear component of response of the head to trunk yaw is that of a low pass filter with a tendency to resonance at circa 1.5 to 4 Hz. Phase lag is greater than one would expect from a second order system at circa 260° over 0.3 to 3 Hz, implying that there is a cascade of second order subsystems or a 'standing wave' effect of transmission delay (Barnes and Rance, 1975, Gresty, 1987).

If subjects are presented with the task of keeping their heads stable in space (i.e. pointing in the same direction), then at low frequencies, below circa 0.5 Hz, head and trunk movement may also be almost entirely uncoupled within the range of neck mobility so that the head remains relatively stationary in space above the turning trunk. For random oscillation this is shown by low Coherence between head in space and trunk in space signals

(Gresty, 1987) [Coherence is degree of covariance and therefore of linear relationship between two signals and expressed at each component frequency]. Subject to passive properties alone the head would always follow the trunk motion after a delay and so an important for the independence of head movement in this low frequency range is the vestibular-collic reflex which stabilizes the head, much as the vestibular-ocular reflex stabilizes the eyes but with a more restricted, dynamic (Guitton et al., 1986, Bronstein, 1988). For low frequency stimuli with essentially sinusoidal envelopes the coherence between head on trunk and trunk in space motion is high which demonstrates an active compensatory mechanism operating fairly linearly (Bronstein, 1988).

Above circa 0.5 Hz, head inertia and the visco-elastic properties of neck tissue dominate the head's response to trunk rotation (Barnes, 1975, Gresty, 1987, 1989). Coherence between trunk and head rotations is high showing a linear relationship which is largely determined by passive properties. The head is underdamped with damping ratio of circa 0.3 to 0.7 with a natural resonant frequency in the region 1.5 to 4 Hz. There is little ability to independently stabilize the head at these higher frequencies, particularly in response to random perturbations.

An instruction to attempt to keep the head aligned with the trunk in normal subjects seems to result in their initiating co-contraction of the neck muscle to increase overall viscosity: a tactic which is successful but only to a point since the high head inertia results in some oscillation at high frequencies of stimulation despite vigorous attempts to stiffen the neck. Undoubtedly the vestibular-collic reflex can be overridden easily in this scenario. It is obscure as to what contribution cervico-collic reflexes may lend.

Response of Labyrinthine Defective Subjects to Yaw Rotation

Patients who have lost labyrinthine function show a markedly defective ability to stabilize their heads in space. In contrast to normals who have a phase lead (possibly predictive) of head stabilization they show head lag and the movements they make in attempts to achieve stability are less well organized then normals (Bronstein, 1988).

Pitch and Roll

The power demands on apparatus capable of whole body oscillation in pitch or roll are formidable so it is not surprising that few normal physiological studies (outside aerospace/military sector) have been attempted. The basic response of the head neck system to oscillation of the trunk in a vertical plane about earth horizontal axes is that of a low pass filter with a slight tendency to resonance at circa 1.5 to 4 Hz (Figure 1). Head resonance is not so pronounced as for yaw and phase lag of 180° over circa 0.3 to 3 Hz is consistent with a second order lag (Barnes and Rance, 1975, Gresty, 1989).

For lower frequencies of oscillation, below circa 1 Hz, normal head movement responses to unpredictable sustained oscillation pitch and roll show decoupling from the body, as with yaw, and some ability to maintain the head stable in space and the presence of visual context enhances stabilization (Gresty, 1987; Gresty and Bronstein, 1992). Coherence between head in space signals is low, underlining the caution with which the response relationships should be couched in terms of a linear system (Figure 1). Within these bounds of caution phase between head and trunk lags typically by 180° over 0.1 to 1.0 Hz for pitch and is more variable for roll (Figure 1).

Overall head in space stability as measured by absolute amount of head movement provoked by similar stimulus profiles is marginally better for roll than pitch. A reason for this may be that the damping of the head is higher for roll than pitch as might be expected from the greater restrictions on spinal articulation in this plane of movement. The higher damping means that the head is more fixed to the trunk and does not 'flop' so much. Although there is greater freedom of voluntary movement in pitch the head is not necessarily stabilized so well in this plane because of its lower damping and tendency thus to oscillate.

Barnes and Rance (1975) examined the higher frequency response of head movement in roll and pitch using sinusoidal stimuli and concluded that the head in these planes was approximately critically damped, i.e. a damping ratio circa 1.0. Our own observations suggest that the damping ratio for pitch varies normally between circa 0.7 and 1.0, whereas for roll it is kept closer to unity.

Surge

Pure linear motion is difficult to manipulate for it requires transporting the whole body at high levels of controlled acceleration. The only study of surge of which we are aware examined head movement responses to velocity steps and to random oscillation with the subject seated upright, torso restrained on a bogie powered by linear induction. The testing scenario is interesting because the imperative is always to keep the head upright on the trunk and therefore in space when shunted backwards and forwards in this fashion. When the bogie moves the head lags and tilts from upright - uncomfortably as with a threatened whip-lash. In the face of this sudden head tilt both the vestibular-collic and neck stretch reflexes are synergistic in bringing the head to the vertical in realignment with the trunk.

Bilateral labyrinthine defective subjects differ little from normals in the stability of their heads in the face of surge perturbation. Normal damping can be critical (circa 1.0) with the defective subjects showing a slight tendency to underdamping. The stimulation to the neck in this situation is so strong that cervico-collic reflexes (stretch) may be the most important determining factors in head control and these would be similar for normal and labyrinthine defective subjects (Gresty, 1989).

Cross Axis Coupling

Because of the complexity of neck anatomy significant cross axis motion may induced although stimuli delivered are restricted to a particular plane. In particular, yaw tends to induce pitch and roll head motion and roll induces yaw and pitch with accompanying frequency distortions, as one might expect (Barnes and Rance, 1975). Cross axis effects are greater for higher frequency stimuli and of such magnitude that they must be taken into account when examining active muscle responses to high frequency perturbation, for substantial components of the response may be related to planes of motion other than that of the stimulus.

Figure 1. Transfer function analysis of normal head movement responses of 4 subjects to random velocity oscillation of the trunk about a head centre axis comparing responses obtained for pitch with roll in the presence of a visual frame of reference. The continuous lines of the transfer functions are plotted where the coherence was above 0.5. The lowest figures show examples of coherence for an individual subject. The data is based on the average of 6 linear spectra calculated from a continuous time record of responses to random oscillation. Note gain-phase plots are more orderly for pitch data approximating a lag. Gain-phase plots for roll data indicate significant non-linearity.

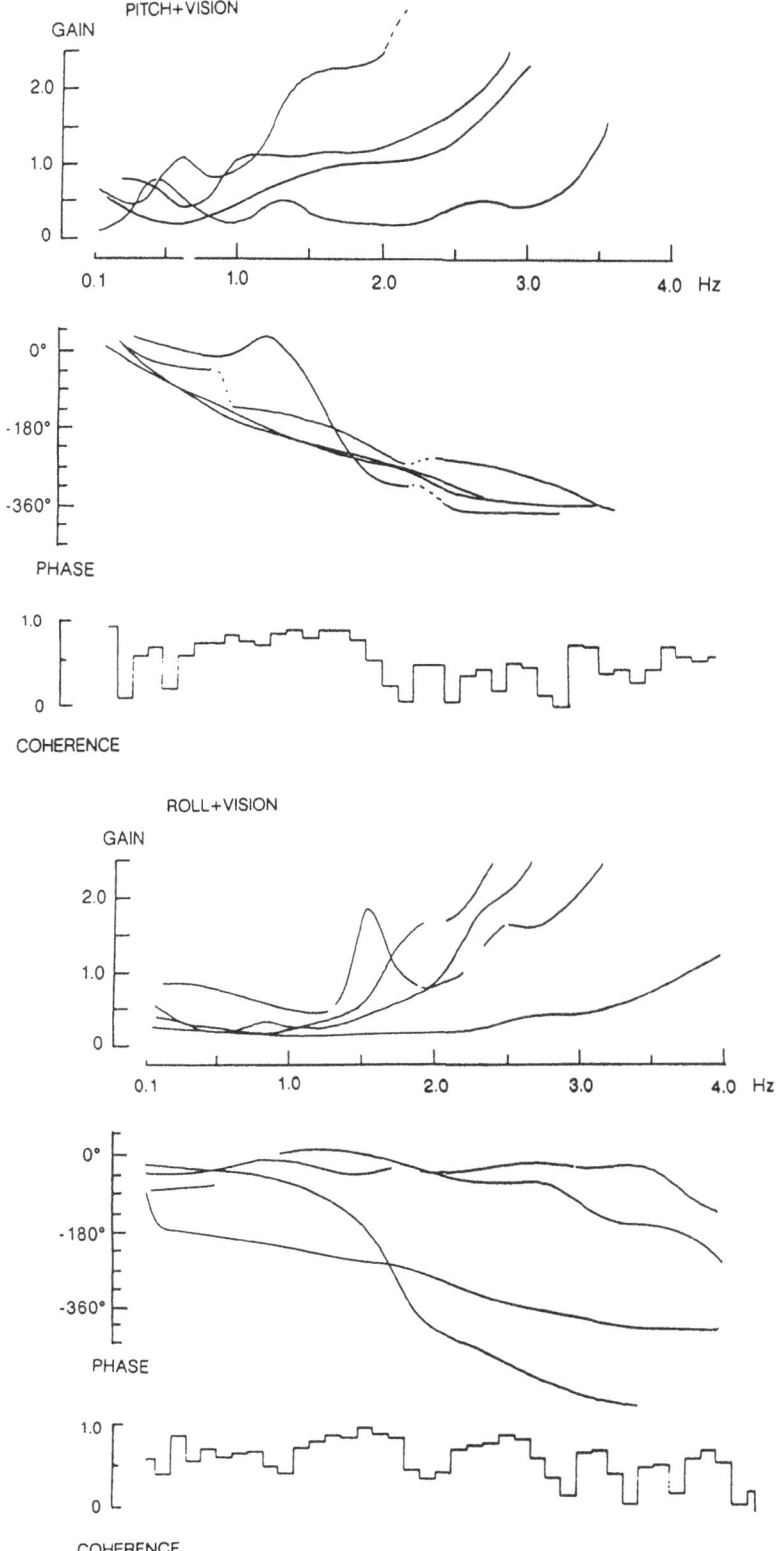

RESPONSE PREDICTABILITY AND POSTURAL 'SET'

Obviously one can 'brace' the neck against impacts of quite high magnitude so that the head oscillates little in response to a sudden jarring. In mechanical terms this is adjusting the suspension characteristics (by increasing muscle co-contractive tone) of the head to maintain a critical damping. The increase in tone is effectively changes the passive characteristics of the system with no active component. When taken unawares however, the head may oscillate, as may be seen in everyday life during sharp, unexpected movement, in which case one would guess that damping was less than critical: prediction of the general movement conditions that a subject may encounter in any experimental situation must be important in determining the kind of results that may be obtained.

An alternative tactic to simply bracing the head against a particular level of disturbance and separate from automatic vestibular and muscle stabilizing reflexes is to predict the stimulus characteristics. Hitherto, predictability has been neglected as a factor in head postural control but is of undoubted importance, particularly for lower frequency stimuli with more sinusoidal profiles (Bronstein, 1988). The presence of predictability is probably best detected in advanced compensatory phase relationships and augmented gain which lend an improvement to task performance; i.e., better stabilization in space or on the trunk which cannot be matched by the automatic reflexes operating with inherent delays.

Some characteristics of predictability in normal human head movement have recently been explored (e.g., Pozzo et al., 1991; Assaiante and Ambelard, 1993) for natural motion situations but the full dynamic capability and restrictions of prediction in postural control of the head have not yet been investigated.

'AUTOMATIC' RESPONSES TO TRUNK PERTURBATION: ORIGINS AND LATENCIES

Despite a wealth of biophysical information and much speculation on the action of presumed vestibular and cervical mechanisms there is so little actually known about the physiology of human head control at the level of receptor-neuronal-effector circuitry that we have taken opportunity to summarize, in Table 1, the few scraps of data available about specific neuronal loops. The table distinguishes those loop identified through natural versus artificial stimulation. It should be borne in mind that in the case of receptors such as the vestibular canals the lag caused by head inertia is likely to delay the reflex neuromuscular effects of natural stimulation so that the path latencies will appear much longer than those revealed by artificial stimulation.

In an attempt to identify some of the control signals responsible for head posture we have recently examined the head movement and neck emg responses to abrupt tilts of the body from upright. The advantage of such a discrete stimulus over sustained motion being that some idea of latencies, and thereby pathways, of the active response components can be ascertained.

Figure 2 shows recordings from a normal subject and from a well-adapted subject with bilaterally absent vestibular function undergoing similar abrupt and unexpected, backwards tilts of the trunk with subsequent return to upright. The overall tilt trajectory is approximately cosinusoidal. The key features of the head movement response are similar for the two subjects and are as follows: there is an initial lag of the head due to inertia which fortuitously maintains the head in a more earth-upright position. The head then tends to fall backwards but is arrested by a vigorous upwards movement tending to maintain earth-uprightness. At the end of tilt when the trunk is repositioned vertically, the head oscillates briefly about upright. Corresponding emg in the early stages of tilt show an initial burst in

Table 1.

Loop	Stimulus/site	Latency/response	Reference
Artificial stimulation			
Vestibular-collic	Sound click normal ear	8 ms inhibition in ipsilateral neck approx 20 ms excitation in contralateral neck	Colebatch (1993,1994)
Vestibular-collic Tullio effect	Sound/mechanical	55-65 ms excitation 12 ms excitation	Dieterich (1989) Bronstein (communication)
Natural stimulation			
Cervico-collic	Stretch to neck due to trunk surgery (in labyrinthine defectives)	25 ms in stretched muscle	Gresty (1989)
Vestibular-collic + proprioceptive	Trunk tilt	35 ms excitation in righting muscles, 25 ms inhibition in antagonists	See text
collic + proprioceptive	Trunk tilt (in labyrinthine defectives)	35 ms excitation in righting muscles25 ms inhibition in antagonists	See text
vestibular-collic	'startle'	55 ms excitation	Stell (1990), Bisdorff (1994)
	tilt	53 ms excitation	Keshner (1988)
	bob/tilt	59 ms excitation, 47 ms inhibition	Mazzini (1993).

the sternomastoid muscles which is appropriate for head righting towards upright followed by a larger packet of activity on the same muscles which maintains the upright alignment. It should be noted that the muscles supporting the earliest activity are not stretched at this point in time but unloaded so any proprioceptively mediated, automatic response is likely to be related to muscle or joint unloading or be a functionally organized righting response (Traub et al 1980). This type of activity has not been described previously for neck muscle.

The source of the control signals for active head righting in the normal subject may be a combination of vestibular and proprioceptive stimulation. In the labyrinthine defective subject one must conclude that proprioception is capable of serving head righting to a near normal level of performance.

MODELING ACTIVE POSTURAL CONTROL

It is instructive to explore the possibility of modeling active postural control of the head as an inroad into determining what type of signals may drive the emg responses seen in neck muscle during active righting in the discrete tilt scenario described above (Figure 2). The performance of the model, shown in Figure 3, was explored in "Simulink" (The Maths Work Co Inc). The simulations are given in Figure 4 and computational details in the appendix (Figure 5).

As a starting point the passive response of the head-neck system was assumed to approximate to a second order lag with constant elasticity and viscosity coefficients through tilt, a damping ratio variable between circa 0.7 and 1.2 and a theoretical undamped natural frequency in the rage 1.5 to 4 Hz.

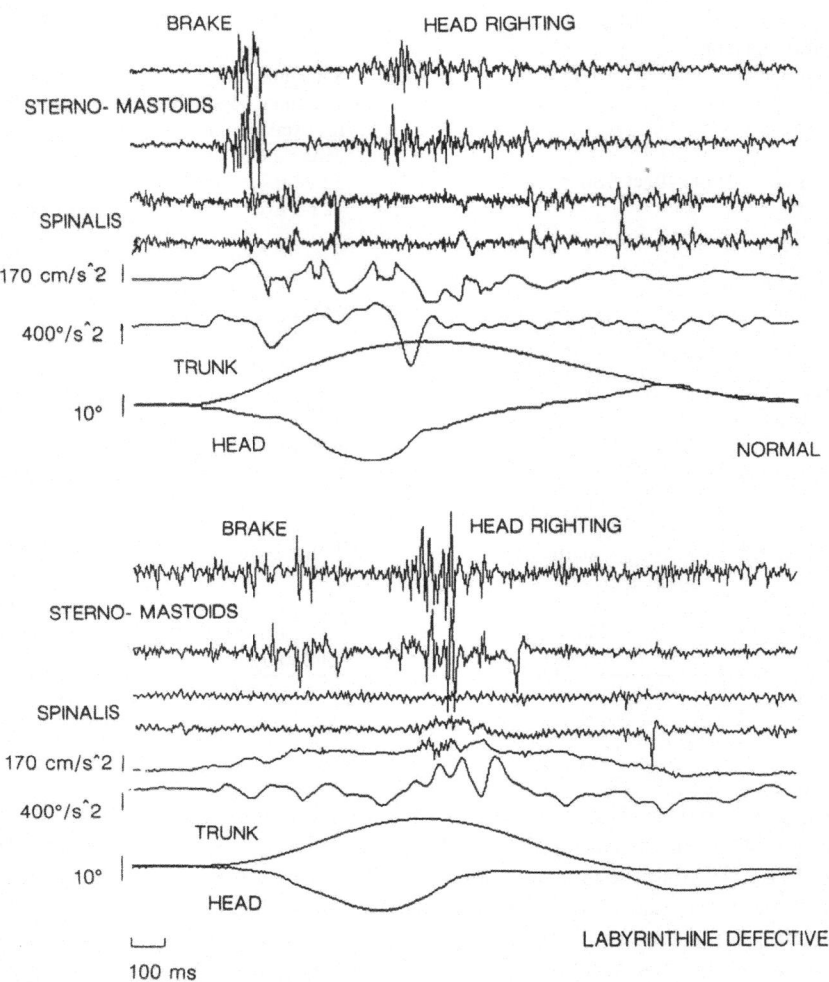

Figure 2. Example of head movement response and associated emg to sudden rotational tilt of the trunk backwards about an axis approximately head centre comparing a normal subject with an adapted patient having bilateral loss of vestibular function.

As tilt progresses the weight of the head exerts a significant torque proportional to the head weight, the sine of tilt and leverage: these parameters cannot be assessed directly but this force term may be manipulated to 'tune up' the performance of the simulation. Head weight acts as positive feedback exacerbating the tendency to fall.

The initial burst of emg was assumed to be a 'one-off' occurrence triggered by sudden tilt of the head from upright; a braking or unloading reaction. This burst is a short sharp activity with latency in the region of 30 ms (earliest) to 70 ms. On averaging we have also observed inhibitory periods in the antagonist muscles at latencies of 20 to 25 ms. The mechanical signals with the profile of a short latency peak which would be available as a source of motor control are: i) relative acceleration between head and trunk which is a high peak at the onset of motion and could be transduced through neck receptors; and ii) initial acceleration, or rate of change of acceleration, of the head in space, which also peaks at the

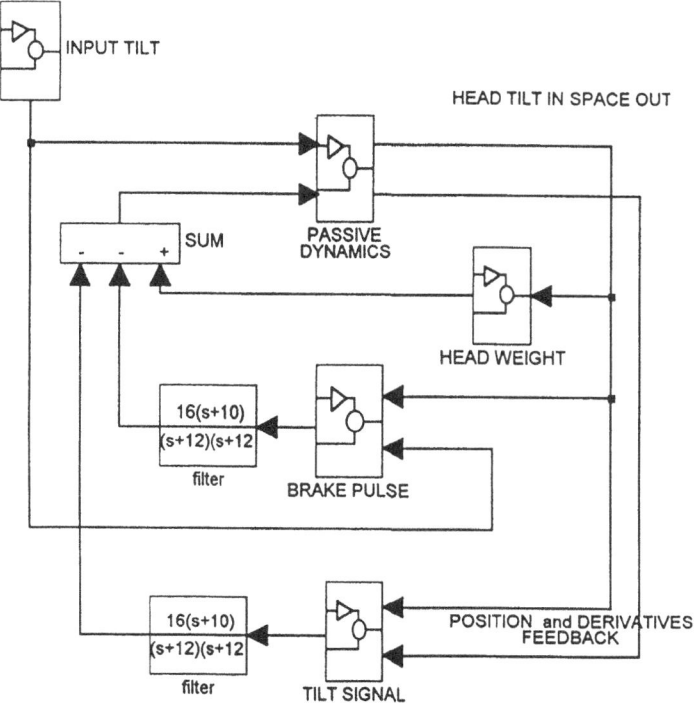

Figure 3. Block diagram of model of postural head control.

onset of motion and is effectively signaled by the otolithic regular and irregular units respectively; iii) touch pressure receptors, particularly in the skin about the shoulders which must be sensitive to the forces tilting the trunk. Relative acceleration of the head to the trunk was selected as the source of the initial emg burst for this signal would be available to both normal and the labyrinthine defective subjects and little is known about dermal control of movement. These signals provide negative control feedback.

Emg bursts occurring later in the motion do a reasonable job of maintaining the head earth upright and thus should be related to head motion in space. The obvious source of a signal of tilt displacement from upright in space is that transduced by regular otolith units; particularly those of the utricles. In labyrinthine defective subjects this signal would have to be derived from combining somatic cues to upright with a signal of head to trunk position derived from neck receptors. This signal provides negative feedback compensating for the trunk tilt, similar to the vestibular-ocular reflex compensation for head rotation.

It was assumed that both control signals were delayed by central processing; about 20 ms for the initial, burst and circa 100 ms for the head in space signal to match the timing of observed emg bursts. Both signals were filtered to approximate the emg envelope and muscle contraction dynamics giving a net dominant time constant of 0.1 s.

Comparison of some normal trajectories of head response to tilt with the simulations achieved with this model are shown in Figure 4. The variety of trajectories were achieved by small variations in the feedback gain, damping ratio or resonant frequency. If significant head-in-space velocity or acceleration negative feedback was introduced, as one might attempt with a mechanical controller to speed up response, higher frequency oscillatory instabilities occurred which were not seen in actual data records. Comparison of actual records with simulations shows the modeling to have promise and to some extent justifies

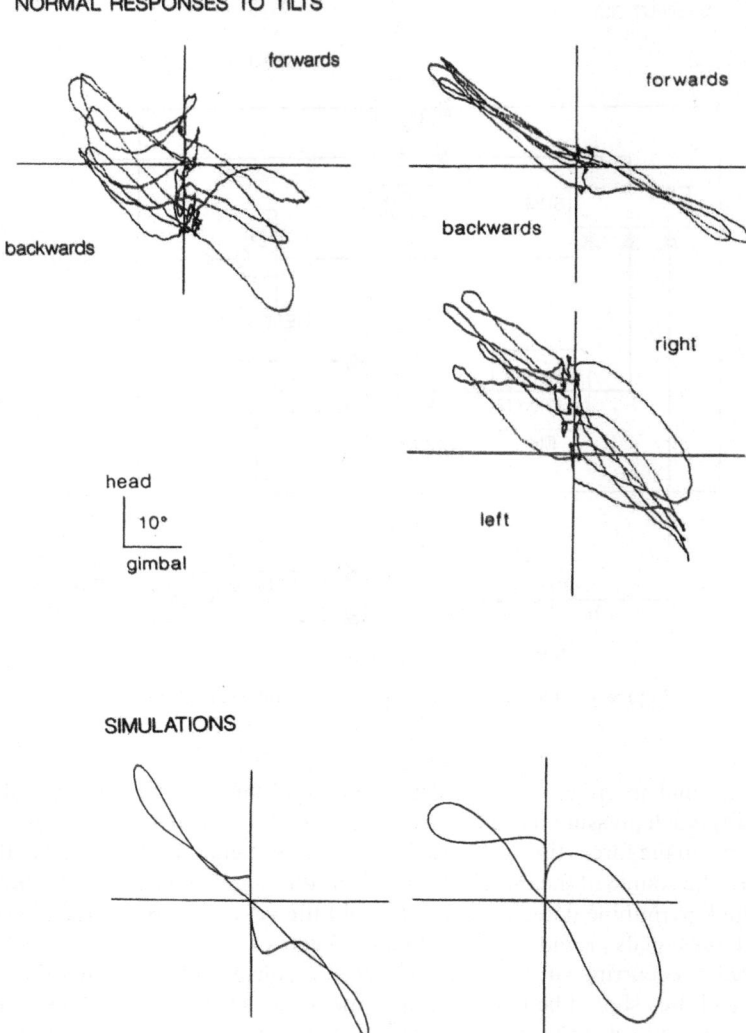

Figure 4. Comparisons of actual data and simulated head movement trajectories in response to discrete tilts of the trunk. Head and trunk relationships are plotted as Lissajous figures for ease of comparison. As a guide to interpretation a Lissajous trajectory approximating a straight line with gradient -1 would indicate perfect compensation: the head moving with equal magnitude and in the opposite direction to the trunk.

the assumptions we have made about the characteristics of the neural control signals underlying active postural responses.

APPENDIX

Figure 5 shows the actual 'Simulink' realization of the sub-block used in the model of head control shown in figure 3.

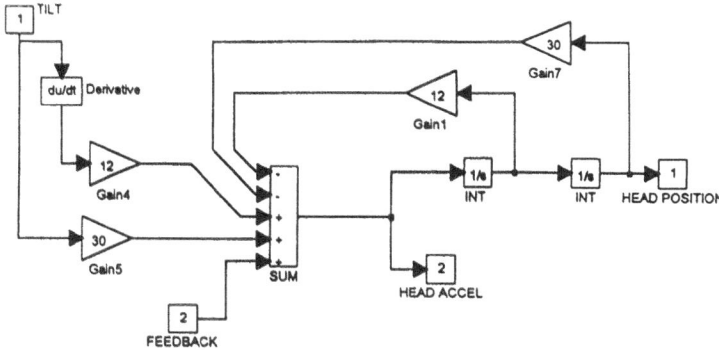

PASSIVE RESPONSE

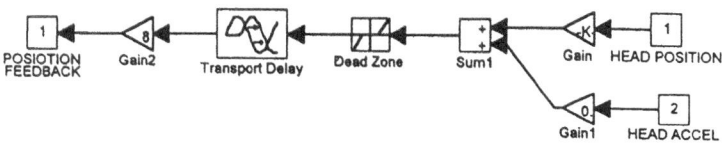

HEAD IN SPACE POSITION SIGNAL

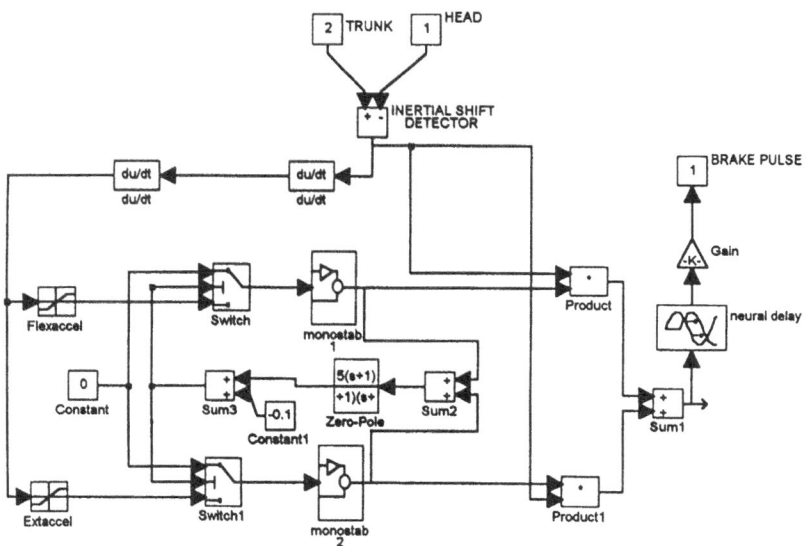

BRAKE PULSE TRIGGER CIRCUIT

Figure 5. Decomposition of the computational blocks used in the simulink model with typical gain values. The passive model block is deliberately 'exploded' into simple terms to show the relationships between the components of head and trunk movements. For the 'head in space position' signal the ratio of position to acceleration feedback was typically 400:0.5. The monostable pulse generated to provide the basis of the 'brake' signal was typically 150 ms long and through the 'product' block made proportional to the size of the initial signal of relative acceleration of the head to the trunk.

ACKNOWLEDGMENTS

This work was in part supported by the EEC Biomed Program Concerted Action 'EHO' (Eye Head and Orientation).

REFERENCES

Assaiante, C., and Ambelard, B., 1993, Ontogenesis of head stabilization in space during locomotion in children: influence of visual cues, Exp. Brain Res. 93:499-515.

Barnes, G., and Rance B., 1975, Head movement induced by angular oscillation of the body in the pitch and roll axes, Aviation Space and Environmental Medicine 46:987-993.

Bisdorff, A. R., Bronstein, A. M., and Gresty, M., 1994, Response in neck and facial muscle to sudden free fall and a startling auditory stimulus, EEG Clin Neurophysiol. 93:409-416.

Bronstein, A., 1988, Evidence for a vestibular input contributing to dynamic head stabilization in man, Acta Otolaryngol (Stockh) 105:1-6.

Colebatch, J., Halmagyi, G., and Skuse, N., 1994, Myographic potentials generated by a click-evoked vestibulocollic reflex, J Neurol. Neurosurg. Psychiat. 57:190-197.

Colebatch, J., and Rothwell, J. 1993, Vestibular evoked EMG responses in human neck muscles, J Physiol. 473:18.

Dieterich, M., Brandt, Th., and Fries, W., 1989, Otolith function in man, Brain 112:1377-1392.

Findley, L., and Gresty, M., 1988, Head, Facial and Voice Tremor. in; Advances in Neurology 48: Facial Dyskinesias. ed: J. Jankovic and E. Tolosa. Raven Press, N.Y. 239-253.

Gresty, M., 1989, Stability of the head in pitch (neck flexion-extension): Studies in normal subjects and patients with axial rigidit, Movement Disorders 4:233-248.

Gresty, M., 1987, Stability of the Head: Studies in normal subjects and in patients with labyrinthine disease, head tremor and dystonia Movement Disorders 2:165-185.

Gresty, M., and Bronstein, A., 1992, Visually controlled spatial stabilization of the human head: compensation for the eye's limited ability to roll, Neuroscience Letters 140:63-66.

Guitton, D., Kearney, R., Wereley, N., and Peterson, B., 1986, Visual, vestibular and voluntary contributions to human head stabilization, Exp. Brain Res. 64:59-69.

Keshner, E., Woollacott, M., and Debu, B., 1988, Neck, trunk and limb responses during postural perturbations in standing human, Exp. Brain Res. 69:455-466.

Mazzini, L., and Schieppati, M., 1993, Short latency neck muscle responses to vertical body tilt in normal subjects and in patients with spasmodic torticolli, EEG Clin Neurophysiol. 108:1-11.

Pozzo, Th., Berthoz, A., Lefot, L., and Vitte, E., 1991, Head stabilization during various locomotor tasks in humans, Exp. Brain Res. 85:208-217.

Stell, R., Gresty, M., Metcalfe, T., and Bronstein, A., 1990, The neurology of short latency of muscles in neck and leg response to sudden onset free fall. In: Brandt Th., Paulus, W., Bles, W., Dieterich, M., Krafczyk, S., Straube, A. (eds.). Disorders of posture and gait. New York: Georg Thieme Verlag Stuttgart, 214-217.

Traub M., Rothwell, J., and Marsden, C., 1980, A grab reflex in the human hand, Brain 103:869-884.

Viviani, P., and Berthoz, A., 1975, Dynamics of the head-neck system in response to small perturbations: Analysis and modeling in the frequency domain, Biol. Cybernetics 19:19-37.

Zangenmeister, W., Arlt, A., and Lehman, S., 1992, Sensitivity coefficients, trajectories and graphs of a human head-movement model, J. Biomed. Eng. 14:451-458.

35

SIMPLE MODEL OF SENSORY INTERACTION IN HUMAN POSTURAL CONTROL

R. J. Peterka

Clinical Vestibular Laboratory and
    R.S. Dow Neurological Sciences Institute
Legacy Good Samaritan Hospital and Medical Center
1040 NW 22nd Avenue, N010
Portland, Oregon 97210

INTRODUCTION

Upright body stance is inherently unstable. Active feedback control utilizing motion cues from various sensory systems is necessary in order to maintain this upright position. Visual, somatosensory, and vestibular sensory systems are the main contributors to this feedback control. In many environments, accurate information is available from all three of these sensory systems. In other environments, motion information from one or more sensory systems may be absent or inaccurate leading to poor performance (increased body sway) or falls in extreme cases. However simple observations indicate that the motion information from the three sensory systems is highly redundant in that upright stance can be maintained when orientation cues are absent or inaccurate in two of the three sensory systems. For example, a normal subject can stand with eyes closed on a foam pad suggesting that vestibular cues alone are sufficient to maintain upright stance. Also, a subject with a bilateral vestibular loss can stand on a flat surface with eyes closed indicating that somatosensory cues alone provide sufficient feedback information for postural control.

Although the influences of sensory cues on postural sway have been clearly demonstrated (Lestienne et al., 1977; Diener et al., 1984; Dichgans and Diener, 1989), very little is known about how visual, somatosensory, and vestibular motion information is combined and processed by the nervous system for postural control. There are many relevant questions. (1) How does the postural control system deal with conflicting sensory orientation information? The postural control system might be capable of assessing the accuracy of available sensory cues and excluding information which is judged to be inaccurate. Alternatively, information from all three sensory systems might be used continuously, but with postural control system parameters set to minimize postural disturbances due to inaccurate sensory information. (2) Are the sensory system interactions linear in nature? Mergner and coworkers (Mergner et al., 1991) have demonstrated that the perception of combined head and body movements can be explained by an essentially linear interaction of vestibular and somato-sensory motion cues. Perhaps the postural control system makes use of the same or similar neural mechanisms for combining motion cues. (3) What types of sensory information processing are necessary for postural control?

The purpose of the present work was to develop a control system model-based approach to the study of sensory interactions in postural control. The models serve as an explicit hypothesis regarding the functional mechanisms involved in the processing of information for postural control. An important goal of the modeling work is to develop an intuitive understanding of how the entire system functions and how individual components influence the overall system's responses.

MODELING ASSUMPTIONS

There are many complications in the application of control systems modeling to the postural control problem. First, it is known that various components of the postural system are nonlinear. Nonlinearities can arise from several sources including body mechanics, sensory system response properties, central processing of sensory signals, time delays in neural processing and transmission, and muscle activation properties. Second, if the body is accurately modeled as a multi-link structure, the equations of motion become very complex (Koozekanani et al., 1980). This inherent complexity can defeat the goal of obtaining intuitive insight into the functioning of the postural control system. Third, limited data are available for the validation of any model.

To overcome these problems, our first attempt at modeling and experimental valida-tion of a model was to simplify the system as much as possible. This was accomplished in several ways. First, the vestibular contribution to postural control was not included. Experi-mental results for model validation came from subjects with a bilateral loss of vestibular function. Second, the body was modeled as an inverted pendulum (single body segment) with rotational motions occurring about the ankle joints in only one plane (anterior-posterior body sway). An inverted pendulum is described by a nonlinear equation of motion. This equation was linearized assuming small angles of rotation about the ankle joint (Ishida and Miyazaki, 1987). Third, the model only attempted to describe the small amplitude responses of the system. Many nonlinear systems effectively behave in a linear manner for small amplitude perturbations. (Exceptions include systems with significant dead-zone and thresh-old properties.) Fourth, the somatosensory and visual information used for postural control were assumed to be processed independently of one another and then combined linearly. Fifth, the central processing of sensory "error" signals were assumed to be by a simple controller mechanism which generates corrective body torque in proportion to a linear combination of position error, velocity error, and the integral of position error. This type of control processing is known as a PID controller (proportional, integral, differential control) and has been used previously to model human postural control properties (Johansson et al., 1988).

MODEL DESCRIPTION

The block diagram model for the postural control system is shown in Figure 1. It consists of two feedback loops, one associated with the processing of somatosensory orientation cues, and one associated with visual orientation cues. There are two inputs to the model. One is the rotation angle, θ_p, of the platform (support surface) upon which the subject stands. The second is the rotation angle, θ_v, of a visual surround which encloses the test

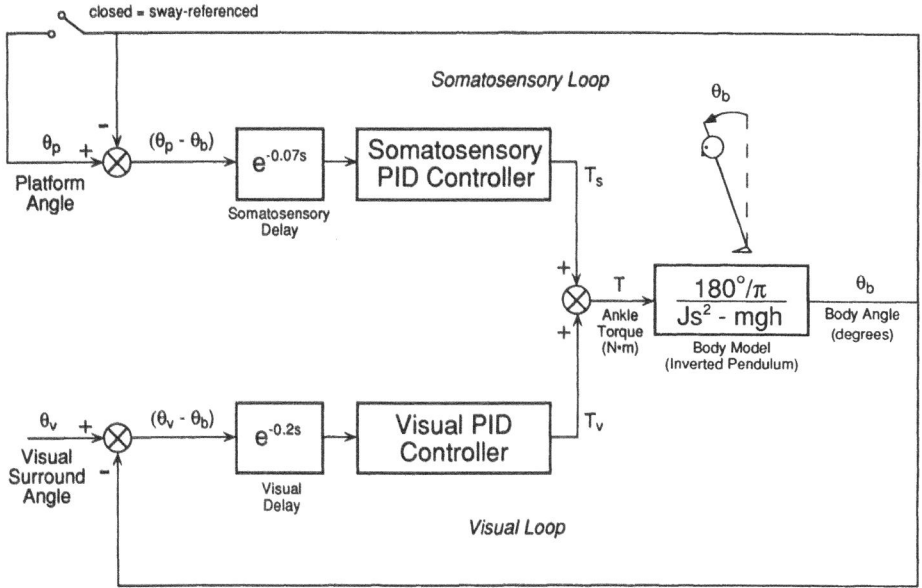

Figure 1. Control system model of human postural control which includes only somatosensory and visual feedback.

subject. The rotation axes of the visual surround and the support surface are assumed to pass through the subject's ankle joint axis.

The subject's body is assumed to be an inverted pendulum, that is, a rigid structure which rotates only in an anterior-posterior direction about the ankle joint axis. The differential equation describing this system is:

$$J\frac{d^2\theta_b}{dt^2} = mgh \sin\theta_b - T \tag{1}$$

$$\approx mgh\theta_b - T$$

where θ_b is the angular deviation of the body away from vertical, J is the body's moment of inertia about the ankle joint, m is body mass, h is the height of the center-of-gravity above the ankle joint, g is the acceleration due to gravity, and T is torque exerted about the ankle joint to maintain stability. The equation is nonlinear due to the $\sin \theta_b$ term, but it can be easily linearized by approximating $\sin\theta_b$ with θ_b for small values of θ_b. If no corrective torque is applied ($T = 0$), the inverted pendulum model is inherently unstable since any small deviation of the body from upright, for example in a positive direction, results in a positive acceleration of the body in the same positive direction. The transfer function equation of the linearized differential equation is given by:

$$\frac{\theta_b(s)}{-T(s)} = \frac{1}{Js^2-mgh} \tag{2}$$

where s is the Laplace transform variable.

The corrective ankle torque, T, is generated in proportion to "error" signals associated with changes in θ_b. The error signal from somatosensory cues is proportional to the difference between the support surface position and the body sway angle (θ_p - θ_b). Similarly, the error signal from visual cues is proportional to (θ_v - θ_b). The error signals are processed through time delay elements representing neural processing, transmission, and muscle activation delays. In addition the error signals are transformed by PID controllers and used to generate corrective torques about the ankle joint. The equation representing PID function for the visual feedback loop is given by:

$$T_v(t) = K_{pv}(\theta_v - \theta_b) + K_{iv} \int_0^t (\theta_v - \theta_b)dt + K_{dv}\frac{d(\theta_v - \theta_b)}{dt}$$

(3)

where T_v is the corrective torque generated by the visual feedback loop, and K_{pv}, K_{iv}, and K_{dv} are constants. Similarly, the somatosensory PID controller generates the corrective torque, T_s, with PID controller constants K_{ps}, K_{is}, and K_{ds}. The two corrective torques are summed to produce the total corrective torque acting on the body to maintain stability.

Finally, a switch is shown in the somatosensory feedback loop. This switch represents an experimental test condition referred to as "sway-referencing." When this switch is closed, the body sway angle measure, θ_b, is used to drive a servo-control system that rotates the support surface in direct proportion to θ_b. This effectively maintains the somatosensory error signal (θ_p - θ_b) at a very low value. If, for modeling purposes, we assume that (θ_p - θ_b) is zero during sway-referencing, then T_s is also zero, and the somatosensory loop is no longer contributing corrective torque for postural control. Therefore the model assumes that only visual motion cues contribute to balance control in the sway-referenced test condition for vestibular loss subjects.

MODEL TUNING AND PREDICTIONS

The dynamic properties of the model are easily summarized by computing the model's overall transfer function. The simplest possible example is for the sway-referenced condition. In addition, assume that the visual PID controller only has proportional and derivative components (i.e. K_{iv} = 0). This assumption is initially made because it is known that only proportional and derivative components are necessary for the stable control of an inverted pendulum (Johansson et al., 1988). The transfer function equation for this system is given by:

$$\frac{\theta_b(s)}{\theta_v(s)} = \frac{(K_{dv}s + K_{pv})e^{-\tau_{dv}s}}{Js^2 + K_{dv}e^{-\tau_{dv}s}s + K_{pv}e^{-\tau_{dv}s} - mgh}$$

(4)

where τ_{dv} is the time delay associated with visual motion processing. This transfer function can be used to predict the amplitude and timing (phase) of θ_b in response to a sinusoidal visual surround motion stimulus at different stimulus frequencies. The gain is defined as the response amplitude divided by stimulus amplitude. The DC gain of this transfer function (obtained by setting s=0) is equal to K_{pv} / (K_{pv} - mgh) and is therefore greater than unity. That is, the model predicts that DC or very low frequency sinusoidal tilting motions of the visual surround will evoke body sways that are greater than the visual surround motion itself! If the integral component is added back into the PID controller, then the overall transfer function is given by:

$$\frac{\theta_b(s)}{\theta_v(s)} = \frac{(K_{dv}s^2 + K_{pv}s + K_{iv})e^{-\tau_{dv}s}}{Js^3 + K_{dv}e^{-\tau_{dv}s}s^2 + (K_{pv}e^{-\tau_{dv}s} - mgh)s + K_{iv}e^{-\tau_{dv}s}} \qquad (5)$$

This transfer function has unity gain at DC since the function of integral control action is to reduce steady state error ($\theta_v - \theta_b$) to zero. However at higher stimulus frequencies, the gain of the transfer function can still be greater than unity. This can be seen in Figure 2 which shows a family of curves, each one graphing gain as a function of stimulus frequency. In all of these curves, the gains remain above unity except for frequencies above about 1.5 Hz. The shape of the various gain curves depends on the values of the PID control parameters. Furthermore, the range of PID parameter values consistent with stable operation of the overall control system is limited. The gain curves in Figure 2 were calculated by keeping visual controller parameters K_{dv} and K_{iv} at fixed values while varying K_{pv} over its entire range of stable operation. As K_{pv} decreases, a sharp resonant peak at about 0.2 Hz develops. The overall system becomes unstable for values of K_{pv} less than about 11 N-m/deg. As K_{pv} increases, a sharp resonant peak at about 0.8 Hz develops, and the system is unstable if K_{pv} is greater than about 19 N-m/deg. A reasonable prediction is that the nervous system would select controller parameters which are near the middle of the stable range, and would therefore avoid the strong resonance created when a parameter is near a stability limit.

If the posture platform support surface is kept in a fixed position (open switch in Figure 1 with $\theta_p = 0$), then ($\theta_p - \theta_b$) changes with body sway and the somatosensory controller generates torque, T_s, which contributes to postural control in addition to T_v. A transfer function equation relating body sway to visual surround motion can be written for the entire system. This transfer function equation is more complicated than the one for the visual loop alone. The transfer function gain curves for the combined somatosensory-visual loop system show similar properties to those in Figure 2 in that resonant peaks emerge as somatosensory control parameters approach their stability limits. However, in contrast to the visual loop transfer functions in which gains are generally greater than unity, the gains of the combined somatosensory-visual loop system are generally less than unity.

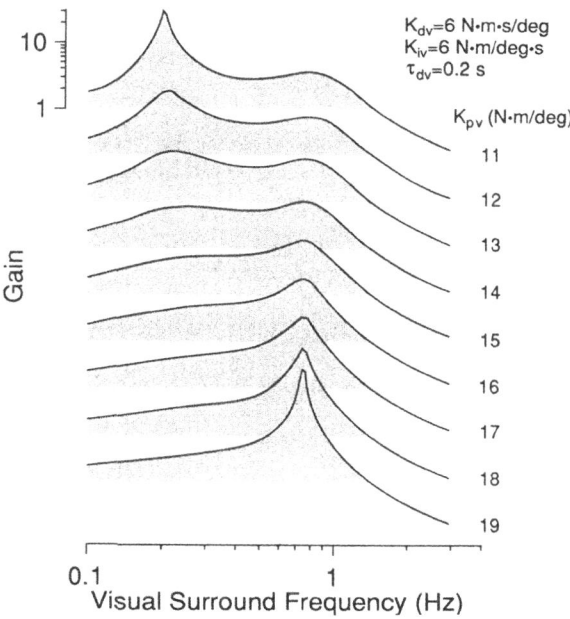

Figure 2. Predicted transfer function gain (body sway amplitude divided by visual surround amplitude) as a function of visual surround stimulus frequency for tests performed under sway-referenced conditions. The family of curves are calculated from Equation 5. The curves show the results of varying the visual PID controller parameter, K_{pv}, over a range consistent with stable operation while other PID parameters remain fixed. The gain scale applies to the upper gain curve and other curves are offset from one another by a gain factor of 5. The shaded areas indicate portions of the curves where gains are greater than unity.

COMPARISON WITH EXPERIMENTAL DATA

One might anticipate that all of the assumptions, simplifications, and approximations used in the modeling of postural responses would leave little chance that the modeling results could correspond with actual experimental observations. However this is not the case. Figure 3 shows gain and phase data at 0.1, 0.2, and 0.5 Hz obtained from 3 subjects with bilateral vestibular losses. The gain and phase data were calculated from center-of-gravity body sway angle responses to visual surround rotations at 0.1, 0.2, and 0.5 Hz. The data are only from responses to low amplitude visual motion stimuli (±0.2° and ±0.5°) in order to avoid nonlinear responses associated with larger stimuli. The data in the right column of Figure 3 are from tests performed with a sway-referenced support surface, and in the left column with a fixed support surface. For the sway-referenced condition, experimental data have gain values between 2 and 3, and there is a phase lead of body sway relative to visual surround position at 0.1 and 0.2 Hz. With the visual time delay parameter set to 0.2 s, the visual loop PID parameters in Equation 5 can be found which provide a reasonable fit to the available gain and phase data. The transfer function given by Equation 5 produces a phase lead required to match the actual gain and phase data from vestibular loss subjects only if the integral component is present in the PID controller. This can be taken as evidence that integral control action is present in the postural control system.

For the fixed support surface condition, the gains were all less than unity but were increasing with increasing frequency. A phase lead was present at 0.1 Hz and a phase lag at 0.5 Hz. However the magnitude of this lead and lag were smaller in the fixed condition compared to the sway-referenced condition. With the visual loop PID parameters remaining fixed at the values that provided a good fit to the sway-referenced condition data, a reasonable fit to the fixed condition experimental data was found by adjusting the somatosensory loop

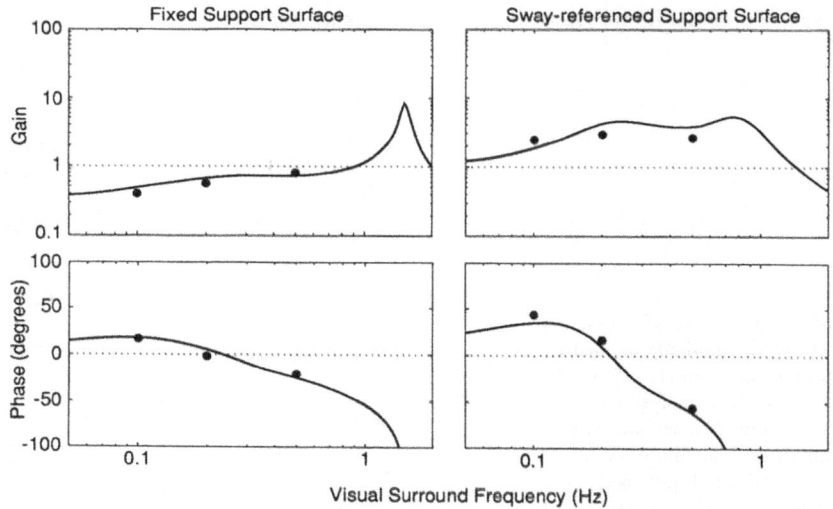

Figure 3. Gain and phase of body sway evoked by visual surround motions of varying frequencies and in conditions with a fixed (left column) or sway-referenced (right column) support surface. Solid dots represent average data from 3 subjects with bilaterally absent vestibular function. The solid lines are fits to the data by the model shown in Figure 1. PID parameters for the visual loop in both the fixed and sway-referenced conditions are K_{pv} = 14 N-m/deg, K_{iv} = 6 N-m/deg-s, K_{dv} = 6 N-m-s/deg. Parameters for the somatosensory loop are K_{ps} = 18 N-m/deg, K_{is} = 12 N-m/deg-s, K_{ds} = 7 N-m-s/deg. The time delay parameters are 0.2 s for the visual loop and 0.07 s for the somatosensory loop, and J = 62 kg-m^2 and mgh = 530 kg-m^2/s^2.

PID parameters. In this case the somatosensory loop time delay was set to 0.07 s. The fixed condition model curve fit is less satisfying than the sway-referenced condition fit in that a resonance peak at about 1.5 Hz is present. While the fit closely approximates the available gain and phase data, the presence of the peak at 1.5 Hz indicates that the system is not far from instability. We currently have no data to confirm the prediction of this resonance peak.

SUMMARY

A simple feedback control model for the control of upright stance in subjects without vestibular function was developed. The model assumes that somatosensory and visual orientation cues are processed independently and then add linearly to produce the corrective torques necessary to maintain upright stance. Different processing time delays were assumed for the somatosensory and visual systems, but no attempt was made to align the timing of signals from the two systems.

The model is consistent with experimental data from vestibular loss subjects whose posture was perturbed by low amplitude visual surround motion. The model provides some insight into the central processing of sensory motion information since it suggests that a mathematical integration of body sway position occurs in postural control, even though this integration is not strictly required for the maintenance of upright stance (Johansson et al., 1988). The model also demonstrates the potential for resonance effects in postural control with the resonance properties largely dependent upon the processing of sensory orientation information.

An obvious direct extension of this model is to include vestibular feedback by adding a third feedback loop which generates corrective torque that sums linearly with the somato-sensory and visual loop torques. However there is some evidence that vestibular information below a frequency-dependent threshold level does not contribute to postural control (Peterka and Benolken, 1992). Since a threshold nonlinearity cannot be linearized in order to calculate transfer function equations, simulations would be required for comparisons with experimental data.

ACKNOWLEDGMENTS

Work supported by NASA grants NAG 9-117, NAGW-3782, and NIH grant P60 DC02072.

REFERENCES

Dichgans, J., and Diener, H.C., 1989, The contribution of vestibulo-spinal mechanisms to the maintenance of human upright posture, *Acta Otolaryngol (Stockh)* 107:338-345.
Diener, H.C., Dichgans, J., Guschlbauer B., and Mau H., 1984, The significance of proprioception on postural stabilization as assessed by ischemia, *Brain Research* 296:103-109.
Ishida, A., and Miyazaki, S., 1987, Maximum likelihood identification of a posture control system, *IEEE Trans Biomed Eng* 34:1-5.
Johansson R., Magnusson, M., and Åkesson M., 1988, Identification of Human Postural Dynamics, *IEEE Trans Biomed Eng* 35:858-869.
Lestienne, F., Soechting, J., and Berthoz, A., 1977, Postural readjustments induced by linear motion of visual scenes, *Exp Brain Res* 28:363-384.
Mergner, T., Siebold, C., Schweigart, G., and Becker, W., 1991, Human perception of horizontal trunk and head rotation in space during vestibular and neck stimulation, *Exp Brain Res* 85:389-404.

Peterka, R.J., and Benolken, M.S., 1992, Role of somatosensory and vestibular cues in attenuating visually-induced human postural sway, In: Woollacott M, Horak F (eds.) Posture and gait: Control mechanisms, vol. 1, University of Oregon Books, pp 272-275.

Koozekanani, S.H., Stockwell, C.W., McGhee, R.B., and Firoozmand, F., 1980, On the role of dynamic models in quantitative posturography, *IEEE Trans Biomed Eng* 27:605-609.

36

EFFORTS TO QUANTIFY ADAPTATION IN MODELING OF POSTURAL CONTROL

M. Magnusson, R. Johansson and P.-A. Fransson

University Hospital of Lund
Department of Oto-rhino-laryngology
22185 Lund
Sweden

INTRODUCTION

The ability to maintain stability and to withstand the effects of gravity on stance and motion is important to a large variety of human activities. It is assumed that adaptation is a necessary quality in human postural control. Studies of adaptation generally include evaluation of gain or amplitude of various parameters of the responses to repeated identical perturbations (Nashner 1976, Nashner and McCollum, 1985, Roos et al., 1988). Such a procedure will inevitably include some possibility of anticipatory action of the subject and one may have some doubts on the generality of the results. Moreover, the evaluation will concentrate on the reduction of responses more than on the time span required to achieve such a reduction. There is a multitude of studies on adaptation of the vestibular ocular reflex where one observes diminishing responses to repeated stimuli, but also over time to repeated sessions of exposure to a stimulus (c.f. Bock et al., 1979; Wilson and Melville-Jones, 1979). Furthermore, the observation of the compensatory process after acute vestibular lesions as well as of reduced symptoms of motion sickness with the adaptation to a moving surrounding or microgravity also exemplifies the conditioning of vestibular responses over time (Wilson and Melville-Jones, 1979).

There are several studies demonstrating alterations in support surfaces to change the postural behavior (Forssberg and Nashner, 1982, Nashner et al., 1982). These studies, however, again focus on habituation to repeated exposures. If one considers the requirements on postural control in every day life, where a subject may be walking down from a firm support surface on to a boat deck or any unstable or soft surface, it is evident that pre-determined and rigid postural or vestibulo-spinal responses or strategies would be inappropriate. The reconditioning or adaptation of the postural control under such circumstances will occur as a continuous process in response to the altering requirements of the changing surrounding. It is therefore appropriate to consider such a process as one of adaptation.

It is feasible to assume that the ability to continuously adapt to different environmental requirements on postural control with short latencies must be of utmost importance to human mobility and to survival in the pre-civilized era. The mechanisms of such conditioning of postural control may be an effect of cognition or selecting different 'strategies' from experience. If so it should vary between different subjects. The conditioning might, however, depend on an adaptive processing of information of perhaps basal importance.

From previous studies we do not know the characteristics of adaptation in human postural control or the time required to develop it. We can postulate, however, that if such a uniform adaptive mechanism exists it should be of paramount importance to human postural control and postural deficits. Therefore the aim of the present studies was to investigate the time span of adaptation in postural control for normal subjects. If this can be quantified, it would allow the development of methods by which the adaptive behavior in human postural control of subjects with postural deficits can be evaluated. In the present report we focus our interest on postural adaptation to a vestibular disturbance in the lateral plane.

METHODS

The present material comprised twelve naive human subjects (mean age, 30.9 years) without any history of vertigo, central nervous disorder, ear disease, or previous injury to the lower extremities. At the time of the investigation, no subject was on any form of medication or had consumed alcoholic beverages for at least 48 hours.

The experimental procedure has been detailed elsewhere (Johansson et al., 1995). The equipment consisted of a square force platform developed at the Institute of Occupational Health, Helsinki, Finland, and the ENT Clinic at University Hospital, Lund, Sweden. Measurements obtained from the force platform were recorded and sampled by a computer at a rate of 20 Hz, which was chosen to permit sufficiently rapid variations in the galvanic stimulus. The subject stood with heels together on the platform while staring at a spot on the opposite wall. Carbon electrodes (Cefar AB, Lund, Sweden) were attached symmetrically on the mastoid process behind each ear and electric stimulation was produced by a constant current generator at 1 mA with opposite polarity of the two electrodes. The polarity of the galvanic stimulation was changed pseudorandomly (PRBS) to produce a spectrum up to 5-10 Hz according to a computer-driven program. The stimulus and the recording of sway forces were carefully synchronized by means of real-time software. During the test, the subject stood erect with arms across the chest either with closed or open eyes, as instructed, and the recording was started. First, spontaneous sway was recorded for 30 s. The stimulus was then started and maintained at a current of 1.00 mA, during which period the stimulus changed polarity over the following 183.6 s.

DATA ANALYSIS

Basic data analysis was made by means of system identification methodology using autospectra, cross spectra, and coherence spectra of the input (i.e., galvanic stimulus) and the outputs (sway force responses). It was verified that the signal levels were adequate and that the stimulus spectrum covered the relevant spectral ranges of biological interest in vestibular research i.e. from below 0.1 Hz up to 10 Hz.

For modeling we used a multi-link inverted pendulum model with coordinates of the links and an associated control model. Briefly, Matlab™ was used for time-series analysis. Since the data exhibited clear nonstationary properties with trends and time-varying behavior, it was not possible to use standard time-series analysis based on stationary stochastic models. To solve the identification problem, we designed a pseudolinear regression that successfully fitted the data to the model structure (Johansson et al., 1995). A detailed account of the modeling assumptions and the evidence that a linearized transfer function from the stimulus to the torque responses for a two-link stabilized inverted pendulum is found in the study of Johansson et al. (1995).

Hence, the identification problem was solved with a pseudolinear regression that fitted the data. The model considered the stimulus (the polarity shifts of the galvanic current), the step input (considering the start of the complete train of galvanic pulses as one stimulus), ramp input (assuming a decay of the strength, or rather the impact, of each pulse). Determination of the estimated model of a suitable model order was supported by model validation criteria (Johansson 1993, Ljung, 1987) such as statistical evaluation of the loss function, the Akaike information criterion (AIC), the final prediction criterion (FPE), and residual analysis (tests of auto-correlation and cross correlation between stimulus and residuals) (Johansson, 1993, Ljung, 1987). The goal of identification was to determine a time-invariant model with the minimum residual power and a sequence of zero-mean non-correlated residuals (,,white noise" in engineering terminology). The correspondence between the resulting ARMAX model and the biomechanical continuous-time model was done by standard methodology (Johansson, 1993).

Responses to the step and ramp inputs and the fitting of the exponential to the residual sequence served to characterize the adaptive properties. Adaptation-related properties were then quantified by extracting the information from sequences of squared residuals, i.e. the sequences of squared non-correlated zero-mean residuals, and in the residual variance in response to the onset of stimulus, and by denoting the time constant of the attenuation of residual power from the peak value at the start of the stimulus. The time constants of the residual attenuation were calculated as shown in Eq. [1]

$$y(t) = a^2 (Ke^{-t/T} + 1) \tag{1}$$

where a is the variance, t is time, T a time constant, K a constant.

Calculation of time constants was done both in the lateral and the anterior-posterior plane.

In the present report we will focus on the analysis of the adaptive properties.

RESULTS

Already from visual inspection of the recordings it was evident that there was a decay of the responses. Calculating the time constant of the decay yielded abundant values for all subjects tested (Table 1). This was observed both for sway in the lateral plane, in line with the effect of vestibular stimulation, and for the concomitant anterior-posterior sway.

Table 1. Time constants (s) of the adaptive decline of postural responses during vestibular stimulation in the lateral and the anterior-posterior (a-p) planes.

Subject	1	2	3	4	5	6	7	8	9	10	11	12	mean	S.D.	S.E.M
$T_{lateral}$	40.8	56.8	47.2	47.4	42.7	40.8	43.3	45.0	208.3	43.0	47.9	36.6	58.3	47.5	13.7
T_{a-p}	46.9	44.4	42.9	48.1	41.7	41.0	40.3	41.7	45.0	42.1	36.2	44.6	42.9	3.18	0.92

DISCUSSION

The present study demonstrate an abundant decline of postural responses to a galvanic vestibular stimulus consisting of right-left poleshifts. This decline can be quantified by means of defining a 'postural adaptive time constant'.

It is well known that a bipolar binaural galvanic stimulus induces vestibular and postural responses. Vestibular nystagmus can be elicited with the fast phase directed toward the cathode at current intensities of several mA, which generally will cause some pain to the subject (Brantberg and Magnusson, 1990). A less powerful stimulus causes asymmetry of vestibulary mediated eye motor reflexes (Nashner and P. Wolfson 1974). Currents of less than 0.4 mA induce postural sway in the direction of the inter-aural axis but does not evoke discomfort (Watanabe et al., 1989). A bipolar binaural galvanic stimulus of 1 mA causes an asymmetric activation of the soleus muscles increasing EMG activity on the side of the cathode and decreasing it on the side of the anode with a latency of approximately 100 ms (Tokita et al., 1989). Thus, a galvanic stimulus to the vestibular nerve as used in the present experiments induce postural movements from the neck down and in the direction of the anode with short latencies. In the present study a constant current stimulator was used; we therefore assume that the galvanic stimuli to the vestibular nerve was of the same magnitude throughout each experiment. Therefore, the decline of the responses can be assumed to result from an adaptation of the interpretation or of the induced output. This might suggest that it results from an adjustment of the neural control of posture. The decline of the responses might be caused by cognition or a conscious learning processes, or by a selection of more appropriate postural strategies. It is felt, however, that, if the decline would be merely based on such processes, one would encounter larger inter-individual variations and perhaps stepwise shifts, rather than a continuous decline of responses. It is noteworthy that the decline of responses is of similar magnitude in the lateral plane (that of the evoked perturbations) and in the anterior-posterior plane. Although one cannot exclude the possibilities of a biomechanical crosstalk, the data suggest that the adaptation represents a central nervous rather than biomechanical phenomenon. Furthermore, we have also observed a similar magnitude of the decline of responses to perturbations invoked by vibratory stimulation of calf and neck muscles, further supporting a unified mechanism for adaptation to perturbations (Fransson et al., 1995)

We have used the term adaptation rather than habituation because the decline of responses is a continuous decline in response to a continuous period of perturbations. One might argue that the different pulses would represent repeated stimuli. However, these pulses were administered according to a PRBS schedule and the subject is considered to be in a stimulated state throughout the experiment. Therefore, the term adaptation and hence, 'postural adaptive time constant', seems appropriate.

The ability to adapt postural control to changes in environmental constraints, as they may take place under natural conditions, should be of major importance in every day life. The observation of a decline of postural responses with a time constant in the magnitude of 40-50 s suggests further, that postural measurements of shorter duration will not reflect deficits in the ability to adapt to disturbances. The present approach however, allows an estimation of such capabilities and may, when applied to patients, contribute to a better understanding of postural deficits as well as of effects of training programs.

SUMMARY

Adaptation to a continuous vestibular disturbance was studied in 12 normal subjects. Perturbations in mainly the lateral plane, corresponding to the intra-aural axis, were evoked by galvanic vestibular stimulation with an amplitude of 1.0 mA. Forces actuated by the feet against the support surface were recorded with a force plate and sampled by computer. The dynamic control of posture was modeled and found to correspond to a fourth order model which could be interpreted as a linked double inverted pendulum. Subjects demonstrated a time-dependent decline of the postural responses to stimuli. The decline is assumed to depend on adaptation and the time span of the adaptation was estimated in the lateral and anterior-posterior planes by means of an exponential function. The calculated 'postural adaptive time constant' was in the range of 40-60 s. The present approach may allow estimation of adaptive behavior in human postural control.

REFERENCES

Bock, O., von Koschitzky, H., and Zangemeister, W.H., 1979, Vestibular adaptation to long-term stimuli, *Biol Cybern* 33:77-9.

Brantberg, K., and Magnusson,M., 1990, Galvanically induced asymmetric optokinetic after-nystagmus, *Acta Otolaryngol* 110:189-195.

Forssberg, H., and Nashner, L.M., 1982, Ontogenetic development of postural control in man: adaptation to altered support and visual conditions during stance, *J Neurosci*. 2(5):545-52.

Fransson, P.-A., Magnusson M., and Johansson, R., 1995, Analysis of adaptation in anteroposterior dynamics of human postural control (Submitted).

Johansson, R., System Modeling and Identification, Prentice-Hall, Englewood Cliffs, N.J., 1993.

Johansson, R, Magnusson, M., and Fransson, P.-A. Galvanic vestibular stimulation for analysis of postural adaptation and stability. Determination of a Postural Adaptation Time Constant. *IEEE Trans Biomed Eng* 1995, 282-292.

Ljung, L., System identification - Theory for the user, Prentice-Hall, Englewood Cliffs, NJ, 1987.

Nashner, L.M., and Wolfson, P., 1974, Influence of head position and proprioceptive cues on short latency postural reflexes evoked by galvanic stimulation of the human labyrinth, *Brain Research* 67:255-268.

Nashner, L.M., Adapting reflexes controlling the human posture. *Exp Brain Res* 26(1):58-72.

Nashner, L.M., Black, F.O., and Wall, C., 1982, 3d. Adaptation to altered support and visual conditions during stance: patients with vestibular deficits, *J Neurosci*. 2:536-44.

Nashner, L.M., and McCollum, G.,1985, The organization of human postural movements. A formal basis and experimental synthesis. *The Behavioral and Brain Sciences*, Vol. 8:135-172.

Roos, H., Bles, W., and Bos, J.E., Postural control after repeated exposure to a tilting room. In: Posture and gait: Development, adaptation and modulation. B. Amblard, A. Berthoz, and F. Clarac (Eds.), Elsevier, Amsterdam, pp. 137-144, 1988.

Tokita, Y., Ito, Y., and Takagi, K, 1989, Modulation by head and trunk positions of the vestibulo-spinal reflexes evoked by galvanic stimulation of the labyrinth. Observations by labyrinth evoked EMG. *Acta Otolaryngol*, 107:327-332.

Watanabe, K., Mizukoshi, H., Ohi, K., Ysuda, N., Ohashi, N., and Kobayashi, H., 1989, Retro-labyrinthine disorder detected by galvanic body sway responses in routine equilibrium examinations, *Acta Otolaryngol Suppl*. 468:343-348.

Wilson, J., and Melville-Jones, G., Mammalian Vestibular Physiology, Plenum New York, 1979.

37

CENTRAL VESTIBULAR DISORDERS OF THE ROLL PLANE

M. Dieterich and Th. Brandt

Ludwig-Maximilians-University Munich
Department of Neurology
Klinikum Grosshadern
Marchioninistrasse 15
81377 München
Germany

INTRODUCTION

It is a well established fact that several postural and ocular motor disorders are secondary to distinct and separate lesions of central vestibular pathways from the vestibular nuclei to ocular motor nuclei. A generally established classification of central vestibular disorders of the brainstem does not exist. One of us (Brandt, 1991) has proposed a first hypothetical and speculative attempt to classify vestibular brainstem disorders according to a lesional tone imbalance in one of the three major planes of action of the VOR: (I) horizontal head rotation about a vertical z-axis = yaw, (II) vertical head extension or flexion about a horizontal binaural y-axis = pitch, (III) lateral head tilt about a horizontal (line of sight) x-axis = roll.

The clinical signs, perceptual and motor, of a vestibular tone imbalance in the roll plane are ocular tilt reaction, skew deviation, ocular torsion, and tilts of the perceived vertical. A unilateral lesion of graviceptive pathways is causative (Brandt and Dieterich, 1994). These pathways travel from the vestibular nuclei, crossing midline at pontine level, towards the contralateral medial longitudinal fasciculus (MLF) to reach the oculomotor nuclei and the interstitial nucleus of Cajal (INC). The major diagnostic rule is based on the crossing of graviceptive pathways for vestibular function in roll: Unilateral pontomedullary lesions cause ipsiversive perceptual, ocular motor, and head tilts, whereas pontomesencephalic lesions cause contraversive tilts. Unilateral vestibular thalamus or cortex lesions manifest as perceptual tilts only without the ocular motor abnormalities typical for the corresponding brainstem lesions. They occur as either contraversive or ipsiversive tilts of perceived vertical in lesions of the vestibular thalamic nuclei and as contraversive tilts of perceived vertical in lesions of the (parieto-insular) vestibular cortex.

Multisensory Control of Posture, Edited by T. Mergner and F. Hlavačka
Plenum Press, New York, 1995

SIGNS AND SYMPTOMS OF VESTIBULAR DYSFUNCTION IN THE ROLL PLANE

The "graviceptive" input from the otoliths converge with that from the vertical semicircular canals at the level of the vestibular nuclei (Angelaki et al., 1993) and the ocular motor nuclei (Baker et al., 1973; Schwindt et al. 1973) to subserve static and dynamic vestibular function in pitch (up and down in the saggital plane) and roll (lateral tilt in the frontal plane). In the "normal" position in the roll plane, the subjective visual vertical (SVV) is aligned with the gravitational vertical, and the axes of the eyes and the head are horizontal and directed straight ahead.

Signs and symptoms of a vestibular dysfunction in the roll plane can be derived from deviations from normal function. A tonic vestibular tone imbalance due to a lesion should result in a syndrome consisting of a perceptual tilt, head and body tilt, vertical misalignment of the visual axes (skew deviation) and ocular torsion (Fig. 1). This has been demonstrated in animal experiments by unilateral stimulation of the utricular nerve (cat: Suzuki et al., 1969), the utricular macula (guinea pig: Curthoys, 1987) or vertical canal nerves (cat: Cohen et al., 1964; Tokumasu et al., 1971) which resulted in either a complete ocular tilt reaction (OTR) - the triad of head tilt, skew deviation and ocular torsion - or in its single components. In humans, inadvertent damage to one utricle (Halmagyi et al., 1979) or inappropriate stimulation of the otoliths in patients with a Tullio phenomenon (Dieterich et al., 1989) also caused OTR, the pattern of which was first described by Westheimer and Blair in monkeys (Westheimer and Blair, 1975 a,b). Single cases of OTR were also described in mesencephalic disorders in humans manifesting in paroxysms (Rabinovitch et al., 1977; Hedges and Hoyt, 1982). Systematic analysis of unilateral brainstem infarctions (Table 1) exhibited that OTR represents a fundamental pattern of eye-head coordination in the roll plane and can be observed not only in patients with peripheral vestibular dysfunction but also in those with

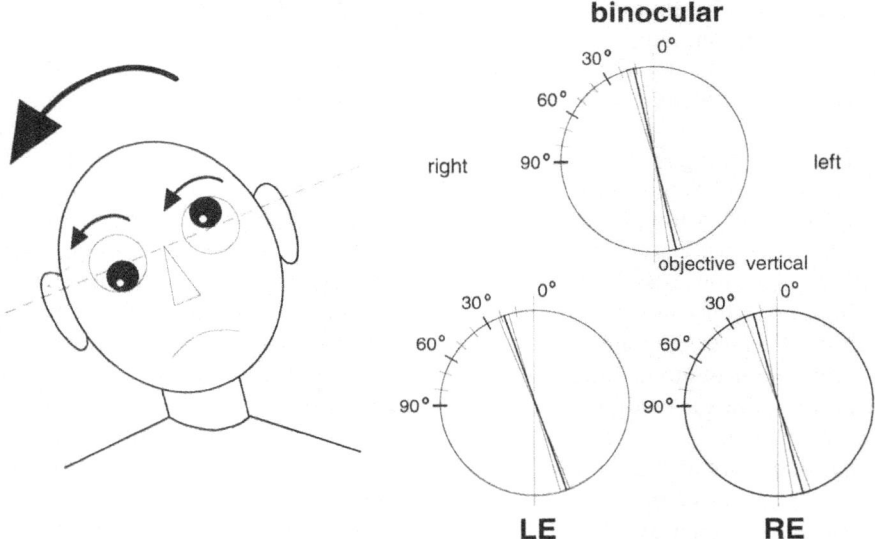

Figure 1. Left: Schematic drawing of ocular tilt reaction to the right with rightward head tilt, skew deviation (with right eye undermost) and binocular torsion. Right: Adjustments of the SVV under binocular or monocular viewing conditions. (Means and standard deviations.) The direction of SVV tilt corresponds to head tilt. RE, right eye; LE, left eye. (Reprinted from: Brandt and Dieterich, 1994)

Table 1. Frequency (in %) of SVV tilt, skew deviation, ocular torsion, and OTR in acute unilateral brainstem and thalamic infarctions

Lesion	Patients (n)	SVV tilt	Ocular torsion		Skew	OTR
			Monocular	Binocular		
Mesodiencephalic						
paramedian thalamic	14	64%	29%[*]	43%[*]	57%	57%
posterolateral thalamic	17	65%	13%[#]	20%[#]	0	0
anterior polar thalamic	4	0	0	0	0	0
Mesencephalic	16	94%	54%	38%	37.5%	25%
Pontomesencephalic	12	92%	64%	18%	25%	25%
Pontine	34	91%	47%	33%	26.5%	12%
Pontomedullary	13	100%	60%	20%	23%	7.7%
Medullary	36	94%	27%	55%	44%	33%
(Wallenberg's S.)						
Total	111	94%	47%	36%	31%	20%

Reprinted from: Brandt and Dieterich [1994]
*Additional third nerve palsy.
#Slight torsion of 2.8°.

lesions of the graviceptive pathways, which run from the medulla to the mesencephalon (Brandt and Dieterich, 1987; Halmagyi et al., 1990; Dieterich and Brandt, 1992; Dieterich and Brandt, 1993 a,b; Brandt and Dieterich, 1993a,b). It is not only the complete synkinesis of OTR which indicates a dysfunction in the roll plane but either one of its components, postural or perceived tilt, skew deviation or ocular torsion, indicates tonic imbalance.

There are clinical rules which may be proven exceptionally incorrect but are helpful for basic understanding. Skew deviation, a supranuclear vertical divergence of the eyes, is always associated with ocular torsion (skew torsion sign) and tilt of perceived vertical towards the undermost eye (Brandt and Dieterich, 1993b), i.e. the only difference between the ocular skew torsion sign and OTR is head tilt. Whereas skew deviation does not manifest itself without ocular torsion, monocular or binocular torsion is frequently seen without concurrent skew deviation. Finally, perceived vertical may be tilted with or without concurrent skew deviation, ocular torsion or head tilt (Fig. 2, Table 1).

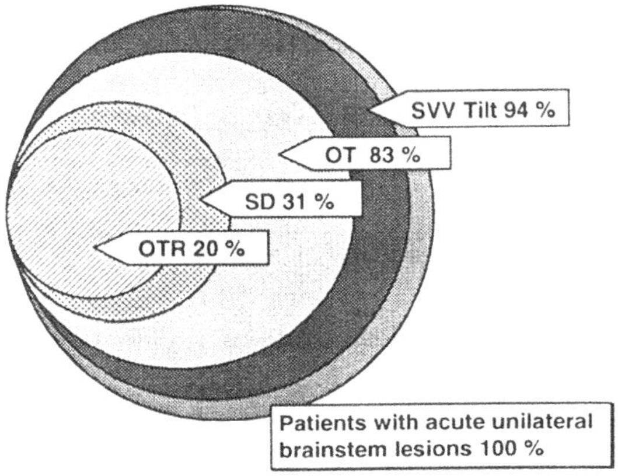

Figure 2. Frequency (in %) of different signs of vestibular dysfunction in the roll plane as determined in a total of 111 patients with acute unilateral brainstem infarctions. SVV tilts are the most sensitive sign, followed by monocular or binocular torsion (OT).

SVV Tilt 94 %
OT 83 %
SD 31 %
OTR 20 %
Patients with acute unilateral brainstem lesions 100 %

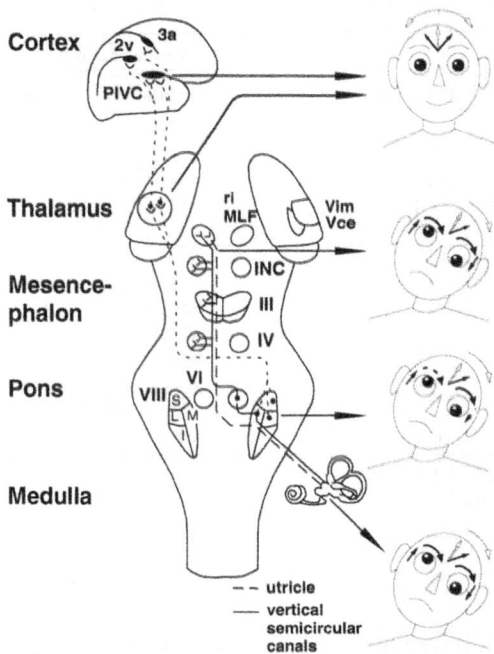

Cortex

Thalamus

Mesence-
phalon

Pons

Medulla

-- utricle
— vertical
   semicircular
   canals

Figure 3. Graviceptive pathways from the otoliths and vertical semicircular canals mediating the vestibular reactions in the roll plane. The projections from the otoliths and the vertical semicircular canals to the ocular motor nuclei (trochlear nucleus IV, oculomotor nucleus III, abducens nucleus VI) and the supranuclear centers of the INC, and the rostral interstitial nucleus of the MLF (riMLF) are shown. They subserve VOR in three planes. The VOR is part of a more complex vestibular reaction which also involves vestibulospinal connections via the medial and lateral vestibulospinal tracts for head and body posture control. Furthermore, connections to the assumed vestibular cortex (areas 2v and 3a and the parieto-insular vestibular cortex , PIVC) via the vestibular nuclei of the thalamus (Vim, Vce) are depicted. Graviceptive vestibular pathways for the roll plane cross at pontine level. OTR is depicted schematically on the right in relation to the level of the lesion: ipsiversive OTR with peripheral and pontomedullary lesions; contraversive OTR with pontomesencephalic lesions. In vestibular thalamus lesions the tilts of SVV may be contraversive or ipsiversive, in vestibular cortex lesions they are preferably contraversive. OTR is not induced by supratentorial lesions above the level of INC. (Reprinted from: Brandt and Dieterich, 1994).

There is convincing evidence that all these signs and symptoms reflect vestibular dysfunction in the roll plane. They may be found in combination or as single components at all brainstem levels. A systematic study of 111 patients with acute unilateral brainstem infarctions revealed that pathological tilts of SVV (94%) and ocular torsion (83%) are the most sensitive signs. Skew deviation was found in one-third and a complete OTR in one-fifth of these patients (Table 1, Fig. 2). Clinical evaluation of vestibular function in roll therefore includes psychophysical adjustments of SVV, determination of the vertical divergence of the visual axes by use of prisms, and determination of ocular torsion by means of fundus photographs (for methods see: Dieterich and Brandt, 1993 a). The presentation of vestibular signs and symptoms in roll plane follows the course of the ascending graviceptive pathways from the vestibular nuclei to the cortex. Two criteria may be useful for topographic diagnosis, the pattern of motor or perceptual dysfunction and the direction of tilt with respect to the side and the level of dysfunction.

RULES FOR TOPOGRAPHIC BRAINSTEM, THALAMUS AND CORTEX CORRELATIONS WITH VASCULAR LESIONS

Current clinical data allow the following preliminary diagnostic rules for vestibular signs in roll (Brandt and Dieterich, 1994) (see Fig. 3):

1. The fundamental pattern of body-eye-head-tilt in roll - either complete OTR or skew torsion without head tilt - indicates an unilateral peripheral deficit of otolith input or a unilateral lesion of graviceptive brainstem pathways from the vestibular nuclei (crossing midline at lower pontine level) to the INC in the rostral midbrain.

2. SVV tilts occur with peripheral or central vestibular lesions from the labyrinth to the vestibular cortex and represent the most sensitive sign of a tonic vestibular tone imbalance in roll.

3. All tilt effects, perceptual, ocular motor and postural, are ipsiversive (ipsilateral eye lowermost) with unilateral peripheral or ponto-medullary lesions below the crossing of the gravceptive pathways. They indicate involvement of medial and/or superior vestibular nuclei, mainly supplied by the vertebral artery (Fig. 4).

4. All tilt effects in unilateral ponto-mesencephalic brainstem lesions are contraversive (contralateral eye lowermost) and indicate involvement of the MLF (paramedian arteries arising from basilar artery; Fig. 4) or INC and riMLF (paramedian superior mesencephalic arteries arising from basilar artery).

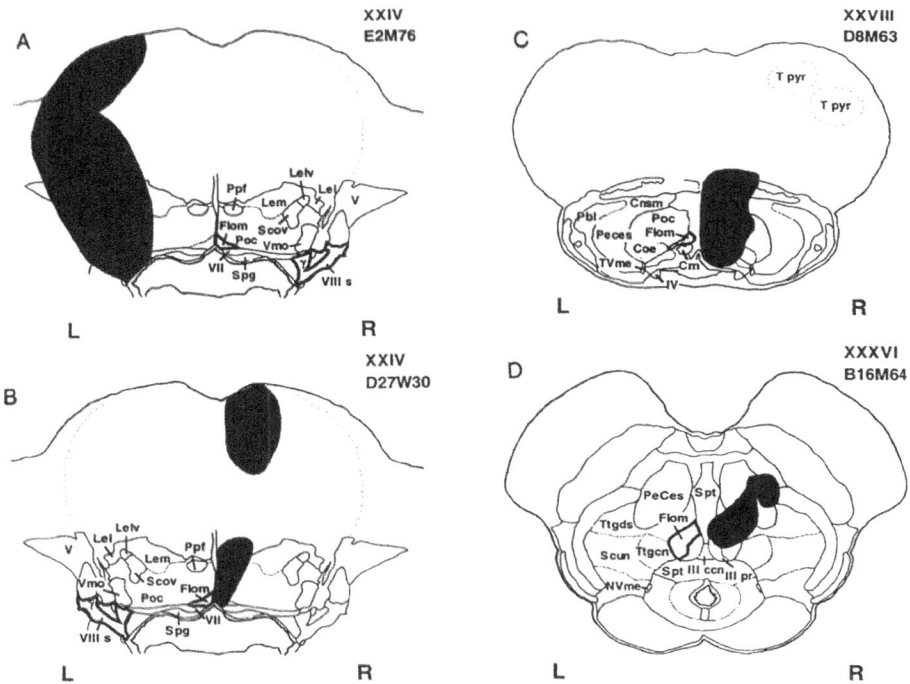

Figure 4. Typical lesioned areas in 4 patients with unilateral pontomedullary infarctions showing ocular skew torsion sign to the left (left eye undermost). Ischemic areas were taken from MRI scans and projected onto the appropriate transverse sections of the stereotaxic brainstem atlas of Olszewski and Baxter (1982). In medullary lesions involving the superior vestibular nuclei (VIIIs in section XXIV) skew torsion was ipsiversive (A). Ocular skew torsion was contraversive if the medial longitudinal fasciculus (Flom) was involved (B, C, D).

5. Unilateral lesions of vestibular structures rostral to the INC typically manifest with deviations of perceived vertical without concurrent eye-head-tilt.
6. OTR in unilateral paramedian thalamic infarctions (paramedian thalamic arteries from basilar artery) indicates simultaneous ischemia of the paramedian rostral midbrain including the INC.
7. Unilateral lesions of the posterolateral thalamus can cause thalamic astasia and moderate ipsiversive or contraversive SVV tilts, thereby indicating involvement of the "vestibular" thalamic subnuclei (thalamogeniculate arteries).
8. Unilateral lesions of the parieto-insular vestibular cortex cause moderate, mostly contraversive SVV tilts (temporal branches of the middle cerebral artery or deep perforators).
9. A SVV tilt found with monocular but not with binocular viewing is typical for a trochlear or oculomotor palsy rather than a supranuclear graviceptive brainstem lesion.

Thus, all clinical signs of vestibular dysfunction in roll plane may be helpful to determine particularly the level as well as the side of the brainstem lesion. If the level of the damage is known from the clinical syndrome, than it indicates the more severely affected side. If the side of the damage is clear from the clinical syndrome, the level of the brainstem is indicated by the direction of OTR, skew deviation and SVV tilt. Their diagnostic topographic value is similar to that of cranial nerve lesions, but is more sensitive. The discussed effects can be explained by a lesional vestibular tone imbalance caused by decreased resting activity of the graviceptive pathways or INC neurons, whereas paroxysmal effects have been described (Rabinovitch et al., 1977; Hedges and Hoyt, 1982; Lueck et al., 1991) of some structures due to transiently increased activity or electrical stimulation. The directions of the effects caused by paroxysmal activation are opposite to that caused by lesional inhibition such as unilateral infarction (Halmagyi et al., 1990).

PRELIMINARY CLASSIFICATION OF VESTIBULAR BRAINSTEM DISORDERS ACCORDING TO THE THREE MAJOR PLANES OF ACTION OF THE VOR

Pathways that mediate the VOR in either one of the three major planes indeed travel separate from each other within the ipsilateral or contralateral MLF, the brachium conjunctivum, or the ventral tegmental tract. Such a classification is helpful for the clinician in terms of pathophysiological thinking and topologic diagnostics. A rigid three-plane-scheme is, however, to simple to cover current knowledge of the complex vestibular neuronal network (Leigh and Brandt, 1993).

Not only do we know critical pathways and structures for dysfunction in the roll plane but also some critical structures which, if lesioned, cause dysfunction in the horizontal yaw or vertical pitch plane. Small infarctions in the territory of the AICA or the PICA or demyelinating plaques in multiple sclerosis have been repeatedly reported to mimic unilateral peripheral labyrinthine disorders (Duncan et al., 1975; Rubinstein et al., 1980; Brandt and Dieterich, 1993a). We are not aware of central vestibular disorders in yaw elicited from an area other than lateral ponto-medullary region involving the vestibular nuclei. Downbeat and upbeat nystagmus are the directional counterparts of a vestibular tone imbalance in pitch plane and are caused by bilateral paramedian lesions. This is why they are missing in unilateral brainstem infarctions. Downbeat nystagmus occurs either with structural lesions which interrupt the commissural fibers between the vestibular nuclei at the floor of the fourth ventricle or with bilateral lesions of the flocculus (Baloh and Spooner, 1981). Upbeat nystagmus arises from two separate intra-axial brainstem lesions in the tegmentum of the ponto-mesencephalic or ponto-medullary junction (Fisher et al., 1983; Ranally and Sharpe, 1988). The close proximity of the areas causing either upbeat or downbeat nystagmus in the medulla fits clinical observations of an occasional transition from upbeating to downbeating and vice versa.

ACKNOWLEDGMENTS

This work was supported by the Deutsche Forschungsgemeinschaft SFB 220, D6, D13, and the Wilhelm-Sander-Stiftung.

REFERENCES

Angelaki D.E., Bush, G.A., and Perachio, A.A., 1993, Two-dimensional spatiotemporal coding of linear acceleration in vestibular nuclei neurons, *J Neurosci.* 13:1403-1417.

Baker, R., Precht, W., and Berthoz, A., 1973, Synaptic connections to trochlear motoneurons determined by individual vestibular nerve branch stimulation in the cat, *Brain Res.* 64:402-406.

Baloh, R.W., and Spooner, J.W., 1981, Downbeat nystagmus. A type of central vestibular nystagmus, *Neurology* 31:304-310.

Brandt, Th., and Dieterich, M., 1987, Pathological eye-head coordination in roll: tonic ocular tilt reaction in mesencephalic and medullary lesions, *Brain* 110:649-666.

Brandt, Th., and Dieterich, M., 1993a, Preliminary classification of vestibular brain-stem disorders, In: Caplan LR, Hopf HC, eds. Brain-stem localization and function, Berlin, Heidelberg. Springer, pp. 79-91.

Brandt, Th., and Dieterich, M.,1993b, Skew deviation with ocular torsion: A vestibular sign of topographic diagnostic value, *Ann Neurol* 33:528-534.

Brandt. Th, and Dieterich M., 1994, Vestibular syndromes in the roll plane: Topographic diagnosis from brainstem to cortex, *Ann Neurol* 36:337-347.

Brandt. Th., 1991, Vertigo. Its multisensory syndromes. London. Springer.

Cohen, B., Suzuki, J.I., and Bender, M.B., 1964, Eye movements from semicircular canal nerve stimulation in the cat, *Ann Otol (St Louis)* 73:153-169.

Curthoys, I.S., 1987, Eye movements produced by utricular and saccular stimulation, *Aviat Space Environ Med* 58:Suppl A192-A197.

Dieterich, M., and Brandt, Th., 1992, Wallenberg's syndrome: Lateropulsion, cyclorotation, and subjective visual vertical in thirty-six patients, *Ann Neurol* 31:399-408.

Dieterich, M., and Brandt, Th., 1993a, Ocular torsion and tilt of subjective visual vertical are sensitive brainstem signs, *Ann Neurol* 33:292-299.

Dieterich, M., and Brandt, Th., 1993b, Thalamic infarctions: Differential effects on vestibular function in the roll plane (35 patients) *Neurology* 43:1732-1740.

Dieterich, M., Brandt, Th., and Fries, W., 1989, Otolith function in man. Results from a case of otolith Tullio phenomenon, *Brain* 112:1377-1392.

Duncan, G.W., Parker, S.W., and Fisher, C.M., 1975, Acute cerebellar infarction of the PICA territory, *Arch Neurol* 32:364-368.

Fisher, A., Gresty, M.A., Chambers, B., and Rudge, P.,1983, Primary position upbeating nystagmus: a variety of central positional nystagmus, *Brain* 106:949-964.

Halmagyi, G.M., Brandt, Th., Dieterich, M., Curthoys, I.S., Stark, R.J., and Hoyt, W.F., 1990, Tonic contraversive ocular tilt reaction due to unilateral meso-diencephalic lesion, *Neurology* 40:1503-1509.

Halmagyi, G.M., Gresty, M.A., and Gibson, W.P.R., 1979, Ocular tilt reaction with peripheral vestibular lesion, *Ann Neurol* 6: 80-83.

Hedges, T.R., and Hoyt, W.F., 1982, Ocular tilt reaction due to an upper brainstem lesion: paroxysmal skew deviation, torsion, and oscillation of the eyes with head tilt, *Ann Neurol* 11:537-540.

Leigh, R.J., and Brandt, Th., 1993, A re-evaluation of the vestibulo-ocular reflex: new ideas of its purpose, properties, neural substrate, and disorders, *Neurology* 43:1288-1295.

Lueck, C.J., Hamlyn, P., Crawford, T.J., Levy, I.S., Brindley, G.S., Watkins, E.S., and Kennard, C., 1991, A case of ocular tilt reaction and torsional nystagmus due to direct stimulation of the midbrain in man, *Brain* 114:2069-2079.

Olszewski, J., and Baxter, D.,1982, Cytoarchitecture of the Human Brain Stem, 2nd edit. Karger, Basel, New York.

Rabinovitch, H.E., Sharpe, J.A., and Sylvester, T.O., 1977, The ocular tilt reaction. A paroxysmal dyskinesia associated with elliptical nystagmus, *Arch Ophthalmol* 95:1395-1398.

Ranally, R.J., and Sharpe, J.A., 1988, Upbeat nystagmus and the ventral tegmental pathway of the upward vestibulo-ocular reflex, *Neurology* 38:1329-1330.

Rubinstein, R.L., Norman, D.M., Schindler, R.A., and Kaseff, L., 1980, Cerebellar infarction. A presentation of vertigo, *Laryngoscope* 90:505-514.

Schwindt, P.C., Richter, A, and Precht, W.,1973, Short latency utricular and canal input to ipsilateral abducens motoneurons, *Brain Res* 60:259-262.

Suzuki, J.I., Tokumasu, K., and Goto, K.,1969, Eye movements from single utricular nerve stimulation in the cat, *Acta Otolaryngol* 68:350-362.

Tokumasu, K., Suzuki, J.I., and Goto, K., 1971, A study of the current spread on electrical stimulation of the individual utricular and ampullary nerves, *Acta Otolaryng (Stockh)* 71:313-318.

Westheimer, G., and Blair S.M., 1975b, The ocular tilt reaction - a brain-stem oculomotor routine, *Invest Ophthalmol* 14:833-839.

Westheimer, G., and Blair, S. M., 1975a, Synkinese der Augen- und Kopfbewegungen bei Hirnstammreizungen am wachen Macacus-Affen, *Exp Brain Res* 24:89-95.

38

POSTURE MAINTENANCE FOLLOWING SENSORY STIMULATION IN SUBJECTS WITH NORMAL AND DEFECTIVE VESTIBULAR FUNCTION

V. Grigorova,[1] K. Stambolieva,[1] and I. Ivanov[2]

[1] Institute of Physiology
Bulgarian Academy of Sciences
[2] Faculty Clinic of Neurology and Psychiatry
Sofia
Bulgaria

INTRODUCTION

Experimental activation of one or more sensory systems controlling body balance is a widely used method in experimental otoneurology. Such experiments have been used to clarify the mechanisms of sensory interactions in posture maintenance. In turn, they can contribute to clinical diagnosis and therapy. Some manipulation used for sensory stimulations, e.g. active head movements for vestibular stimulation, can induce mechanical sway which may considerably modify the sensory stimulation one originally intended (Straube et al., 1989). Adopting the "reafference principle" of von Holst and Mittelstaedt (1950) and Groen·s hypothesis of a pattern center in the CNS (Groen, 1957), we assume that immediately after stimulation postural stabilization should be the same as during stimulation. At that time, however, nonspecific effects of the stimulation would be absent. Although this hypothesis is based mainly on prolonged stimulation, we believe that it holds true also for stimuli of short duration. The difference between the short and long lasting stimulation would be related to the time that is required for the copy of motion pattern to fade away. This time is determined by the mode of stimulation, by the sensory interactions that occur during the stimulation, and by the functional state of presumed central adaptive mechanisms.

The present study was aimed at estimating after-effects of (i) head movements in the horizontal and in the frontal plane, and (ii) a combination of central optokinetic stimulation and yaw head movements on body sway during stance with open and with closed eyes, both in healthy subjects and in patients with peripheral and central vestibular disorders.

Multisensory Control of Posture, Edited by T. Mergner and F. Hlavačka
Plenum Press, New York, 1995

MATERIAL AND METHODS

Sixteen healthy persons (aged 20-44 years) without history of disequilibrium (N) and forty six patients (aged 22-54 years) were investigated. The patients were divided into three groups: 10 patients with clinically diagnosed central cerebral disorders (PC), including tumors in the brainstem or at supratentorial sites (5), post-stroke state (3), arachnoiditis (1) and post commotion state (1); 15 patients with peripheral vestibular diseases in the phase of active vestibular dysfunction (PPD); 11 patients with peripheral vestibular disorders in a compensated state (PPC). The main disequilibrium symptoms in the first two groups of patients were vertigo or dizziness and unstability of walking, while the patients of the third group only had headache or slight dizziness.

The experiments were performed with the help of a posturographic system (Vankov et al.,1989) which consisted of a triangular force plate, a microcomputer and a 12-bit A/D converter. Optokinetic stimulation (OKS) was carried out by another microcomputer with specially designed software. The subjects stood on the force plate 40 cm in front of the computer display. The stimulated visual field subtended 30°, the velocity of OKS was 12°/s. The investigation was conducted in 3 sessions (S), each consisting of 6 trials.

The first two and the last two trials were identical in each session, i.e. upright stance with eyes open, followed by stance with eyes closed. In the third and fourth trials of the 1st session, subjects performed head movements in yaw with eyes open and with eyes closed (VSR) whereas in the 2nd session head movements were frontal tilts with open and closed eyes. The third and fourth trials in the 3rd session were combined VSR and leftward and rightward OKS (OVS). The trial duration was 25 s. The head movements reached a frequency of up to 0.13 Hz, prompted by a computer generated sound signal. The intersession intervals lasted about 3-4 minutes. The first two trials of the 1st session (with open and closed eyes) were considered as baseline of the body sway. The body sway in the last two trials of each session was taken as the early stimulation after-effect (AR, AT and AOV) while the first two trials of the second and the third session and the two trials, carried out 3-4 minutes after the third session, were taken as late after-effect of the previous stimulation (AAR, AAT and AAOV).

The estimation of body sway was based on the sway path (SP) expressed as a change of the center of pressure in cm/s. To avoid large inter- and intra-individual variability, the sway path shift (ΔSP) was calculated, separately for the L-R and A-P planes and for the early and late after-effects as follows:

$$\Delta SP = (SP_{aSn} - SP_{bS1})/SP_{bS1}$$

where SP_{aSn} was the sway after each session or the corresponding pause, and SP_{bS1} was the baseline value of the sway with open and closed eyes, respectively.

Three-way analysis of variance (MANOVA) was used to estimate the significance of factors Group (A) , Stimulation (B) and Visual Input (C), separately for the L-R and A-P planes of body sway. The difference between each group after VSR, VST and OVS under the different experimental conditions was calculated by one-way ANOVA.

RESULTS

The main effects Group (A), Stimulation (B) and Visual Input (C) were significant in the L-R plane, while in the A-P plane the effect of C was not significant (Table 1). Post-hoc analysis showed that patients with compensated vestibular peripheral disorders did not differ from the controls. The significance of Stimulation was mainly due to the difference between

Table 1. Three-way ANOVA for ΔSP in the L-R and A-P planes with the factors Group (A), Stimulation (B) and Visual Input (C), and post-hoc comparisons

		L-R	A-P
Variable			
F(A)		**19.33** (p<0.01)	**14.45** (p<0.01)
F(B)		**11.4** (p<0.01)	**2.6** (p<0.05)
F(C)		**4.77** (p<0.05)	**2.5** (N.S.)
Post-hoc mean comparisons			
F(A)	N vs PC	*	*
	Ń vs PPC	*	N.S.
	N vs PPD	N.S.	*
	PC vs PPC	*	*
	PC vs PPD	*	N.S.
	PPC vs PPD	N.S.	*
F(B)	AR vs AOV	*	N.S.
	AAR vs AAT	*	*
F(C)	O.E. vs C.E.	*	N.S.

* p < 0.05; N.S., no significant difference.

the early after-effects of head movements in yaw and opto-vestibular stimulation in the L-R plane as well as between the late after-effects of yaw head movements and head tilts in the L-R and A-P planes (Table 1).

After-Effect of Head Movements in Yaw

The sway path of the N group showed a decrease. This was significantly different from the PC and PPD groups, which showed a trend towards an increase in the A-P plane, both in the early and the late after-effects during open eyes stance. The PPD group showed a significant difference from the N group in the L-R and A-P planes with respect to the early after-effect (Fig. 1A). In the closed eyes condition the sway path of the N group was reduced in the A-P plane in both after-effects. In the PPC group the sway path tended to decrease, while in the PPD group it increased in the A-P plane for the early and late after-effects (PPD > PPC, p < 0.05) (Fig. 1B). The after-effect of yaw head movements in the PC group was almost negligible on body sway in the eyes closed condition (Fig. 1B).

After-Effect of Head Tilt

A significant decrease in sway path in the A-P plane was found in the N group for the early after-effect during stance with open eyes, while the three groups of patients showed a trend towards a sway increase. A similar trend was observed in all groups for the late after-effect. Particularly pronounced was the sway increase in the PC group, which was significantly different (p < 0.05) from the PPC and PPD groups (Fig. 2A). In the closed eyes stance condition, only the PPD group showed an increased sway in the the A-P plane (PPD

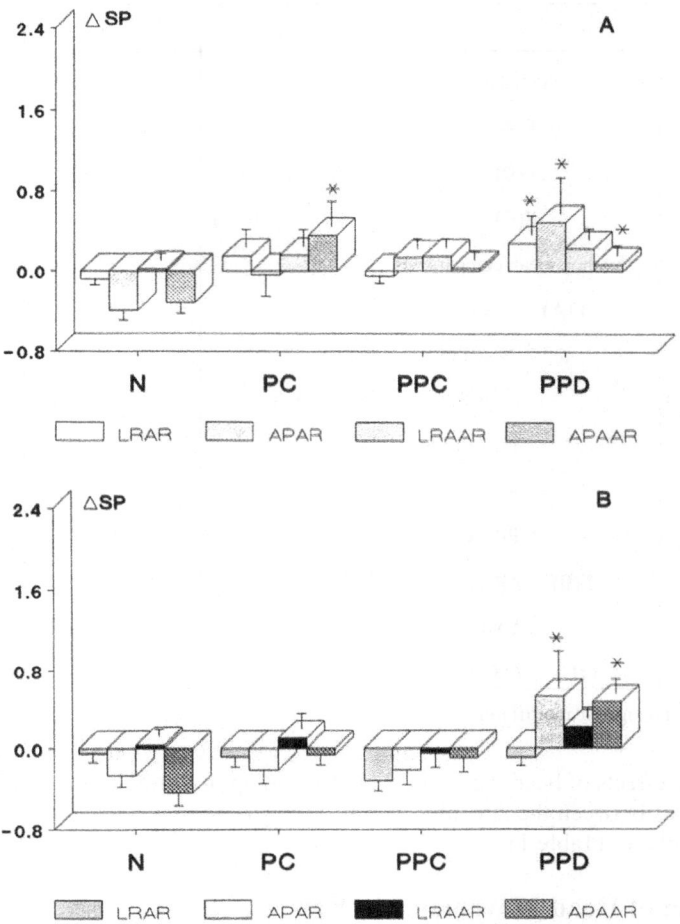

Figure 1. The early and the late after-effects of yaw head movements on body sway in the L-R and A-P planes (LRAR and APAR; LRAAR and APAAR). (A) Eyes open condition. (B) Eyes closed condition. * significant difference (p < 0.05) between N group and indicated patient groups.

> PPC and PC, p < 0.05) for the early after-effect, while for the late after-effect the sway of the PC group increased in both planes (PC > PPC and PPD, p < 0.05 (Fig. 2B).

After-Effect of Combined Optokinetic and Vestibular Stimulation

The effect of OVS consisted of a trend of sway amplitude to increase during stance with eyes open, which was most pronounced in the PC group followed by the N group and the PPD groups (Fig. 3A). The sway increase in the A-P plane of the PC group significantly differed from that of the other groups (p < 0.05) with respect to both the early and the late after-effect in the L-R plane. Almost the same was true for stance with eyes closed (PC > PPC, PPD and N, p < 0.05) where only the PPC group showed a trend towards sway reduction with respect to the early and late after-effects in both planes (Fig. 3B).

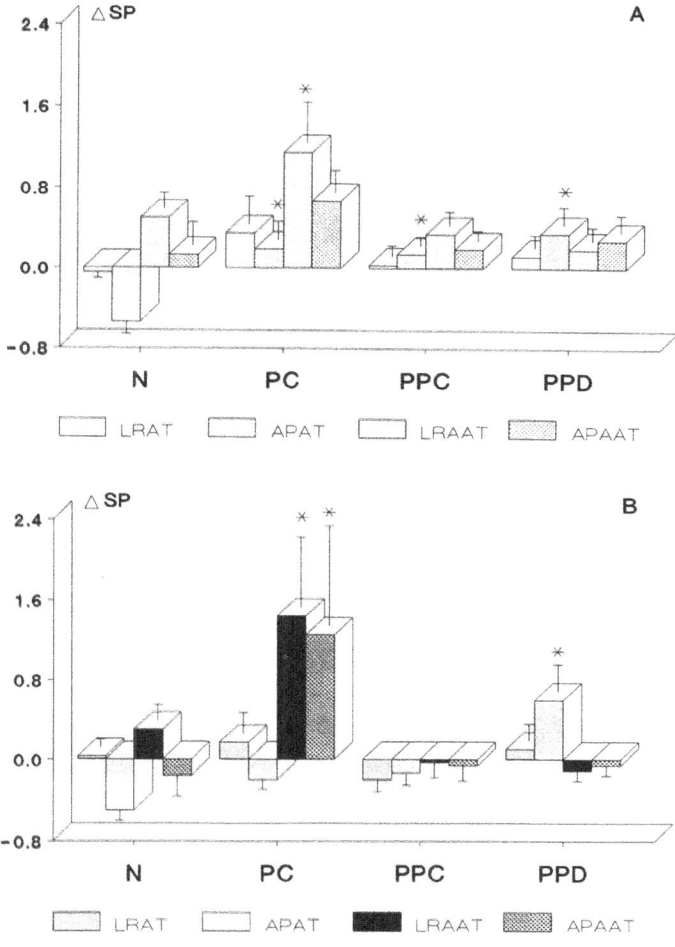

Figure 2. The early and the late after-effect of head tilt for body sway in the L-R and A-P planes (LRAT and APAT; LRAAT and APAAT). (A) Eyes open condition. (B) Eyes closed condition.

DISCUSSION

The interpretation of the present results is based on the assumption that the body sway in the early after-effect is similar to that during the stimulation. If the stimulation is of short duration, as it was in our experiments, a time period of 3-4 min after stimulus end is enough for the copy of motion pattern to fade away and to bring to the light the real after-effect, which in our investigations resulted from the sensory interaction or a profound distortion in the functional state of the central adaptive mechanisms.

During quiet stance the vestibular system participates in postural stabilization mainly via the tonic vestibulo-spinal reflex (VSR). Stimulation of the system during head movements evokes in addition a vestibulo-ocular reflex (VOR). A well functioning VOR helps the VSR in postural stabilization, as it appeared to be the case in the healthy subjects, whose body sway decreased during and after yaw head movements. This even applied to the patients with central lesions, in whom the lesion does not effect the proper functioning of the VOR.

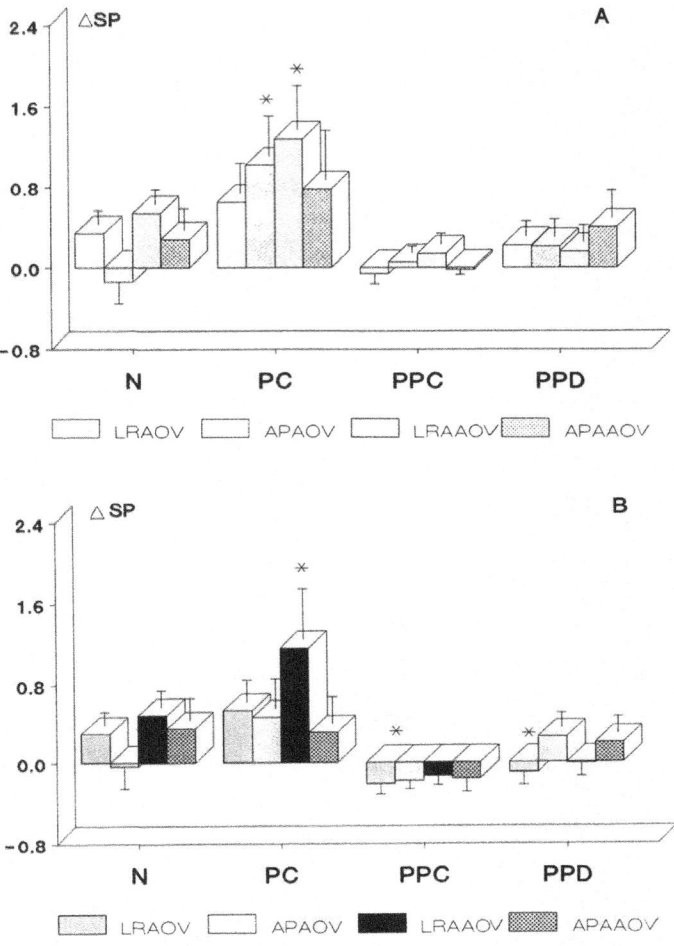

Figure 3. The early and the late after-effects of combined horizontal central OKS and yaw head movements on body sway in the L-R and A-P planes (LRAOV and APAOV; LRAAOV and APAAOV) with eyes open (A) and with eyes closed (B).

In contrast, one would expect an increase in body sway in patients with impaired peripheral vestibular function, resulting from an impairment of both, VSR and VOR.

In our previous work (Grigorova et al., 1992) we have shown an increase of body sway during head tilts in healthy subjects. According to Straube et al. (1989) this increase is due to an additional mechanical sway from the head that adds to the spontaneous body sway movements. If this was the whole truth, there should be no increase in body sway after the stimulation. However, we found a trend towards a sway increase in all of our patient groups with respect to the early after-effect. Brandt (1988) suggested that postural imbalance may result from a mismatch between sensory inputs, or between the actual and expected patterns of reafferent input during voluntary movements, the parameters of the expected input being determined by earlier sensorimotor experience. During head tilts, not only the horizontal semicircular canals but also the vertical canals and the otoliths are stimulated, leading to a complex combination of VOR and motion pattern copies and resulting in a pronounced body destabilization that overlasts the stimulus. A considerable postural desta-bilization has been found after combined central optokinetic stimulation and yaw head movements, although the after-effect to each of these stimuli alone was negligible or even reduced body sway (Grigorova et al., 1992,1993). The sensory conflict that results from the discordant combination of an OKR evoked by an OKS in a given direction (right or left) with a VOR that alternates its direction during the sinusoidal head rotation represent a considerable challenge to higher CNS functions. This may explain the pronounced destabi-lization observed for the after-effect of VST and OVS in the patients with central disorders. We assume that in patients with peripheral vestibular disorders central adaptive mechanisms are well trained by the continuous effort to compensate the vestibular deficit. This notion is based on our finding that the patients with compensated peripheral lesions were more stable than our normal controls after VST and OVS. Possibly, adaptive mechanisms in normals are more "sluggish" and require more time in order to compensate the sensory mismatch.

Vision is of major importance for postural stabilization, and this the more in patients with not fully compensated peripheral vestibular disorders. This may explain why, after stimulation of impaired vestibular input and eyes closing, the body sway increased so clearly in these patients. The difference we observed between the compensated and the decompen-sated patients with peripheral vestibular disorders, established during stance with eyes closed in the early after-effect, might be of relevance for clinicians dealing with the diagnosis and therapy of such patients.

REFERENCES

Brandt, Th., 1988, Sensory function and posture, In: Amblard B., Berthoz, A., and Clarac, F. (ed), Posture and gait: Development, adaptation and modulation, Elsevier pp. 127-136.

Grigorova, V., Ganchev, G., and Vankov, A., 1992, Effect of separated and simultaneous stimulation of visual and vestibular inputs on the postural balance in humans, Satellite Symposium of XVIIth BARANY SOCIETY meeting, Smolenice castle, Slovakia, pp. 12.

Grigorova, V., Stambolieva, K., and Ivanov, I., 1993, Post-effect of sensory stimulation on posture stability in normals and patients with peripheral vestibular disorders, Proc. VII Int. Sym. Motor Control, Sofia, Bulgaria (in press).

Groen, J., 1957, The semicircular canal system of organs of equilibrium. II, *Physiol. Med. Biol.* 1(3):225-242.

Straube, A., Paulus, W., Quintern, J., and Brandt, Th., 1988, Visual ataxia induced by eye movements: posturographic measurements in normals and patients with ocular motor disorders, *Clin. Vision Sci.* 4(2):107-113.

Vankov, A., Dunev, S., and Videnov, S., 1990, Force platform for stabilographic studies, *Acta physiologica et pharmacologica bulgarica* 16:63-68.

v. Holst, E., and Mittelstaedt, H., 1950, Das Reafferenzprinzip, *Naturwissenschaften* 37:464-476.

39

LONG-TERM ADAPTATION OF DYNAMIC VISUAL ACUITY TO TELESCOPIC SPECTACLES BY LOW VISION PATIENTS

Jefim Goldberg,[1]* Franklin I. Porter,[2] Janis M. White,[2] Ann Koval,[2] and Kim A. Schmidt[1]

[1] Center for Balance Disorders
Baylor College of Medicine
[2] College of Optometry
University of Houston
Houston, Texas

INTRODUCTION

Telescopic spectacles (TS) magnify images of objects on the retina. Magnification improves visual acuity, provided the magnified images are sufficiently stable on the retina. Image motion across the retina with velocity greater than 2-3°/s causes a loss of visual acuity (Westheimer and McKee, 1975; Demer and Amjadi, 1993). When vision is not magnified, retinal image motion is minimized for a wide range of head movements by the vestibulo-ocular reflex (VOR) augmented by visually induced tracking. Magnification places an extraordinary burden on these mechanisms when viewing is attempted through high-power headborne magnifiers such as TS (Demer et al., 1991b). The result is that even small, involuntary head oscillations accompanying static postures (Demer et al., 1991a; Goldberg, 1992) induce retinal image motion of sufficient velocity to reduce visual acuity achieved with TS (Porter et al., 1989). Head oscillations of moderate velocities, similar to those observed during walking (Grossman et al., 1988; Demer et al., 1991a) induce acuity losses that can exceed any gains due to image magnification (Demer et al., 1988).

Dynamic visual acuity (DVA) is generally defined as the ability of an observer to resolve spatial detail during relative motion between the observer and an object of regard. DVA will be considered below in the specific context of acuity measured while the observer's head is moving and the object is stationary. Static visual acuity (SVA) is generally defined as the acuity measured while both the head and the target is stationary. SVA also represents the best acuity that can be obtained with TS or the highest level DVA can reach in the special case of zero head velocity. In most practical applications of TS, this level of acuity is not

* Address correspondence to Jefim Goldberg, PhD, OTO Department, Baylor College of Medicine, One Baylor Plaza, Houston, TX 77030. Tel.: (713)798-3226; fax.: (713)798-8658; e-mail: jefimg@bcm.tmc.edu.

achieved and DVA is reduced relative to SVA due to head motion. The amount of the reduction may defined as the dynamic visual acuity loss (DVA loss). DVA can be then considered as the difference between its static and dynamic component:

$$DVA = SVA\text{-}DVA \text{ loss} \tag{1}$$

with the acuities expressed in suitable units.[*]

DVA loss is determined by retinal image velocity, which increases with head velocity and magnification (Demer and Amjadi, 1993). Given measurements of SVA and DVA, DVA loss can be calculated using formula (1) and used as a simple measure of sensitivity of visual acuity to head movement with TS.

It is estimated that millions of people in the USA have severe visual impairments, although the majority of them have useful residual vision (Tielsch et al., 1990). The goal of visual rehabilitation is to maximize the use of the remaining vision and thus enhance the individual's capacity for independent living (Browning and Jose, 1984). Patients whose visual acuity cannot be increased to better than 20/70 with ordinary corrective lenses are clinically considered to have low vision. Telescopic spectacles are routinely used in low-vision rehabilitation to address the patients' needs for distance vision, since telescopic magnification can potentially bring their acuities closer to the 20/20 level. Spectacle-mounted telescopes range from 2x to 10x in magnification, with 4x being the most common. TS are prescribed to motivated low-vision patients whose SVA is improved when TS are tried in the clinic. However, many of these patients fail to benefit from their prescribed TS. Among patients without congenital nystagmus, rehabilitative failure with TS occurs at a rate of almost 50% (Porter et al., 1993) and is related to high DVA loss and excessive involuntary head oscillations (Demer et al., 1989b, 1991c). These findings confirm that dynamic acuity is a critical factor in TS use by low-vision patients and partially explain the high failure rate of patients who are clinically considered to be good candidates for TS. Since clinical decisions to prescribe TS are based on the measurement of SVA rather than DVA, only the beneficial effect of magnification on static acuity in formula (1) enters clinical judgment, whereas its detrimental effect on dynamic acuity, reflected by DVA loss, is not considered.

Static and dynamic acuity measurements obtained under laboratory conditions before any attempted use of TS exhibit considerable variations between patients (Demer et al., 1991c). Ranges of DVA and DVA loss values of patients who later do succeed with TS overlap the corresponding ranges of those who fail. Since the successful patients wear TS regularly, head movement combined with TS magnification may be expected to induce adaptive changes affecting their DVA loss and DVA over time. The stimulus for adaptation would be expected to be the strongest in TS users with the worst acuity loss. DVA loss improvements observed in normal subjects after only 15 minutes of imposed head motion with TS are consistent with these expectations (Demer et al., 1988). The present study addresses these issues by monitoring DVA loss changes in low-vision patients who are first-time TS users. The results have been reported in a preliminary form (Goldberg et al., 1993).

---

[*] Clinical acuity measurements are often recorded as Snellen fractions, relative to the smallest letters that can be read by an average normal observer at a distance of 20 feet (equivalent to 1 minute of arc resolution). For example, 20/40 Snellen acuity means that the size of the smallest letter the patient can read at 20 feet is the same as a normal observer would read at 40 feet, i.e. the patient is capable of only half the normal spatial resolution. Snellen acuity may be converted to the linear units of log minimum angle resolvable (logMAR) by taking the logarithm to the base of 10 of the Snellen fraction and negating it. Thus 20/20 and 20/40 Snellen acuities are equivalent to 0.00 and 0.30 logMAR, respectively. With SVA and DVA in logMAR units, the relationship between static and dynamic visual acuity is conveniently expressed by formula (1).

METHODS

Subjects were recruited among low-vision patients who received a prescription for TS at cooperating clinics. Only patients who had never used TS and did not have clinically detectable spontaneous nystagmus were considered. The 59 study participants ranged in age from 14 to 84 years (mean±standard deviation: 53.0±18.9); 18 were females and 41 were males. Each patient's static and dynamic visual acuity with TS was tested in the laboratory on two occasions. The "initial" acuity was tested before the patient's prescribed TS was dispensed to him or her, i.e., before any use of TS. The "final" acuity was tested after the patient had the TS for a minimum of 70 days. Several additional acuity measurements were obtained at more closely spaced intervals in three of the patients. All 59 patients at least tried to use their TS in their normal living environments. However, 12 of them stopped using TS, most of them within a few weeks of dispense, and were considered to be unsuccessful with the TS. The other 47 patients continued using their TS and were considered successful. This classification is based on the judgment of independent evaluators who observed the patients in their normal environment, as previously described (Demer et al., 1991c).

Visual acuities were tested while the patient was seated in a computer-controlled rotatory chair with the head strapped to a headrest. Each patient wore the same laboratory, binocular TS with 4x power and viewed letter optotypes projected onto a screen 3m in front of the patient (Demer et al., 1988). The TS were focused and aligned on the optotypes to ensure binocularity, when applicable. Each telescope provided a central visual field 10° in diameter, with the periphery kept dark by occluders attached to the spectacle frame. For the static visual acuity (SVA) measurement, the chair was stationary. For the dynamic visual acuity (DVA) measurement, the chair was rotated sinusoidally at 1 Hz, 20°/s velocity amplitude (± 3.2° peak excursion) about its vertical axis. Dynamic visual acuity loss (DVA loss) due to the imposed head rotation was calculated as SVA minus DVA after converting the SVA and DVA measurements from Snellen to logMAR units. A negative DVA loss value indicates that acuity is decreased by the head rotation.

Since DVA loss was determined under the same test conditions for all subjects and on both test occasions, it could be used as an index of sensitivity of visual acuity to head motion. Thus a difference between DVA loss values corresponding to the initial and final acuity measurement of each patient reflects a change in the sensitivity. A negative value of the difference (initial minus final) reflects a smaller DVA loss on the final than on the initial test and a reduction in the sensitivity with time. Student's t-test on the difference averaged over the subjects was used to test the statistical significance of these differences. In addition, the differences were plotted against initial DVA loss values and Pearson product moment coefficient computed to test whether these variables were linearly related.

RESULTS

Regular, successful use of the prescribed TS was confirmed in 47 of the 59 study participants, as described in Methods. The other 12 tried TS but were ultimately unsuccessful. Multiple sequential measurements of SVA and DVA obtained in three patients were used to characterize changes in acuity as a function of time and TS use. The time course of DVA loss, reflecting changes in sensitivity of visual acuity to head motion, is illustrated in Fig 1 for two successful and one unsuccessful patient. Patient 3 started with a large loss which was reduced rapidly and substantially in the first month of TS use. The small initial loss of patient 1 was reduced more slowly to zero. Patient 2 tried using TS on several occasions a few minutes at a time, but gave up in the first two weeks. These data suggest that TS use

Short-Term DVAloss Adaptation

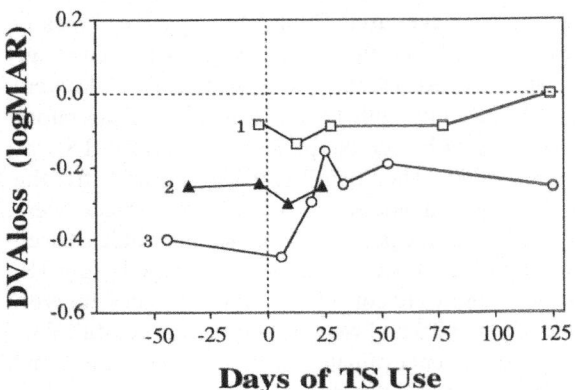

Figure 1. Changes in DVA loss of three representative patients. Dynamic visual acuity loss (DVAloss=SVA-DVA) due to head motion is plotted for two TS users (open symbols) and one non-user (filled symbols) against time. Prescribed TS were dispensed to the patients at the time indicated by the dashed line. Negative numbers on the vertical axis indicate dynamic acuity loss; -0.1 logMAR is equivalent to a loss of one Snellen line. The dashed horizontal line indicates ideal performance. Changes directed up toward this line reflect improvement.

reduces DVA loss, thus improving DVA. The size and rate of this improvement appears to be related to initial DVA loss value.

The notion that DVA loss changes adaptively in proportion to its initial value is supported by acuity data of the 47 successful TS users. Note that at the time of the final acuity measurements, the successful patients had been using their TS regularly for at least 70 days. Initial SVA (obtained before any TS use) averaged 0.34±0.24 logMAR. Final SVA had a nearly identical average of 0.33±0.26 logMAR. DVA averaged 0.55±0.23 logMAR initially, improving to 0.48±0.24 logMAR at the time of the final measurement. Since the corresponding change in the SVA was negligible, nearly all of this improvement was due to a reduction in DVA loss, reflecting reduced sensitivity of acuity to head motion. DVA loss was computed for each of the 47 successful TS users and averaged. The reduction in DVA loss between the initial and final tests amounted to -0.06±0.16 logMAR, which was statistically significant (p<0.02). Furthermore, TS users with the worst initial DVA loss tended to have the largest DVA loss reduction, as illustrated by the open squares in Fig 2. The magnitude of the reduction was significantly correlated with initial DVA loss (r=0.57, p<0.001) in the 47 TS users.

DVAloss Adaptation vs DVAloss

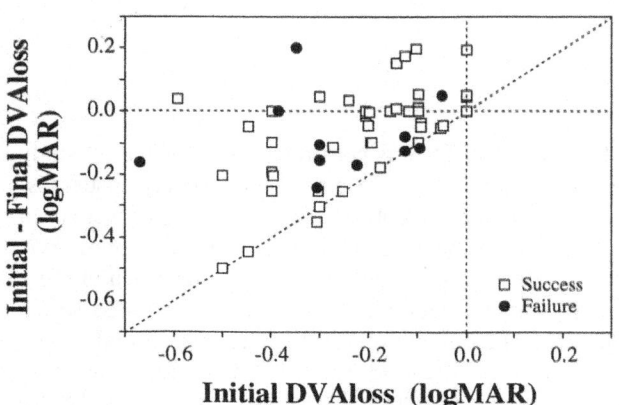

Figure 2. Changes in DVA loss as a function of initial DVA loss. Differences between the initial and final DVA loss values are plotted against initial DVA loss for 47 successful and 12 unsuccessful TS users. Points that lie below the dashed horizontal line correspond to patients whose final DVA loss was reduced relative to the initial one. Points falling on the diagonal correspond to patients who adapted exactly by the amount needed to reduce their initial DVA loss to zero. Most patients show some adaptation and are between the two lines.

Most of the 12 unsuccessful patients stopped using their TS after in first several weeks after dispense, i.e., well before the final acuity test. Filled symbols in Fig 2 suggest that DVA loss of these was reduced by their limited experience with TS. However, neither the average DVA loss reduction nor its tendency to be related to initial DVA loss was statistically significant.

DISCUSSION

Dynamic visual acuity achieved with TS depends on static acuity and on sensitivity of TS acuity to head motion. Sensitivity changes associated with long-term TS use were quantified using dynamic acuity loss (DVA loss) induced by controlled head rotation. A significant reduction in DVA loss averaging -0.06 logMAR was observed in the 47 successful TS users. A similar reduction was reported previously for 24 of these subjects (Demer et al., 1991c) but was not significant, most likely due to the small sample size.

When a factor has a graded effect on different strata of a heterogeneous study population, the strength of the effect may become obscured by averaging over the whole population. Although the -0.06 logMAR reduction in DVA loss represents only a small adaptive improvement in dynamic acuity, it is an average that lumps all patients together regardless of the size of their initial DVA loss. A large initial DVA loss might be expected to provide a strong stimulus for adaptation and produce a large adaptive change, whereas a small one might not produce any change. The data shown in Fig 2 support this hypothesis and contain several cases of very large adaptive changes. An analogous relationship between improvement in performance and its initial level has been also demonstrated in normal subjects trained on a dynamic acuity task with moving optotypes (Long and Burke, 1989; Long and Riggs, 1991).

The correlation found between the reduction in DVA loss and its initial value implies that the amount of acuity loss a patient experiences during attempted TS use is a factor affecting the degree and speed of adaptation. Initial DVA loss depends on the magnitude and frequency content of head movement, as well as on the power of the TS. Other factors likely to affect TS adaptation are the frequency and duration of TS use. Most of these variables were not controlled in the present study and thus may have contributed to the inter-individual variability seen in Fig 2. However, large inter-individual variability was seen in the amount of initial DVA loss and its short-term adaptation to TS induced under controlled conditions in normal subjects (Demer et al., 1988). The data illustrated in figures 1 and 2 of that study also indicate that the amount of short-term DVA loss adaptation is related to its initial value.

The mechanisms subserving the adaptation of dynamic acuity to TS remain elusive. A significant contribution of VOR plasticity is unlikely because of the optical configuration of TS dispensed to low-vision patients. The central field of view provided by telescopes is very restricted. To allow orientation and mobility, the telescopes are mounted on the spectacle frame so that the unmagnified visual periphery is not occluded. During head motion, the unmagnified retinal slip in the periphery inhibits short-term VOR gain increase induced by the central magnification (Demer et al., 1989a). However, one cannot rule out the possibility that long-term TS use produces a high-gain VOR that is switched on when gaze is shifted to the magnified field (Shelhamer et al., 1992). TS magnification causes an immediate enhancement in the gain of visually-augmented VOR, or VVOR. For very low head velocities, the increased VVOR gain may even reach the power of the TS and thus prevent retinal image slip and acuity loss. However, the VVOR gain enhancement is reduced substantially by the unmagnified periphery (Demer et al., 1991b). It remains to be determined whether the long-term TS use reduces these limitations and improves VVOR performance.

In contrast to its effect on VVOR, unoccluded periphery improves dynamic acuity with TS in normal subjects (Demer et al., 1988). This seemingly paradoxical effect suggests that the contribution of high-level visual processing mechanisms should be also considered. It may be hypothesized that TS users become adept at "capturing" visual information during brief low-velocity segments of head or target movement trajectories. Low-vision patients with congenital nystagmus (CN) develop strategies of this kind to minimize effects of retinal image slip induced by involuntary eye movements (Del'Osso et al., 1975). Acuity measurements, obtained under the same conditions as described above, indicate that these patients' DVA loss is only about one third of the loss found in patients without nystagmus (White et al., 1994). A lifetime of adaptation to retinal slip serves CN patients well when they try to use TS. They succeed at a nearly 100% rate.

ACKNOWLEDGMENTS

Supported by National Institutes of Health grant EY06394, The Clayton Foundation for Research and the Feinbloom Legacy for Vision

REFERENCES

Bedell, H.E., and Loshin, D. S., 1991, Interrelations between measures of visual acuity and parameters of eye movement in congenital nystagmus, *Invest Ophthalmol Vis Sci.* 32:416-421.

Browning, R., and Jose, R., 1984, The low vision triangle, *Rehabilitative Optometry* 2:2-12.

Del'Osso, L.F., and Daroff, R.B., 1975, Congenital nystagmus waveform and foveation strategy, *Doc Ophtalmol.* 39:155-182.

Demer, J.L., Porter, F.I., Goldberg, J., Jenkins, H.A., and Schmidt, K., 1988, Dynamic visual acuity with telescopic spectacles: improvement with adaptation, *Invest Ophthalmol Vis Sci.* 29(7):1184-1189.

Demer, J.L., Porter, F.I., Goldberg, J., Jenkins, H.A., and Schmidt, K., 1989a, Adaptation to telescopic spectacles: vestibulo-ocular reflex plasticity, *Invest Ophthalmol Vis Sci.* 30(1):159-170.

Demer, J.L., Porter, F.I., Goldberg, J., Jenkins, H.A., Schmidt, K., and Ulrich, I., 1989b, Predictors of functional success in telescopic spectacle use by low vision patients, *Invest Ophthalmol Vis Sci.* 30(7):1652-1665.

Demer, J.L., Goldberg, J. and Porter, F.I., 1991a, Effect of telescopic spectacles on head stability in normal and low vision, *J Vestib Res.* 1(2):109-122.

Demer, J.L., Porter, F.I., Goldberg, J., Jenkins, H.A., and Schmidt, K., 1991b, Visual-vestibular interaction with telescopic spectacles, *J Vestib Res.* 1(3):263-277.

Demer, J.L., Goldberg, J., Porter, F.I., and Schmidt, K., 1991c, Validation of physiologic predictors of successful telescopic spectacle use in low vision, *Invest Ophthalmol Vis Sci.* 32:2826-2834.

Demer, J.L., and Amjadi, F., 1993, Dynamic visual acuity of normal subjects during vertical optotype and head motion, *Invest Ophthalmol Vis Sci.* 34(6):1894-1906.

Goldberg, J., Porter, F.I., White, J., Demer, J.L., and Koval, A., Comparison of clinical and physiological predictors of low vision rehabilitation with telescopic spectacles, (submitted).

Goldberg, J., 1992, Nonlinear dynamics of involuntary head movements, In: The Head-Neck Sensory Motor System. Berthoz, A., Graf, W.M., and Vidal, P.P. (eds). Oxford University Press. pp 400-403.

Goldberg, J., Porter, F.I., White, J., Koval, A., and Schmidt, K. A., 1993, Long-term adaptation of dynamic visual acuity to telescopic spectacles in low vision patients, *Invest Ophthalmol Vis Sci.* 34(4, Suppl):791.

Grossman, G.E., Leigh, R.J., Abel, L.A., Lanska, D.J., and Thurston, S.E., 1988, Frequency and velocity of rotational head perturbations during locomotion, *Exp. Brain Res..* 70:470-476.

Long, G.M., and Rourke, D.A., 1989, Training effects on the resolution of moving targets-dynamic visual acuity, *Human Factors.* 31(4):443-451.

Long, G.M., and Riggs, C.A., 1991, Training effects on dynamic visual acuity with free-head viewing, *Perception.* 20:363-371.

Porter, F.I., Goldberg, J., and Demer, J.L., 1989, Sensitivity of visual acuity to spontaneous head motion while wearing telescopic spectacle low vision aids, *Invest Ophthalmol Vis Sci.* 30(Suppl):399.

Tielsch J.M., Sommer, A., Witt, K., Katz J., and Royall, R. M., 1990, Blindness and visual impairment in an American urban population, *Arch Ophthalmol.* 108:286-290.

Shelhamer, M., Robinson, D.A., and Tan, H.S., 1992, Context-specific gain switching in the human vestibu-loocular reflex, *Ann New York Acad Sci*: 656:889-891.

Westheimer, G., and McKee, S.P., 1975, Visual acuity in the presence of retinal image motion, *J Opt Soc Am.* 65:847.

White, J.M., Goldberg, J., Porter, F I., and Koval, A., 1994, Rehabilitation with telescopic spectacles in low vision patients with nystagmus, *Invest Ophthalmol Vis Sci.* 35(4, Suppl):1554.

40

SUBJECTIVE POSTURAL AND VISUAL VERTICAL IN SPASMODIC TORTICOLLIS

D. Anastasopoulos, A. R. Bisdorff, A. M. Bronstein, and M. A. Gresty

MRC Human Movement and Balance Unit
National Hospital for Neurology and Neurosurgery
Queen Square
London WC1N 3BG
United Kingdom

INTRODUCTION

The pathophysiological basis of the abnormal head posture in spasmodic torticollis, the most common form of focal dystonia, is unknown. The disorder is thought to be one of the extrapyramidal system, although the literature on its neuropathology is still inconclusive (Tarlov, 1970; Fahn et al., 1988). Several studies have suggested an involvement of the central vestibular system in spasmodic torticollis (Bronstein and Rudge, 1986; Diamond et al., 1988). It is not clear, however, whether the abnormalities observed are secondary to the abnormal head posture, or imply a causative disruption of the vestibular brainstem pathways in this disorder.

Normal subjects have a remarkable ability to perceive accurately the gravitational vector, a function which mainly depends on the otolithic and somatosensory inputs. When asked to signal uprightness as they are tilted slowly about the earth vertical, normal subjects can orient with an accuracy to within ±2-3 deg. They are also able, when seated upright and in the absence of a visual framework, to place a line within ±2-3 deg of earth vertical (Guedry, 1974). It has been found that patients with vestibular disturbances can be less sensitive to the gravitational vector (Bisdorff et al., 1994) and that a pathological tilt of the subjective visual vertical can occur after peripheral (Friedmann, 1971) or central vestibular lesions (Dieterich, 1993). In order to explore the possibility that they have a defect of spatial orientation corresponding to their abnormal head posture, the subjective postural and visual vertical were investigated in patients with spasmodic torticollis. The patients were compared to a group of age-matched normal subjects and to a group of avestibular patients.

MATERIAL AND METHODS

In order to obtain estimates of their postural vertical, the subjects were seated in a motorized gimbal with the head and trunk comfortably restrained and their eyes closed. The

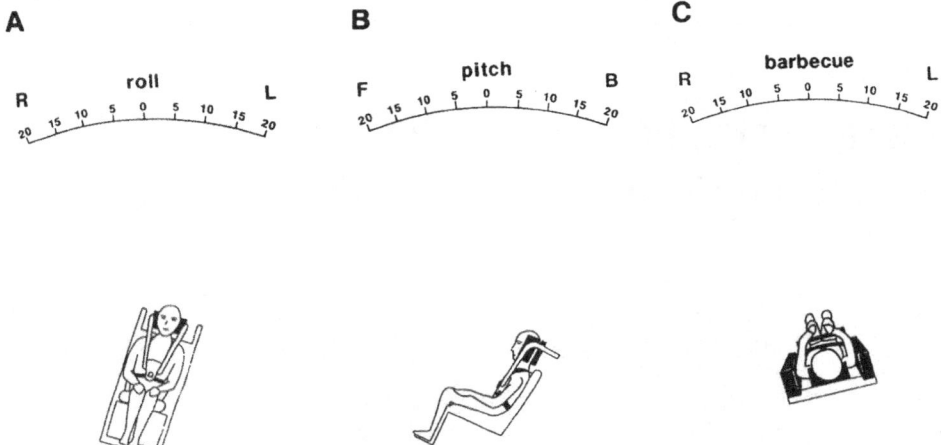

Figure 1. Experimental procedure for the judgment of the postural vertical. The subjects were seated in a padded chair with their torso and head restrained and their eyes closed. The gimbal executed cycles of 15° tilt to either side around the vertical, i.e. about a horizontal axis perpendicular to the plane of the figure, producing rotations in the roll plane (A), in the pitch plane (B) and in barbecue (C).

gimbal executed 10 cycles of tilt around the vertical at 1.5 °/ s in roll (Fig. 1A), pitch (Fig. 1B) and during z-axis oscillation ("barbecue-spit", Fig. 1C). The machine vertical was defined when the plane of the seat was orthogonal to the gravitational vertical. Starting in a 15 ° tilt position to the right or back, subjects had to indicate when they began to feel upright (horizontal for the barbecue paradigm) and again when they began to feel tilted using a 3-position joystick. In each cycle four indications were thus made (Fig. 2), two during each half cycle, signaling entering and leaving upright (resp. horizontal for barbecue). These two indications defined a sector of subjective verticality during motion from right to left (R→L) and left to right (L→R) in roll (Fig. 2, bottom). Means and standard deviations were calculated after rejecting the data of the first cycle. Similar sectors were defined for the motion in pitch and in barbecue. The width of the sectors is considered to be a parameter for the sensitivity of the perception, i.e. the bigger the sector the less the sensitivity to the vector of gravity. The sum of the four values indicated in one plane divided by four is a parameter for the position of the subjective vertical with respect to (machine) vertical.

For the measurement of the perceived visual vertical, the unrestrained patient sat upright in front of a disk at a distance of 30 cm. The disk was covered with a random pattern of colored dots, being devoid of cues as to verticality. The centre of the disk was fixed to the shaft of a DC motor. A smaller circular white target, divided by a central black line, was mounted on a coaxial shaft, just in front of the large disk. It subtended 18° of visual angle, while the large disk covered, at the above distance, almost completely the visual field of the subject. With a circular controlling knob the subject was able to rotate the target and adjust the black line to subjective vertical. The static visual vertical was determined by means of 6 adjustments of the target from a random offset position, the large disk being stationary. The angular deviation from the true vertical in degrees was measured by a potentiometer (+ clockwise deviation, - counter-clockwise deviation) and saved after each positioning. The judgments were made binocularly, with the head directed to face the disk and with eyes in primary gaze position. To determine the dynamic visual vertical, the large disk was rotated around the subject's line of sight either clockwise or counter-clockwise at a constant speed

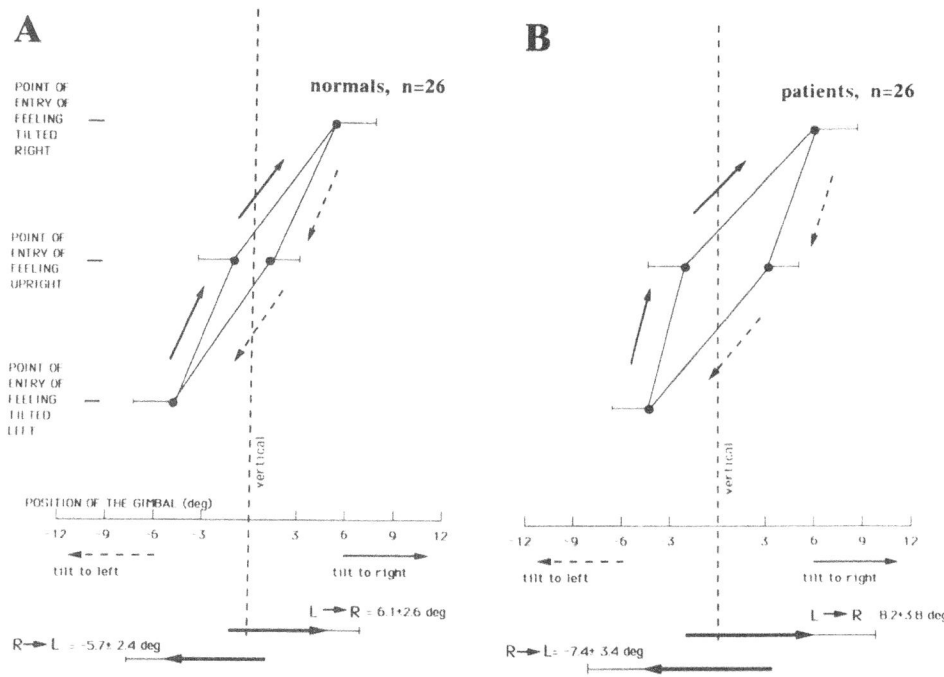

Figure 2. Roll data of the normals (A) and torticollis patients (B). The thin solid and dashed arrows indicate rotations to the right and left respectively. The filled circles correspond to the means of the angle where subjects felt entering or leaving uprightness. The variability of the indications of the individuals during the 9 cycles , i.e. the mean of the standard deviations around the mean constant error is represented by the thin horizontal lines. The thick arrows at the bottom represent the sectors of subjective postural vertical.

of 30°/ s. The stimulation period before beginning with the judgments lasted at least 30 s. Visual orientation to an egocentric reference, i.e. to the plane of symmetry of the head, was investigated after asking the patient to align the line of the target disk with a line passing through the centres of his forehead, nose and chin. To determine the egocentric visual vertical, measurements were taken only once from each patient.

RESULTS

The middle of the sectors indicated by the normal subjects and the patients was in all planes tested shifted to the side of the direction of the motion (Fig. 2). In 26 normal subjects (mean age 49.6 years, range 25-82) there was a significant increase of the width of the sectors with age, i.e. loss of sensitivity. In 26 patients (mean age 50.7 years, range 20-76, all tested immediately or during the first week after botulinum toxin injections) the width of the sector of the subjective vertical in roll was with R→L -7.4 ± 3.4° and L→R 8.2 ± 3.8° significantly bigger than the normal values (R→L -5.7 ± 2.4°, p=0.03 and L→R 6.1 ± 2.6°, p=0.02, Mann-Whitney test). The patients perceived entering subjective vertical at a higher tilt angle during each half cycle compared to normals. The standard deviations i.e. the variability of the judgments of a subject for each of the four values indicated in a plane did not differ significantly between the patients and the normals. Even after splitting the torticollis patients according to the direction of their head tilt, no directional perception bias for the point of entry of feeling upright or for the position of the subjective vertical (i.e. the sum of the four values indicated divided by four, normal value - 0.1 ± 0.9°) was found. Not only the width of the sector of the subjective vertical of the patients in pitch was again with F→B -8.3 ± 4.2° and B→F 9.2 ± 4.5° significantly bigger than the normal values (F→B —6 ± 2.2°, p= 0.01 and B→F 7.1 ± 2.2°, p=0.04), but also their perception of entering the subjective vertical occurred at a higher tilt angle compared to normals. As in the roll plane, splitting the patients into two groups (antero- and retrocollis) did not reveal any directional preponderance for the point of entry of feeling upright or for the position of the subjective vertical (normal value 0.2 ± 1.5°). The mean width of the sectors, within which the patients felt horizontal during the barbecue paradigm did not differ from the normal values (R→L -5.7 ± 2.3°, p=0.08 and L→R 6.1 ± 2.4°, p=0.09). No directional bias was observed for the point of entry of feeling horizontal and for the position of the subjective horizontal (normal -0.2 ± 1.0°) according to the right or left turn of the patients. A mild asymmetry of the sector size in roll and barbecue related to the direction of torticollis was detected, i.e. patients with right tilt indicated larger sectors during roll motion from left to right and vice versa. Those with left turn indicated larger sectors during barbecue rotation from left to right.

Five patients with absent vestibular function had bigger sectors in roll and pitch but not in the barbecue paradigm.

The mean deviation of the visual vertical assessed against a static background for each of 29 patients is depicted in Fig. 3A. For some patients, it deviated significantly from

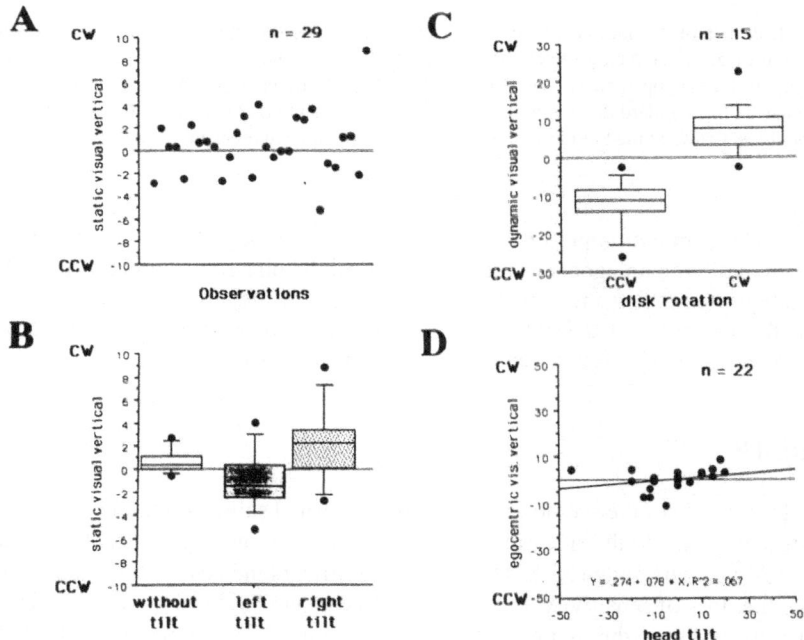

Figure 3. A Each dot represent the mean deviation of the static visual vertical of one patient. B Distribution of the static visual vertical splitted by the direction of the laterocollis. Horizontal lines show the 10th, 25th, 50th, 75th and 90th percentiles of the variable. Outliers are displayed as dots. C Settings during rotating visual background (30 deg/ s) normalized for left laterocollis. Exaggerated deviations to the left were observed during counter-clockwise rotation of the disk. D When asked to adjust the line to the longitudinal axis of the head, the patients usually positioned the line near the true earth vertical, even in cases of marked head tilts.

the gravitational vertical (normal value $0.4 \pm 0.9°$, n=17), but never exceeded 9°. A weak correlation with the direction of torticollis in the frontal plane was observed, i.e. patients tended to set the visual vertical in the same direction as their head tilt (p=0.06 Mann-Whitney test, Fig. 3B, left laterocollis n=11, right laterocollis n=9). Settings during rotation of the visual background amplified this asymmetry (normal values with clockwise stimulation $8 \pm 3.8°$ and $-6.6 \pm 3.5°$ with counter-clockwise stimulation). If the patients without laterocollis are excluded and the settings are normalized as if they were all obtained from patients with left tilt, the dynamic visual vertical showed a higher tilt angle when the visual scene rotated counter-clockwise ($7.5 \pm 6.0°$ compared to $-12.4 \pm 6.3°$, n=15, p=0.03 Wilcoxon signed rank test, Fig. 3C). When asked to adjust the line of the target disk to the longitudinal axis of the head, the patients usually positioned the line near the true earth vertical, even in cases of marked head tilts (Fig. 3D). Under the same experimental conditions, normal subjects (n=13) performing the same task during voluntarily maintained lateral head tilt, tended to slightly overestimate the degree of their head tilt.

DISCUSSION

We examined the possibility that torticollis patients may have a disturbance of the orientation with respect to gravity corresponding to their abnormal head posture, by assessing their subjective postural and visual vertical. The paradigm requiring the subjects to signal when they enter and leave verticality is based on the assumption that the subjective postural vertical is not confined to a single point but to a sector in space. The width of the sectors indicated by avestibular and torticollis patients in roll and pitch was bigger than the normal values obtained from an age-matched group, suggesting that these patients are probably less sensitive to the gravitational vector. No directional bias for either the point of entry of feeling upright or for the position of the subjective vertical was found, suggesting that the pathological motor phenomenon is not a reaction to an abnormal perception of the body in space. The standard deviations for each of the four values indicated in a plane (i.e. the variability of judgments) did not differ significantly between the patients and the normals. The same was found for avestibular patients. In earlier studies, this parameter was thought to be important in making inferences about the sensitivity of the position sense. However, the recorded differences between judgments of observers with absent vestibular function and normals have been not statistically significant (Guedry, 1974). Both patient groups showed no abnormalities in the barbecue paradigm. The interpretation of this finding can be based on the assumption that, in the recumbent position, the maculae of the inner ear are less sensitive to changes of the gravitational force as compared to the upright position, explaining the fact that the influence of somatosensory cues on orientation becomes more important when the axis of rotation is horizontal (Lackner and Graybiel, 1978). It is worth mentioning that the degree of torticollis decreases in many patients in the recumbent position.

The static visual vertical deviated significantly in about one third of the patients (38%). In a previous study only 16% of 19 patients were found to have pathological tilts from the true vertical (Straube and Dieterich, 1993). After unilateral labyrinthectomy (Friedmann, 1971), unilateral vestibular neurectomy (Curthoys et al., 1991) and in acute unilateral vascular brainstem lesions (Dieterich, 1993) almost all patients show a pathological tilt of the static visual vertical. This tilt tends to regress rapidly in a few weeks. The visual vertical is usually within the normal limits in patients with chronic vestibular disorders. Our patients tended to set the visual vertical in the same direction as the head tilt in the frontal plane. Patients with an acute dysfunction of the tonic, symmetrical vestibular input can also present with a head tilt and a deviation of the visual vertical towards the same direction, but this deviation is usually greater. Lateral head tilt up to 45° relative to the upright trunk in normal subjects alters the visual orientation; when the subject is allowed to adjust a line to the position that appears vertical, this is on average to the opposite direction of the head tilt (E-effect, Müller, 1916; Wade, 1969). Prolonged head tilt of 30° over several minutes decreases the magnitude of the E-effect from about 6° to 4°, but it does not reverse it (Wade, 1970). Our data cannot thus be interpreted as an adaptation effect due to the prolonged head tilt.

Contrary to the higher tilt angle of the dynamic visual vertical during rotation in the same side as the laterocollis, estimates of the visual vertical obtained under dynamic conditions while normal subjects have their head tilted relative to the trunk, have shown larger deviations when the visual scene is moving opposite to the head inclination (Dichgans et al., 1974).

The most striking pathological finding was the inability of the patients to assess correctly their head position relative to the trunk. They used mainly their trunk as reference during this task, just as in the case of the judgments of the postural vertical. It remains uncertain weather this finding represents an adaptation effect due to the chronic head inclination or a disturbance of the information input from the neck receptors.

ACKNOWLEDGMENT

Financial support from the Human Capital and Mobility Program (EEC) is gratefully acknowledged.

REFERENCES

Bisdorff, A.R., Bronstein, A.M., Gresty, M.A., and Anastasopoulos, D., Subjective postural vertical in peripheral and central vestibular disorders and Parkinson's disease in: Taguchi, K., Igarashi, M., and Mori, S., 1994, Vestibular and neural front, Elsevier Science B.V. New York.

Bronstein, A.M. and Rudge, P.,1986, Vestibular involvement in spasmodic torticollis, *J Neurol Neurosurg. Psychiatry* 49:290-295.

Curthoys, I.S., Dai, M..J, and Halmagyi, G.M., 1991, Human ocular torsional position before and after unilateral vestibular neurectomy, *Exp. Brain Res.* 85:218-225.

Diamond, S.G., Markham, M.D., and Baloh, R.W., 1988, Ocular counterrolling abnormalities in spasmodic torticollis, *Arch Neurol* 45:164-169.

Dichgans, J., Diener, H.C., and Brandt, Th. , 1974, Optokinetic-graviceptive interaction in different head positions, *Acta Otolaryng* 78:391-398.

Dieterich, M., Brandt, Th., 1993, Ocular torsion and tilt of subjective visual vertical are sensitive brainstem signs, *Ann Neurol* 33:292-299.

Fahn, S., Marsden, C.D., and Calne, D.B., 1988, Adavances in neurology. Dystonia 2. Raven Press, New York, Vol. 50.

Friedmann, G., 1971, The influence of unilateral labyrinthectomy on orientation in space, *Acta Otolaryng* 71:289-298.

Guedry, F.E., 1974, Psychophysics of vestibular sensation in: Kornhuber, H.H., *Handbook of Sensory Physiology* vol. VI/2 Vestibular System part 2, Springer Verlag, New York: 3-154.

Lackner, J.R., and Graybiel, A., 1978, Some influences of touch and pressure cues on human spatial orientation, *Aviation, Space, and Environmental Medicine* 49:798-804.

Müller, G.E., 1916, Über das Aubertsche Phänomen, *Z. Psychol. Physiol. Sinnesorg.* 49:109-246.

Straube, A., and Dieterich, M., 1993, Neuroophthalmologische und posturographische Untersuchungen bei Patienten mit idiopatischem Tortikollis, *Nervenarzt* 64:787-792.

Tarlov, E., 1970, On the problem of the pathology of spasmodic torticollis in man *J Neurol Neurosurg. Psychiatry* 33:457-463.

Wade, N.J., 1969, The effect of monocular and binocular observation on visual orientation during head tilt, *Am. J. Psychol.* 82:384-388.

Wade, N.J., 1970, Effect of prolonged tilt on visual orientation. *Q. J. exp. Psychol.* 22:423-439.

41

SMOOTH PURSUIT EYE MOVEMENTS IN PATIENTS WITH IMPAIRED VISUAL MOTION PERCEPTION

H. Kimmig, C. Pinnow, T. Mergner, and M. Greenlee

Neurologische Universitätsklinik
Neurozentrum
Breisacher Str. 64
D- 79106 Freiburg
Germany

INTRODUCTION

The smooth pursuit system, like other eye movement systems, has been extensively studied in rhesus monkeys. This work has identified two cortical areas that are important for both visual motion processing and smooth pursuit eye movements - the middle temporal (MT) area and the adjacent medial superior temporal (MST) area in the depth of the superior temporal sulcus (see Wurtz et al., 1990a,b). There is some evidence that a homologue to MT/MST exists in the human cortex, located in the temporo-parieto-occipital region (TPO). Passive visual motion perception leads to activation of the TPO-region in positron emission tomography (PET; Watson et al., 1993). A similar topography was found in a PET study on ocular pursuit (Miezin et al., 1988). Patients with posterior brain damage may exhibit motion perception deficits or a disturbance of smooth pursuit eye movements (Morrow and Sharpe, 1990; see Newcombe and Ratcliff, 1989).

It has recently been shown that patients with a small cortical lesion in the TPO-region, due to brain surgery, show an increased threshold in a velocity discrimination task, even two years after the surgery (Greenlee et al., 1995). The deficits were especially pronounced if the task involved the short-term memory for this stimulus dimension (velocity). Contrast detection thresholds for drifting stimuli, on the other hand, were normal in these patients.

In the present study we asked whether a small cortical lesion in the TPO region, which affects the velocity discrimination threshold, also affects the performance of smooth pursuit eye movements. The close vicinity of MT and MST in monkey might suggest that also in humans the neuronal assemblies dealing with visual motion processing and smooth pursuit eye movements are located in close vicinity to each other. We were able to reinvestigate six of the patients of the previous psychophysical study for their performance of smooth pursuit, assessing their pursuit gain and phase and comparing the results to those of normal controls. The visual target for smooth pursuit was presented either in the dark or on a structured background. It is known that a stationary structured background can degrade smooth pursuit gain up to 10-20% (Collewijn and Tamminga, 1984; Barnes and Crombie, 1985). We asked whether the TPO lesion might affect the interaction between the two visual stimuli, i.e., between background and tracking target.

METHODS

Six patients with a TPO-lesion were investigated. The lesions (Fig.1) resulted from brain surgery of a cavernoma (n=3), angioma (n=1) or astrocytoma (n=2, WHO I or II). The patient group was compared to a group of healthy controls (n=10). Smooth pursuit eye movements were elicited by a horizontally and sinusoidally moving light spot (peak target displacement, ±8 and ±16°; frequency, 0.1 - 1.6 Hz; peak acceleration, 12 - 808°/s²). The target was presented either in complete darkness or against a stationary textured background. Horizontal eye movements were recorded with an infrared light technique. Data were stored on a computer and analyzed off-line. Saccades were removed from the eye position signal using an interactive computer program. This smooth component of the eye tracking signal as well as the target position signal were Fourier transformed. The ratio of the amplitudes of the fundamental component of eye and target positions were taken as smooth pursuit gain.

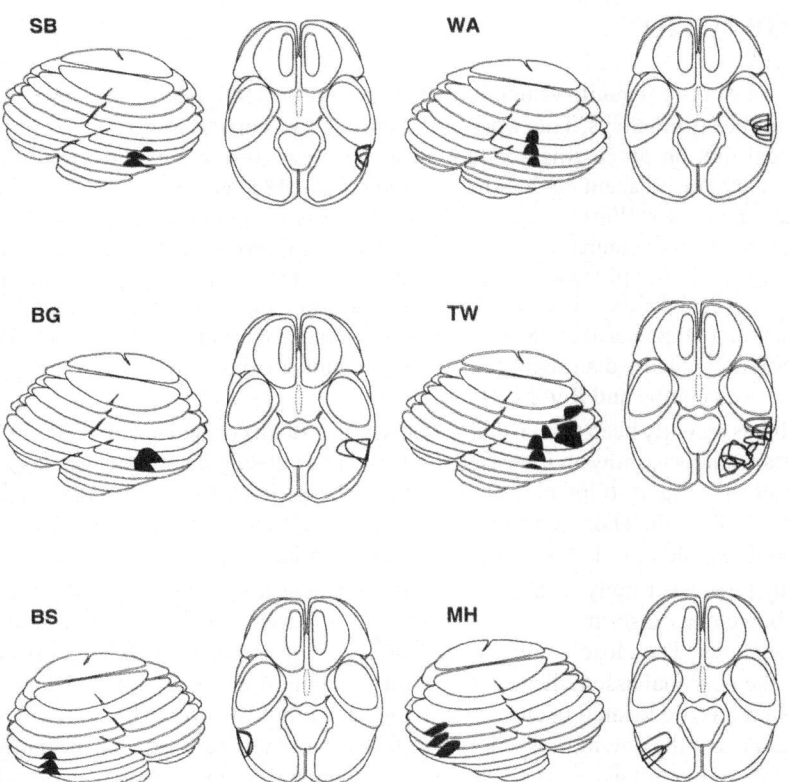

Figure 1. Schematic representations of the CT-scans of six patients with a lesion in the TPO region. Four patients had the lesion on the left side, the other two on the right side.

The temporal shift between eye and target fundamentals was used to calculate the phase difference (in deg).

RESULTS

Figure 2 shows smooth pursuit gain as a function of stimulus frequency of the sinusoidally moving target for both the patients and the normal controls. The thick curves give the results obtained in the dark. At 0.1 Hz there is no statistically significant difference between patient and control groups (gain about 0.85). In contrast, at 0.2 Hz and above the data of the two groups diverge, the gain of the patient group becoming lower than that of the controls by approximately 0.1-0.2. In both groups gain decreases with increasing frequency, reaching 0.4 in the controls and 0.2 in the patients at 1.6 Hz. These differences at 0.2 - 1.6 Hz were statistically significant. Up to 0.4 Hz the eyes were slightly leading the target. Above 0.4 Hz they developed a phase lag that amounted to 23 deg at 1.6 Hz in the normal controls. In the patients the phase lag was somewhat larger at 1.6 Hz (32 deg).

As a further step, we analyzed pursuit gain of the patient group separately for pursuit towards the side of the cerebral lesion and towards the intact side. The gain of ipsilateral pursuit was slightly lower than that of contralateral pursuit (not shown). The difference was in the order of 0.05 at all stimulus frequencies tested and was statistically significant (p2 0.01).

The thin curves in Fig.2 show pursuit gain of both subject groups for target motion on a stationary textured background. In the group of normal controls gain is lower by approximately 0.1 as compared to pursuit in the dark. This difference is about constant over the frequency range tested. In the patient group the gain is reduced by a similar amount as in the normal controls.

DISCUSSION

Before considering the present findings on the presumed visual motion area in man, a short outline of neural responses in the MT and MST areas in monkey is given. The neurons

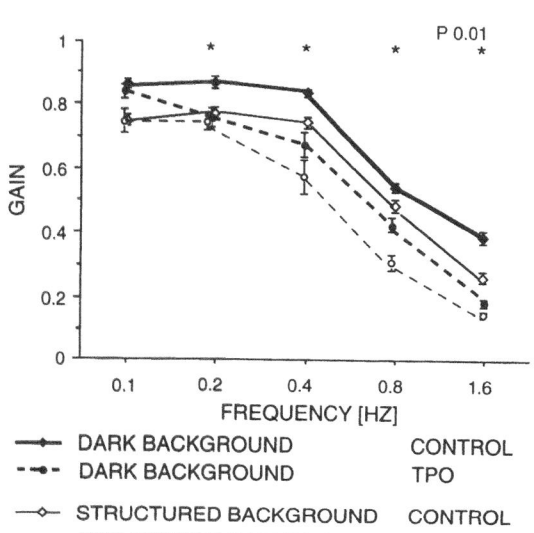

Figure 2. Gain of smooth pursuit eye movements plotted as function of stimulus frequency. Mean values (±S.D.) of patients (dashed curves) and normal controls (solid curves). Tracking in the dark, thick curves; tracking against a stationary textured background, thin curves.

in MT exhibit a directional selectivity for moving visual stimuli (Maunsell and Newsome, 1987). It is assumed that they carry a pure visual signal (Newsome et al., 1988). Punctate ibotenic acid lesions in MT lead to a retinotopic deficit of pursuit initiation, thought to result from a localized inability of motion processing (Newsome et al., 1985). MT neurons code stimulus velocity and the population response seems to be a function of the rate of retinal slip (Maunsell and Van Essen, 1983a). MT projects heavily to the immediately adjacent area MST (Maunsell and Van Essen, 1983b; Ungerleider and Desimone, 1986). The population of neurons in area MT responds to ipsilateral or contralateral motion in the contralateral hemifield.

MST receives input from ipsilateral area MT (ipsilateral movement in the contralateral hemifield) and, via splenium corporis callosi, input from the contralateral MT (ipsilateral movement in the ipsilateral hemifield) (Tusa and Ungerleider, 1988). Thus the population of MST neurons responds to ipsilateral motion regardless of the visual hemifield. Many MST neurons continue to fire during smooth pursuit when the stimulus is blanked for a short moment. This extraretinal activity was related to the pursuit eye movement (Newsome et al., 1988). Since this activity started after the onset of pursuit it cannot be responsible for pursuit initiation, but could well reflect a signal related to the steady state pursuit command (Newsome et al., 1988).

Postmortem studies of the human brain identified heavily myelinated regions in the lateral temporo-parieto-occipital cortex (Brodmann area 19, 39, 37) (Sereno et al., 1988; Clarke and Miklossy, 1990), thought to represent a possible homologue to area MT in monkey. Indeed, patients with lesions of occipito-parietal cortex may exhibit, among neuropsychological disorders like optic ataxia (misreaching), neglect, constructional or gaze apraxia and disorders of spatial cognition, a deficit of visual motion perception (akinotopsia; see Newcombe and Ratcliff, 1989; also Zihl et al. 1983; Vaina 1989; Plant and Nakayama 1993). Watson et al. (1993) showed that passive perception of motion leads to an increase in regional cerebral blood flow (rCBF; measured by PET) in the ascending limb of the inferior temporal sulcus. In a PET study on smooth pursuit visual tracking an activation of similar localization was observed by Miezin et al. (1988). Lesions in this region have been shown to reduce ocular pursuit gain, especially for pursuit towards the lesion side (Morrow and Sharpe, 1990). The lesions in the latter study, however, were relatively large and included substantial portions of the white matter.

Greenlee et al. (1995) selected patients with a relatively small lesion in the TPO region. In their study a deficit of visual motion perception was found for velocity discrimination (threshold for velocity differences when two velocity stimuli were presented). When the two stimuli were presented sequentially, the deficit increased with increasing stimulus interval (i.e., when the short-term memory for velocity was required). Part of these patients were tested in the present study for a possible impairment of smooth pursuit. We found that smooth pursuit gain was reduced in these patients as compared to normal controls, especially for pursuit towards the lesion side. We like to speculate that the pursuit deficit is related to that of the short-term storage of velocity, both being represented in the same cortical region, or in closely adjacent regions. Closed loop (steady state) pursuit depends to a considerable degree on the prediction, or an internal model of the forthcoming (expected) stimulus velocity. Provided that expected velocity is extrapolated from previously experienced velocity, one is in need of a short-term velocity storage. There is a considerable amount of work dealing with velocity stores for smooth pursuit (see Barnes, this volume), one of which could well be represented at cortical levels in the TPO region.

Pursuit of a visual target moving on a structured background has a somewhat reduced gain as compared to tracking in the dark. Since in our study this effect occurred to about the same extent in the patient group and the control group, it cannot be mediated by the TPO region. Possibly, the interaction between pursuit and background occurs on a cortical

processing stage (Kimmig et al., 1992) prior to the TPO region (e.g., V1 or V2), or at some later stage, e.g. in the frontal lobe (see Keller and Heinen, 1991; MacAvoy et al., 1991).

REFERENCES

Barnes, G.R., and Crombie, J.W., 1985, The interaction of conflicting retinal motion stimuli in oculomotor control, *Exp. Brain Res.* 59:548-558.

Clarke, S., and Miklossy, J., 1990, Occipital cortex in man: organization of callosal connections, related myelo- and cytoarchitecture, and putative boundaries of functional visual areas, *J. Comp. Neurol.* 298:188-214.

Collewijn, H., and Tamminga, E.P., 1984, Human smooth and saccadic eye movements during voluntary pursuit of different target motions on different backgrounds, *J. Physiol. (Lond)* 351:217-250.

Greenlee, M.W., Lang, H.-J., Mergner, T., and Seeger, W., 1995, Visual short-term memory of stimulus velocity in patients with unilateral posterior brain damage, *J. Neurosci.* 15 (in press)

Keller, E.L., and Heinen, S.J., 1991, Generation of smooth-pursuit eye movements: neuronal mechanisms and pathway, *Neurosci. Res.* 11:79-107.

Kimmig, H.G., Miles, F.A., and Schwarz, U., 1992, Effects of stationary textured backgrounds on the initiation of pursuit eye movements in monkeys, *J. Neurophysiol.* 68:2147-2164.

MacAvoy, M.G., Gottlieb, J.P., and Bruce, C.J., 1991, Smooth-pursuit eye movement representation in the primate frontal eye field, *Cerebral Cortex* 1:95-102.

Maunsell, J.H.R., and Newsome, W.T., 1987, Visual processing in monkey extrastriate cortex, *Annu. Rev. Neurosci.* 10:363-401.

Maunsell, J.H.R., and Van Essen, D.C., 1983a, Functional properties of neurons in middle temporal visual area of the macaque monkey. I. Selectivity for stimulus direction, speed and orientation, *J. Neurophysiol.*, 49:1127-1147.

Maunsell, J.H.R., and Van Essen, D.C., 1983b, The connections of the middle temporal visual area (MT) and their relationship to a cortical hierarchy in the macaque monkey, *J. Neurosci.* 3:2563-2586.

Miezin, F., Applegate, C., Peterson, S., Fox, P., 1988, Brain regions in humans activated during smooth pursuit visual tracking, *So.c Neurosci. Abstr.* 14:795.

Morrow, M.J., and Sharpe, J.A., 1990, Cerebral hemispheric localization of smooth pursuit asymmetry, *Neurology* 40:284-292.

Newcombe, F., Ratcliff, G., 1989, Disorders of visuospatial analysis. In F. Boller and J. Grafman (Eds.), Handbook of Neuropsychology, vol. 2, Elsevier, Amsterdam, pp. 333-356.

Newsome, W.T., Wurtz, R.H., and Komatsu, H., 1988, Relation of cortical area MT and MST to pursuit eye movements, II. Differentiation of retinal from extraretinal inputs, *J. Neurophysiol.* 60:604-620.

Newsome, W.T., Wurtz, R.H., Dürsteler, M.R., and Mikami, A., 1985, Deficits in visual motion processing following ibotenic acid lesions of the middle temporal visual area of the macaque monkey, *J. Neurosci.* 5:825-840.

Plant, G.T., and Nakayama K., 1993, The characteristics of residual motion perception in the hemifield contralateral to lateral occipital lesions in humans, *Brain* 116:1337-53

Sereno, M.I., McDonald, C.T., Allman, J.M., 1988, Myeloarchitecture of flat-mounted human occipital lobe: possible location of visual area MT, *Soc. Neurosci. Abstr.* 14:1122.

Tusa, R.J., and Ungerleider, L.G., 1988, Fiber pathways of cortical areas mediating smooth pursuit eye movements in monkeys, *An.n Neurol.* 23:174-183.

Ungerleider, L.G., and Desimone, R., 1986, Cortical connections of visual area MT in the macaque, *J. Comp. Neurol.* 248:190-222.

Vaina, L.M., 1989, Selective impairment of visual motion interpretation following lesions the right occipito-parietal area in humans, *Biol. Cybern.* 61:347-359.

Watson, J.D., Myers, R., Frackowiak, R.S.J, Hajnal, J.V., Woods, R.P., Mazziotta, J.C., Shipp, S., and Zeki, S., 1993, Area V5 of the human brain: Evidence from a combined study using positron emission tomography and magnetic resonance imaging, *Cereb. Cortex* 3:79-94.

Wurtz, R.H., Komatsu, H., Yamasaki, D.S., and Dursteler, M.R., 1990a, Cortical visual motion processing for oculomotor control, *Res. Publ. Assoc. Res. Nerv. Ment. Dis.* 67:211-31.

Wurtz, R.H., Yamasaki, D.S., Duffy, C.J., and Roy, J.P., 1990b, Functional specialization for visual motion processing in primate cerebral cortex, *Cold Spring Harb. Symp. Quant. Biol.* 55:717-27.

Zihl, J., Von Cramon, D., and Mai, N., 1983, Selective disturbance of movement vision after bilateral brain damage, *Brain* 106:313-340.

42

INERTIAL CORIOLIS FORCE PERTURBATIONS OF ARM AND HEAD MOVEMENTS REVEAL COMMON, NON-VESTIBULAR MECHANISMS

Paul DiZio and James R. Lackner

Ashton Graybiel Spatial Orientation Laboratory
Brandeis University
Waltham, Massachusetts 02254-9110

INTRODUCTION

Pitch or roll head movements made during passive rotation tend to be nauseogenic and disorienting. Traditionally these effects have been associated exclusively with the unusual pattern of vestibular stimulation, often termed Coriolis, cross-coupled stimulation (CCS), generated by such movements (Johnson, et al., 1951; Graybiel and Johnson, 1963). Models have been developed to explain the contribution of semicircular canal and otolith activation to disorientation and motion sickness (Guedry and Benson, 1978). In the course of experiments on the etiology of space motion sickness, we found that the nauseogenic and disorienting effects of head movements during rotation were highly gravitoinertial force (GIF) dependent, being greatly lessened in 0 g and greatly heightened in 1.8 g relative to 1 g baseline values (Graybiel et al.,1977; Lackner and Graybiel, 1984). This was the case despite maintaining the patterns of semicircular canal activation constant across different GIF levels.

In other experiments, we made detailed measurements of actual head trajectory and perceived head trajectory, rather than assuming what perceived trajectory would be on the basis of theoretical assumptions as has previously been the case (e.g. Guedry and Benson, 1978). These observations indicated that the apparent path of a pitch head movement during rotation was skewed in a fashion that could not be related to semicircular canal or otolith activity. This skewed path is illustrated in Figure 1. When a subject rotating counterclockwise tilts his head forward he feels it deviate rightward and then return somewhat toward the midline as the movement is completed. The subject also experiences a complex pattern of whole body tumbling. This "scalloping" was force level dependent, paralleling the disorienting and nauseogenic potential of the head movements, being less in 0 g and greater in 1.8 g than in 1 g.

We thought this pattern of apparent path deviation might be due to altered motor control of the head owing to the Coriolis forces present in movements made during rotation.

Multisensory Control of Posture, Edited by T. Mergner and F. Hlavačka
Plenum Press, New York, 1995

Figure 1. Illustration of a subject attempting to move arm and head straight down (broken arrows) while rotating counterclockwise. The movements are deviated rightward (solid arrows) which is correctly perceived.

Coriolis forces are proportional to the velocity and mass of the head (or other moved body part such as an arm) when the head is moved during body rotation, $F_{cor} = -2m(\omega \times v)$, where m is the mass of the head, v its linear velocity relative to the torso and ω is the angular velocity of the torso. These Coriolis forces also act on the otolith organs and semicircular canals of the inner ear. It is these latter effects that vestibular theorists have emphasized.

To see whether alterations in sensory-motor control of the musculature was occurring during exposure to Coriolis forces, we had subjects make arm movements during body rotation and indicate the experienced path of the arm. Subjects while rotating counterclockwise made arm movements in a parasaggital plane. When extending the arm outward they felt it deviate rightward and then return toward the intended plane as the movements slowed; flexing the arm resulted in a mirror symmetric apparent deviation. This was precisely the same form of apparent scalloping experienced when head movements rather than arm movements were made. Moreover, the magnitude of illusory curvature of arm movements was also, like that of head movements, proportional to background gravitoinertial force level.

Together these observations signified to us that an understanding of human spatial orientation would require consideration of skeletal musculature control mechanisms as well as vestibular function. As a first step in this direction we wanted to measure precisely the actual and perceived motion of an arm movement perturbed by a Coriolis force. Moreover, we wanted to characterize the time course and form of any adaptive compensations that might develop to correct deviations from the intended movement path or movement terminus. We expected that the understanding of arm movement control and perception so gained would generalize to the control of head movements made in a rotating environment.

PERCEPTION AND CONTROL OF ARM MOVEMENTS DURING CORIOLIS FORCE PERTURBATIONS

A circular rotating room (22 feet in diameter, 7.5 feet in height) that could be completely darkened was the site of the experiment. Subjects sat at the center of the room in a chair with a contoured headrest and a horizontal, waist-high desktop. The surface of the Plexiglas desktop was smooth, providing no tactile clues about the location of light-emitting diodes (LEDs) embedded in it from underneath. These LEDs served as targets the subject could point to. A low profile button switch was placed on the surface, to the right of the subject's torso. Subjects initiated a pointing trial by pressing the button switch with their right hand, which turned on an LED 35 cm straight ahead. Releasing the button during the onset of the pointing movement extinguished the LED and triggered data collection. The position of the fingertip was monitored with a WATSMART motion analysis system to an accuracy and resolution below 2 mm.

Eleven subjects reached in total darkness to the location of the extinguished LED before the room began to rotate, during 10 rpm counterclockwise rotation, and after the room stopped. The pre-rotation reaches established a baseline performance. During rotation, a transient rightward Coriolis force was generated on the arm when it moved forward; otherwise, there were no differences between pre- and per-rotation conditions. The post-rotation period was included to assess the effects on reaching of adaptation retained from the rotation period.

When reaching, subjects were attempting to make a smooth, continuous movement to the target at a comfortable speed. They were told to correct perceived in-flight errors if they could do so without slowing down, stiffening the arm, or stopping. During the pre-rotation period, the room was dark and subjects made 40 reaches to the target, in five sets of eight. Then, with the lights on they mimicked the last movements they felt they had executed in darkness. The room was darkened again and accelerated to $60°/s$ counterclockwise at $1°/s^2$. Subjects remained as motionless as possible for one minute after constant velocity was reached and then made 40 more reaching movements. Accelerating at a low rate, delaying reaching onset, and locating the subject on the axis of rotation prevented vestibular signals, postural reflexes, and visual illusions from influencing the reaches. The same procedure was followed for decelerating the room to a stop. Afterwards, the subjects made 40 reaching movements. At the end of the experiment, with the room lights on, subjects mimicked what the initial per-rotation and initial post-rotation movements had felt like. They had been forewarned that this mimic task would be required. It should be emphasized that during testing the only object ever visible was the target LED which extinguished with the onset of a reach. Consequently, subjects never had visual feedback about the accuracy of their reaches.

Pre-rotation, subjects (N=11) pointed in a straight line toward the target LED. The path of the fingertip deviated only 5 mm left of a straight line path between the start button and the movement endpoint. The average movement endpoint was 3 mm left of the target and 30 mm short of it. Movement time was 660 ms and peak velocity 825 mm/s.

On their first reach during rotation, every subject showed significant lateral deviations of movement path and endpoint in the direction of the Coriolis force generated. Throughout, our criteria for significance will be post-hoc Scheffé tests (p<.05, at least) following ANOVAs that showed a significant main effect for rotation period. Movements reached an average, peak rightward deviation from a straight line path of 15 mm before turning back toward baseline, but still ended 41 mm right of pre-rotation reaches (Figure 2). With additional reaches subjects rapidly adapted to the Coriolis force perturbations: by the last set of eight per-rotary reaches, the movements were straight, deviating only 2 mm right of a straight line path, and had regained pre-rotation endpoint accuracy. These movements also felt straight and accurate.

Every subject's first post-rotation reach was a mirror-image of his or her initial per-rotation reach. The paths of the initial movements deviated 25 mm left of a straight line path and ended 36 mm left of the baseline endpoint, both of which are significant differences from pre-rotation. Re-adaptation to normal conditions occurred at the same rate as adaptation to Coriolis force perturbations had. In the mimetic movements made after re-adaptation was complete, subjects matched the actual post-rotation endpoint errors but underestimated the

REAL INITIAL   REAL FINAL   MIMIC INITIAL

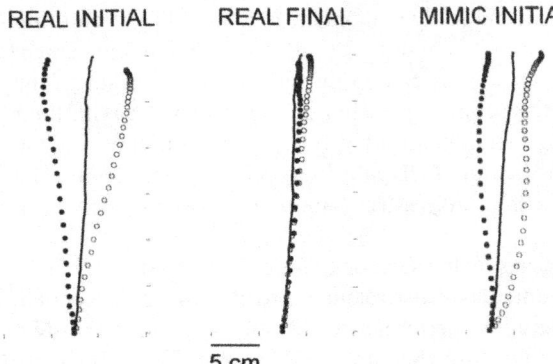

5 cm

Figure 2. Top view of average (N=11) reaching movements made before (solid lines), during (open dots) and after (solid dots) 10 rpm counterclockwise rotation. Represented are the first attempt to reach in darkness for an extinguished visual target (REAL INITIAL) and the fortieth attempt (REAL FINAL), as well as subsequent attempts to mimic what the initial movements had felt like (MIMIC INITIAL).

curvature of the movement paths. The mimetic movements ended 33 mm left of baseline but their maximum deviation from a straight line was only 5 mm leftward. The subjects' mimic of their initial per-rotary movements accurately reproduced the endpoint errors but erred in the shape of the actual movement path. The mimetic movements ended 38 mm right of baseline, which was also their peak deviation from the baseline movement path, and the mimic paths had one more inflection point than the actual movements, first deviating right then bending back toward the center and finally deviating right again.

Thus, our major findings were that transient Coriolis force perturbations initially prevented subjects from reaching the intended endpoint. Additional reaching movements, despite the absence of visual and tactile feedback about reaching accuracy, led to adaptive compensations such that movement paths initially deviated rapidly resumed their original straight-line characteristics and subjects accurately reached to the target position. Moreover, from their mimic movements it was clear that subjects perceived their initial endpoint errors and deviations in movement path. These results indicate that the nervous system must be continuously, monitoring and updating movement trajectory control. They also implicate muscle spindle and efferent signals in the process and perhaps joint and tendon afferents, as well.

These results are directly relevant to theories that address how the central nervous system regulates muscle properties responsible for posture and movement. For example, equilibrium-point theories propose that central nervous system commands directly regulate the spring-like characteristics of muscles (Feldman, 1966a,b, 1986; Bizzi et al., 1976, 1984, 1992).These theories avoid the great computational demands placed on the nervous system by theories that propose muscle innervation commands are generated from inverse dynamic calculations. In equilibrium-point theories, a desired posture is achieved by programming a length/tension relationship that results in muscle lengths and tensions and muscle torques around a joint that balance external loads. Movements are planned as a series of length/tension relationships that would bring about a corresponding series of equilibrium postures if sufficient settling time were allowed.

Proponents of equilibrium-point theories predict that transient perturbations of a movement would not affect the programmed endpoint of a movement. Feldman (1980), for example, taught subjects to move their forearms to well-defined positions in a familiar force field. He then instructed them not to voluntarily intervene if he transiently mechanically perturbed their arms during voluntary movements. Perturbed and unperturbed movements reached the same endpoint. Bizzi et al. (1984) trained monkeys to move a manipulandum to visual targets without sight of their arm. He then deafferented their arms by dorsal rhizotomy and studied their performance on the learned task after they were given additional training with visual feedback. When a torque was applied at movement onset that propelled a monkey's arm beyond its desired endpoint and then removed, the arm went to the desired position, as predicted.

Equilibrium-point theories also predict that transient Coriolis force perturbations such as in our experiments, should not deviate arm movement endpoints because the programmed length/tension relationship and the external load at the end of the movement would remain constant. The significant deviations we observed cannot be accounted for by the idea that movement and posture are the trajectory and equilibrium state of a mass-spring system whose spring-like parameters are set by the nervous system.

The apparent conflict between the endpoint errors caused by Coriolis force perturbations and the endpoint accuracy found by Bizzi, Feldman and others may be explained by the differences between paradigms. In our experiment, inertial Coriolis force perturbations were given without contact, but in the others there were local contact cues from perturbations given by torque motors via a contoured arm rest or manipulandum. Monkeys with arms deafferented by dorsal rhizotomies require significant retraining in the task of moving a manipulandum to a visual target even without perturbations. In the retraining process, they may learn to make use of spatially relevant sensory feedback arising from reaction forces generated in parts of the body with intact sensory innervation. Interestingly, Day and Marsden (1982) found that perturbations delivered via local contact to an unanesthetized thumb do not prevent achievement of learned positions, but with cutaneous anesthesia there are endpoint errors. These considerations show that our new technique for transiently perturbing a limb is ideal for evaluating theories of movement control. The results obtained disconfirm equilibrium-point hypotheses. In fact, methods traditionally used to evaluate equilibrium-point hypotheses are more appropriate for investigation of tool use and object manipulation, e.g. how something in contact with the arm or hand behaves or affects the arm movement.

In our experiment on Coriolis force perturbations of reaching movements, adaptations occurred as additional reaches were made, despite the absence of visual or tactile feedback about reaching accuracy, restoring straight-line trajectories and target accuracy. Post-rotation, when Coriolis forces were first absent there were mirror-image aftereffects to the initial errors made during rotation. Thus, the form of adaptation involved generating a new set of muscle commands that canceled the spatial deviations of the pointing finger from its intended straight-line path.

The pattern of findings indicates that muscle spindle signals during movement are interpreted in relation to motor commands. Under normal conditions, reaching involves generating a set of motorneuronal innervations to accomplish the desired movement. Coriolis forces deviate the arm from the expected path and activate spindle primary and secondary receptors in the unexpectedly stretched muscles of the arm. Spindle primary and secondary outputs code velocity and position, respectively (Matthews, 1972). Unexpectedly, high levels of spindle discharge are interpreted as lengthening of the host muscle and referred to the joint about which the muscle acts (cf. Matthews, 1988). This would largely account for the perceived deviation of an arm exposed to Coriolis force perturbations of its path. How such signals contribute to the adaptive process is an issue we have discussed in detail elsewhere (Lackner and DiZio, 1994; Lackner, 1985).

PERCEPTION AND CONTROL OF HEAD MOVEMENTS AND MOTION SICKNESS DURING CORIOLIS FORCE PERTURBATIONS

Armed with these ideas we returned to our original goal of understanding human perceptual responses to head movements during body rotation. We set out to test whether the control and appreciation of movement path, and possibly the evocation of motion sickness symptoms would be influenced by factors like the non-vestibular, sensory-motor control and calibration mechanisms we observed in reaching movements. The situation is more complex with head movements because of the vestibular stimulation also generated by Coriolis forces. Therefore, we developed a paradigm that permitted the vestibular stimulus to be held constant and the non-vestibular, sensory-motor control component to be experimentally manipulated. This was achieved by comparing the responses to head movements during rotation when subjects tilted their head with no external load versus for the same attempted movement with extra mass added to the head. Adding mass externally has no effect on vestibular stimulation, but it does affect the magnitude of the Coriolis force applied to the head as a whole (mass is a factor in the equation for Coriolis force).

We expected that Coriolis force perturbations of the head/neck system would affect the control and perception of head trajectory relative to the trunk, whereas the vestibular signal would affect perceived movement of the whole body and reflexive control of gaze. In particular, we predicted that larger Coriolis forces with constant vestibular stimulation would produce larger perceptual deviations of the head from its actual path.

Precise descriptions of the perceptual consequences of CCS are difficult for subjects to make because CCS can be confusing and nauseogenic. CCS is usually generated by having subjects make head movements after they have been rotating at constant velocity for some period. Guedry and Benson (1978) found that relatively little disorientation and motion sickness is elicited by CCS when head movements are made immediately after reaching constant velocity. In pilot experiments, we found that subjects could easily discern the effects of head-loading on perceived head trajectory when head movements were made immediately after acceleration to constant velocity was complete. The following describes an experiment using this paradigm for testing non-vestibular effects of head loading on perceived trajectory of the head relative to the torso.

Eight subjects participated in the experiment which involved making head movements in a rotating chair. There was a lap belt and padded stops to limit the extent of pitch head movements but no other restraints. The head could be moved actively between an upright position and 40° pitched forward. The subjects wore a three-axis rate sensor on a light-weight headband (total mass = 88 grams) which was used in preliminary sessions to give feedback about movement velocity and in experimental sessions to measure actual and perceived head trajectory. In one of the two experimental sessions, a 500 gram mass was added to a biteplate held by the subject, in the other, no mass was added. The added mass required the subjects to activate neck extensor muscles beyond their normal levels for maintenance of a static upright or pitched forward head posture and it increased the effective inertia of the head that had to be controlled during movements. Vestibular stimulation was the same in both sessions because head movements always had comparable amplitudes and velocities, but Coriolis force perturbations of the head/neck system were greater in the session with the added mass. CCS was produced by having the subject make a pitch forward head movement immediately after acceleration at $9°/s^2$ to $90°/s^2$ counterclockwise rotation. The chair was then stopped, the subject released the biteplate and, after a two minute pause, we recorded from the rate sensors while the subject mimicked the perceived path of the head movement that had been made during rotation. Subjects also reported motion sickness severity on a 1 to 10 scale. This procedure was repeated eight times on each test day, with enough rest between trials for any disorientation and motion sickness symptoms to abate completely.

Subjects attempted to pitch their head straight forward in a midsagittal plane but perceived a deviation to the right, the direction of the Coriolis force. The perceived deviation was greater and more severe motion sickness symptoms were elicited in the condition with the mass added to the head. Both differences were highly significant statistically. Importantly, the perceived and actual pitch amplitude and velocity of head movements did not differ across the head loaded and unloaded conditions.

Interestingly, neither our baseline nor our head loaded observations give patterns of experienced trajectory that are in accord with traditional vestibular models of orientation. For example, Benson and Guedry's model (1978) fails to predict either the perceived scalloping of head movements or the effects of head weighting on motion sickness and disorientation. It also predicts semicircular canal stimulation that codes leftward roll of the head when in fact our results show that subjects perceived rightward roll.

These findings demonstrate that non-vestibular, sensory-motor mechanisms also contribute to the disorientation and motion sickness elicited by head movements made during body rotation. The Coriolis forces generated in this experiment deviated movements to the right and subjects accurately perceived this direction of deviation. Although the synergies among muscles controlling head movements are quite complex, multi-dimensional frames of reference can be identified (Baker et al., 1985). Such complex but systematic relationships establish the conditions necessary for the nervous system to make use of muscle spindle feedback in perception of head movement trajectory and updating of motor control. Spindle error signals that drive the adaptation process by which straight forward pitch head movements are restored may also be implicated in the evocation of motion sickness.

The involvement of mechanisms that process muscle spindle signals during movement may explain why head movements, of matched kinematic properties, made while rotating generate disorientation and motion sickness whose severity is proportional to GIF level. We have already shown that other phenomena that depend on muscle spindle signals, for example vibratory myesthetic illusions (Lackner et al., 1992) and the ability to reproduce practiced arm movements (Fisk et al., 1993), are GIF dependent. Additional experiments are in progress to further discriminate vestibular and non-vestibular influences on motion sickness susceptibility and head movement control. The results are demonstrating that vestibular signals represent only one of the factors influencing body orientation and that skeletomuscular control affects both the control and appreciation of orientation.

In summary, we have utilized the inertial, non-contacting Coriolis forces that can be generated in a rotating environment to study basic mechanisms of arm and head movement control and perception. We have identified sensory-motor components influencing head movement control that explain aspects of performance that have been misattributed to vestibular function. Our overall findings indicate that it is essential to broaden our theoretical perspectives and not rely only on vestibular based models to explain spatial orientation. The reliance on such models is seriously hindering an adequate understanding of posture, movement, and orientation.

REFERENCES

Baker, J., Goldberg, J., and Peterson, B., 1985, Spatial and temporal response properties of vestibulocollic reflex in decerabrate cats, *J. Neurophysiol.* 54:736-756.

Bizzi, E., Accornero, N., Chapple, W., and Hogan, N., 1984, Posture control and trajectory formation during arm movement, *J. Neurosci.* 4:2738-2744.

Bizzi, E., Polit, A., Morasso, P., 1976, Mechanisms underlying achievement of final head position, *J. Neurophysiol.* 39:435-444.

Bizzi, E., Hogan, N., Mussa-Ivaldi, F.A., and Giszter, S., 1992, Does the nervous system use equilibrium point control to guide single and multiple joint movements? *Behav. Brain Sci.* 15:603-613.

Day, B.L., and Marsden, C.D., 1982, Accurate repositioning of the human thumb against unpredictable dynamic loads is dependent upon peripheral feedback, *J. Physiol.* 327:393-407.

Feldman, A.G., 1986, Once more for the equilibrium point hypothesis (model), *J Motor Behavior* 18:17-54.

Feldman, A.G., 1980, Superposition of motor programs. I. Rhythmic forearm movements in man, *Neurosci.* 5:81-90.

Feldman, A.G., 1966a, Functional tuning of the nervous system during control of movement or maintenance of a steady posture. II. Controllable parameters of the muscle, *Biophysics* 11:498-508.

Feldman, A.G., 1966b, Functional tuning of the nervous system during control of movement or maintenance of a steady posture. III. Mechanographic analysis of the execution by man of the simplest motor task., *Biophysics* 11:766-775.

Fisk, J., Lackner, J.R., and DiZio, P., 1993, Gravitoinertial force level influences arm movement control, *J. Neurophysiol.* 69(2):504-511.

Graybiel, A., and Johnson, W.H., 1963, A comparison of the symptomatology experienced by healthy persons and subjects with loss of labyrinthine function when exposed to unusual patterns of centripetal force in a counter-rotating room, *Annals of Otology, Rhinology and Laryngology* 72:357-373.

Graybiel, A., Miller, E.F., and Homick, J.L., 1977, Experiment M-131, Human vestibular function. In: Johnston, R.S., Dietlein, L.F. (eds.), Biomedical results from Skylab, Sect. II. US Government Printing Office, Washington DC.

Guedry, F.E., and Benson, A.J., 1978, Coriolis cross-coupling effects: disorienting and nauseogenic or not? *Aviation Space and Environmental Medicine* 49:29-35.

Johnson, W.H., Stubbs, R.A., Kelks, G.F., and Franks, 1951, W.R. Stimulus required to produce motion sickness, *J Aviation Medicine* 22:365-374.

Lackner, J.R., and Graybiel, A., 1984, Influence of gravitoinertial force level on apparent magnitude of Coriolis cross-coupled angular accelerations and motion sickness. NATO-AGARD Aerospace Medical Panel Symposium on Motion Sickness: Mechanisms, Prediction, Prevention and Treatment. AGARD-CP-372, 22, 1-7.

Lackner, J.R., and DiZio, P., 1994, Rapid adaptation to Coriolis force perturbations of arm trajectory, *J. Neurophysiol.* 72(1):299-313.

Lackner, J.R., 1985, Human sensory-motor adaptation to the terrestrial force environment. In: D. Ingle, M. Jeannerod and D. Lee (Eds.). Brain Mechanisms and Spatial Vision, Amsterdam: Nijhoff, 175-210.

Lackner, J.R., DiZio, P., and Fisk, John D., 1992, Tonic vibration reflexes and background force level, *Acta Astronautica* 26(2):133-136.

Matthews, P.B.C., 1972, Mammalian Muscle Receptors and their Central Actions, London: Edward Arnold.

Matthews, P.B.C., 1988, Proprioceptors and their contribution to somatosensory mapping: complex messages require complex processing. *Can. J. Physiol. Pharmacol.* 66:430-438.

43

INFLUENCE OF SHORT- AND LONG-TERM EXPOSURE TO REAL MICROGRAVITY ON KINEMATICS OF POINTING ARM MOVEMENTS

M. Berger,[1] S. Mescheriakov,[1] E. Molokanova,[1] S. Lechner,[1]
F. Gerstenbrand,[1] I. Kozlovskaya,[2] B. Babaev,[2] and A. Sokolov[2]

[1] Institute for Space Neurology
Anichstr 35
A-6020 Innsbruck, Austria
[2] Institute for Biomedical Problems
Chorochevskoje Ch. 76a
Moscow 123007, Russia

INTRODUCTION

Movement control in microgravity is complicated by several factors: changed physical properties of the body and extremities, functional proprioceptive deprivation due to reduction of the input from proprioceptors of postural muscles, and increased excitability of the motor control structures. These factors cause complex disturbances of co-ordination of aimed voluntary movements, which are known as hypogravitational ataxia (Kozlovskaya, 1988) and manifest by reduced accuracy and altered movement kinematics.

If movements should be performed in microgravity under high accuracy constraints (as most real tasks), a pronounced increase of movement duration is observed. Reason for that was supposed to be a re-orientation of the motor control system to a more reliable source of information - the visual system. According to this point of view the motor control changes to the visual tracking mode. Movements performed in the visual tracking mode tend to have more than one deceleration, which results in prolongation of the deceleration phase.

Numerous studies of movement organization (for review Jeannerod, 1988), reported that changes of motor control strategy lead to the deformation of normal bell-shaped velocity profile, which becomes asymmetric with prolonged deceleration phase. This is true also for visually controlled movements. Therefore, the symmetry of the velocity profile seems to be an important parameter, providing information about the role of the visual system in movement organization.

The gravitational force is integrated in the organization of every movement performed on the earth surface (Lackner, 1993). As shown recently, movements performed under normal gravity conditions in different directions, with and against gravity, have very similar kinematics. However, the correspondent EMG patterns tend to change in order to compensate the influence of the gravity force (Virji-Babul et al., 1993).

Under earth conditions the body is fixated on the surface by the gravity force and can compensate the impact of the moving limb by additional postural adjustment. In microgravity any rapid muscular contraction evokes a counter-movement of the body if no additional fixation is provided. In this situation the central nervous system should elaborate new control patterns to preserve its ability to organize and conduct temporally and spatially accurate aimed movements, and to reduce the destabilization of the body.

Observations have shown that movement accuracy, defined as final position of the limb at the end of the movement, tends to be preserved in microgravity. Furthermore, cosmonauts are able to perform almost the full repertoire of movements from the beginning of exposure to microgravity. However, movements become slower and their kinematics are altered. Visual input, supposedly, becomes more important for movement organization in the microgravity environment.

Since for single-joint movements the kinematics are proportional (neglecting viscosity) to force production, the kinematic analysis can elucidate the mechanisms underlying the reorganization in motor control during the adaptation to microgravity.

We studied simple single-joint arm movements in order to analyze the alterations of movement kinematics, and to elucidate the role of the visual system in the movement organization in microgravity.

METHOD

Subjects

Two cosmonauts, which were crew members of two different space missions on board the Russian MIR space station performed several pre-flight, in-flight, and post-flight sessions. Cosmonaut A participated in 10-days space flight and performed three pre-flight, two in-flight (on the 2d and 5th day), and two post-flight sessions. Cosmonaut B participated in 132-days space flight and performed two pre-flight, four in-flight (on the 27th, 60th, 70th, and 102d day) and two post-flight sessions.

Hardware

The MONIMIR equipment was used for data collection (Berger et al., 1992). The equipment was developed for investigation of eye-head-arm coordination in microgravity and consists of a matrix of LED's for signal presentation and of an opto-electronic system for three-dimensional recording of the position of infrared light emitting diodes (IR-LED's). Arm movements were recorded by using an arm clamp fixed on the forearm and shoulder to prevent possible elbow flexion. The lamp was equipped with two pairs of IR-LED's allowing reliable definition of the arm position. The opto-electronic system included two IR video cameras that allowed registration of the position with a 25 Hz sampling rate.

Experiment

On earth the subjects were tested in a sitting position. In flight they were fixed to the floor of the station by two belts. They were instructed to perform pointing arm movements towards flashing LED's as accurate and as fast as possible, and to begin the movement as soon as the next flashing LED appears. Importance of accuracy was emphasized. The sequences of flashing LED's were randomized in time and direction. Targets appeared at 4 and 16 angular degrees from the center of the visual field. They required movements in the horizontal and vertical planes. In this paper only horizontal movements at 16 angular degrees are analyzed.

Data Processing

Three-dimensional Cartesian position data were converted into values of arm angular displacement and interpolated using the "two-bells" model of simple single joint arm movements (Mescheriakov et al., 1994). The model was proposed in order to separately estimate the angular acceleration produced by agonist and antagonist muscles (or muscle groups) during simple pre-programmed arm movements (that is one acceleration and one deceleration phase). The basic assumption of the model was that time profiles of the angular acceleration produced separately by agonist and antagonist muscles could be approximately described using Gaussian distribution functions. Two best fitting Gaussian functions were selected using simplex algorithm. Linear combination of these two functions yielded the acceleration-time profile which, when integrated, produced velocity- and position-time profiles. This interpolation procedure showed high fitting ability. The original software MONIGRAF was used for extraction of the kinematic parameters.

Parameters

Three different time profiles were analyzed. (i) Position-time profile: movement amplitude (maximal angular displacement), movement duration, and mean velocity. (ii) Velocity-time profile: peak velocity, acceleration phase (time from the movement onset to the peak velocity), and deceleration phase (time from the peak velocity to the end of movement). (iii) Acceleration-time profile: peak acceleration, peak deceleration, acceleration time (the time from the movement onset to the peak acceleration), "switch" time (the time from the peak acceleration to the peak deceleration), and deceleration time (the time from peak deceleration to the end of movement). Mean values and standard deviations of each parameter were calculated and analyzed.

RESULTS

Constantly high accuracy of angular displacement for movements of 16 angular degrees is observed in all test sessions including in-flight for cosmonauts. Mean deviation from the target in all in-flight sessions did not exceed 5% of the movement amplitude. Movement duration increased in-flight in both short- and long-term flights and was up to 300 ms longer in short-term and up to 200 ms longer in long-term flight. Significant decrease of peak and mean velocities is observed in all in-flight sessions for both cosmonauts.

Velocity-time profile was asymmetric in all sessions due to prolonged deceleration phase. No changes of the symmetry of the velocity-time profile could be detected in in-flight sessions, because both phases increased in-flight by the same factor.

Analysis of the acceleration-time profile shows that for cosmonaut A the acceleration and deceleration times were almost equal in all sessions. The lower curve on the Fig. 2 results from overlapping values of the acceleration and deceleration times. For cosmonaut B the deceleration time was up to 100 ms longer than the acceleration time. The switch time for both cosmonauts constituted about 50 % of the movement duration.

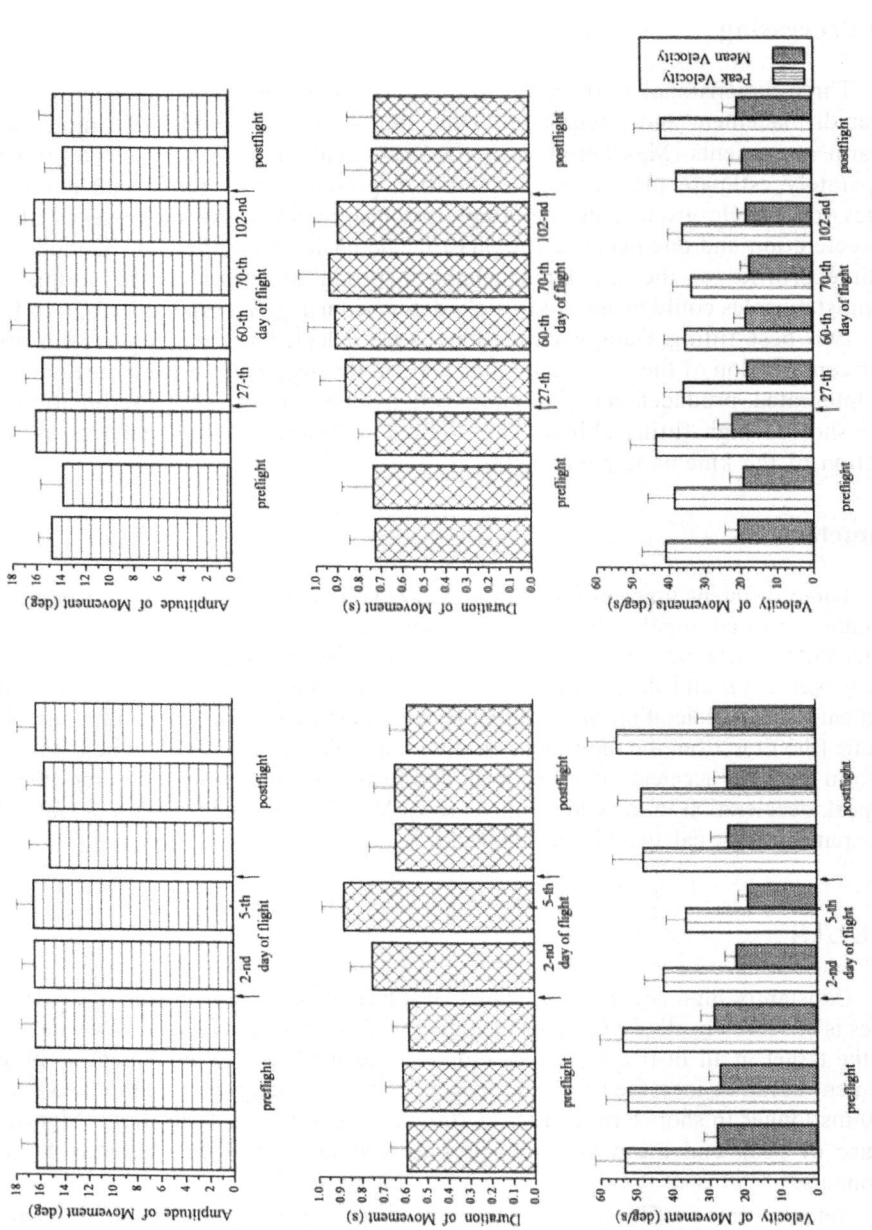

Figure 1. Angular displacement, duration, and peak and mean velocities of pointing arm movements. Left: Cosmonaut A - short-term flight, right: Cosmonaut B - long-term flight. Arrows below the horizontal axis indicate launch and landing. Target angular displacement was 16 degrees. All values are means and S.E.D. per session.

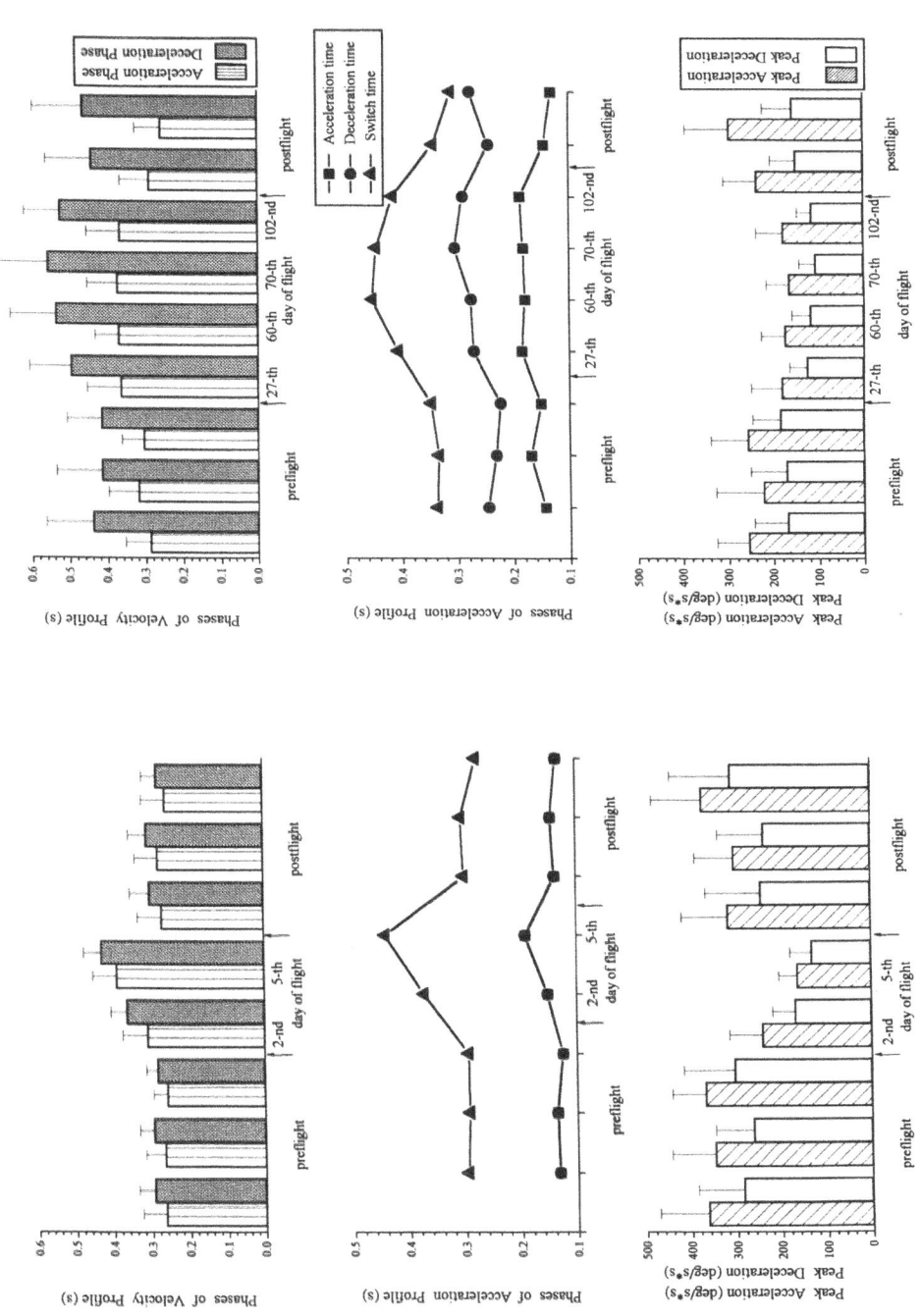

Figure 2. Parameters of the velocity-time and acceleration-time profiles. Left: Cosmonaut A - short-term flight. Right: Cosmonaut B - long-term flight. Arrows below the horizontal axis indicate start and landing. Further explanations in text.

The increase of movement duration in-flight results from lengthening of all three phases of the acceleration-time profile. However, the absolute contribution of the switch time is considerably higher. Mean values of peak acceleration were higher than values of peak deceleration and both considerably decreased in-flight.

DISCUSSION

Our data suggest that the CNS can successfully organize movements and maintain high accuracy in weightlessness and during the re-adaptation to normal gravity. However, it requires additional time for the movement organization under new conditions, which results in prolongation of movement. Since in our task the accuracy constraints were high and cosmonauts were instructed to perform movements as fast as possible, certain decrease of movement duration can be expected when changes in control mechanisms occur.

Pointing to visual targets is obviously controlled by proprioceptive and visual input. Movements seem to be ballistic, that is they are not controlled during the movement. Only terminal visual feedback is possible. The question arises whether movements in microgravity are also executed ballistically or if the visual system intervenes on the late phases of movement. If the latter is true, general undershooting and/or prolonged final phases of movement could be expected. Participation of the visual system in the middle phase of movement, where actual velocity is very large, seems to be impossible because of difficulties in visual perception. The visual system can control movement only in the phases where velocity is small, that is after the deceleration peak.

Since no specific changes in the final part of movement (deceleration phase and deceleration time) were observed, we can not make a conclusion about an increased role of the visual system in control of this type of movements in microgravity during both short and long-term space flights.

The most sensitive part of the acceleration profile, the switch time, seems to be very important for the regulation of movement amplitude in microgravity. Generally, the switch time contributes more to the prolongation of movement than either acceleration or deceleration times. There are strong correlations between the peak velocity and length of the switch time, as well as between peak acceleration and deceleration and switch time.

In general, we observed that lower accelerations and decelerations were characteristic for movements in weightlessness. There are several possible explanations for this effect. The first is that the proprioceptive system, altered in microgravity, needs more time for coordination between the agonist and antagonist activity because of disturbances in proprioceptive control loops. This explains the prominent increase of the time between peak acceleration and peak deceleration (switch time on the Fig. 2). The second explanation could be that lower and prolonged accelerations of extremities reduce the counter-movements of the body in microgravity. The CNS elaborates a strategy of careful and slow movements and therefore avoids the destabilization of the body during the movement. This explains the decrease of peak values of accelerations and decelerations.

Detailed analysis of kinematics of movements in microgravity should be conducted in order to clarify strategies of the motor control system in adaptation to altered force environment.

ACKNOWLEDGMENTS

This work was supported by Austrian Ministry for Science and Research.

REFERENCES

Berger, M., Gerstenbrand, F., Kozlovskaya, I., Holzmueller, G., Hochmair, E., and Steinwender, G., 1992, Eye, head and arm coordination and spinal reflexes in weightlessness - MONIMIR experiment. In: Health from Space Research, Ed: Austrian Society for Aerospace Medicine, Springer-Verlag/Wien, N.Y., pp. 119-135.

Jeannerod, M., 1988, The neural and behavioural organization of goal-directed movements, Clarendon Press, Oxford.

Kozlovskaya, I. B., 1988, Gravitational mechanisms in motor systems. Studies in real and simulated weight-lessness. In: Stance and Motion. Facts and concepts. Ed. by Gurfinkel et al. Plenum Press. NY, London, pp. 37-48.

Lackner, J. R., 1993, Orientation and movement in unusual force environments, *Psychol. Science* 4:144-142.

Mescheriakov, S., Holzmueller, G., Berger, M., and Molokanova, E., 1994, On the coordination of simple pre-planned arm movements: a kinematic model of agonist-antagonist interaction, Suppl. No. 7 to the *European Journal of Neuroscience* Abs. 67-17.

Virji-Babul, N., Cooke, J.D., and Brown, S. H., 1994, Effects of gravitational forces on single joint arm movements in humans, *Experimental Brain Research* 99:338-346.

44

IS POSTURAL STABILITY CHANGED BY AVIATION PRACTICE?

Milos Sázel

Institute of Aviation Medicine
160 60 Prague
Czech Republic

INTRODUCTION

Various methods are used for the control and selection of pilots and pilot candidates. The outcome of these examinations are of considerable importance, not only for flight safety, but also because of the high costs for pilot training. Therefore, many different types of the aviators' reactions are evaluated. Measurements of postural control have repeatedly been used to monitor the state of health, rather rarely, however, in air forces.

Previous results have shown that specific features of postural stability of pilots can be attributed to vestibular afferents (Sázel, 1993). Postural control may reflect the pilot's ability to deal successfully with the stress of spatial disorientation. Therefore measurement of postural stability may be a predictor of certain parameters of flight performance (Kohen-Raz et al., 1994).

METHODS

Postural stability was measured on a stabilometer platform covered with a 10 cm thick foam rubber. Four groups of healthy subjects were tested: 36 jet fighter pilots (37 ±4 years old, 1014 ±472 flight hours), 79 jet pilot students (21 ±1 yrs., 40 ±18 h.), 27 helicopter pilot students (22 ±1 yrs., 26 ±2 h.), and 45 control subjects (27 ± 8 yrs.). Stabilometer data of the blindfolded subjects were collected during 50 s and processed with the help of a computer (Kundrát and Hlavacka, 1989). Power spectrum density analysis of the data also was performed. Statistical results were calculated using ANOVA or, alternatively, the Kruskal-Wallis test and post hoc comparisons (CSS Statsoft).

RESULTS

Significant differences in stabilogram data were observed between groups for the parameter 'amplitude' ($p < 0.05$). The most stable subjects were the jet fighter pilots in the

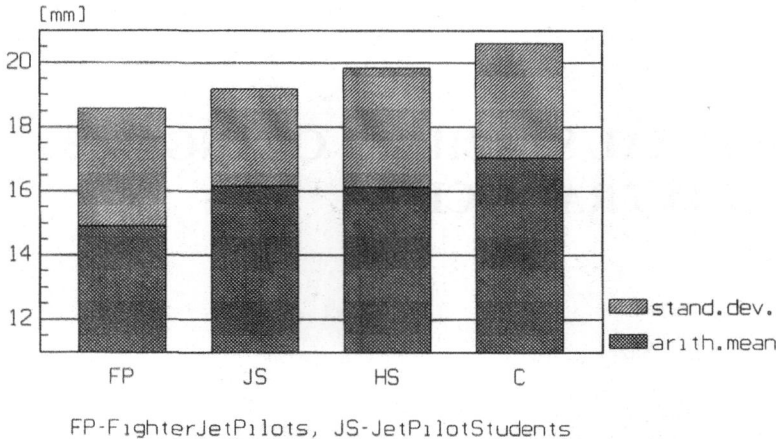

FP-FighterJetPilots, JS-JetPilotStudents
HS-HelicopterPilotStudents, C-Control Ss

Figure 1. Root mean square of statokinesigram for the four groups of subjects. FP, jet fighter pilots; JS, jet pilot students; HS, helicopter pilot students; C, control subjects.

lateral plane (vs. control $p<0.01$, vs. other groups, $p<0.05$). Almost the same results were obtained in the anterior-posterior plane. In contrast, no significant differences were found in the indexes of speed.

Differences between the groups also were observed for the root mean square of statokinesigram ($p<0.05$), as shown in Fig. 1. The differences were significant between controls and jet fighter pilots ($p<0.01$) as well as between jet fighter pilots and jet pilot students ($p<0.05$). No differences were found in line integral and in the total area.

Power spectrum density (PSD) for both axes at 0.25 Hz was higher in the jet fighter pilots than in the other groups ($p<0.05$). At 0.7 Hz and 0.8 Hz, PSD of lateral sway was highest in the two student groups, but this effect was not very pronounced.

DISCUSSION

The results support the hypothesis that a fighter pilot career has some effect on postural stability. It appears that intensive flying improves postural stability. We suppose that the importance of vestibular afferents for spatial orientation increases with aviation practice. This presumption is supported by higher values of the PSD at 0.25 Hz in the jet fighter pilots. The most influencing vestibular stimulation in flight occurs at 0.2 Hz frequency (Sázel 1992). Maximal resonance of this stimulation on a stabilometer also occurred at 0.2 - 0.3 Hz (Kapteyn and Wit, 1972; Bizzo and Baron, 1972).

It appears necessary for further research on this issue to determine in detail the possible role of age, of the type of flight training used, and of the criteria used to select the pilot candidates. A prospective study on the same subjects during their further flight career would be desirable.

SUMMARY

Postural stability was measured using a stabilometer during a time period of 50 s. The stabilometer was covered with 10 cm foam rubber. Jet fighter pilots (n=36), jet pilot and helicopter pilot students (n=79 and n=27, respectively) and controls (45) were examined. The largest sway responses were observed in the control group. Significantly lower amplitudes in both axes of sway were found in the jet fighter pilot group (p<0,01). The latter group also showed higher values in power spectrum density (statistically significant at 0.25 Hz). It is supposed that postural stability is improved with extensive aviation practice, possibly due to the influence of vestibular afferents.

References

Bizzo, G., Baron, J.B., 1972, Aspect cybernétique des déplacements du centre de gravité du corps induits par des stimulations labyrintiques électriques rectangulaires ou sinusoidales. *Agressologie* 113 B:41-50.

Kapteyn, T.S., and Wit, G., 1972, Posturography as an auxiliary in vestibular investigation, *Acta Otolaryng.* 73:104-111.

Kohen-Raz, R., Kohen-Raz, A., Erel, J., et al., 1994, Postural control in pilots and candidates for flight training, *Aviat Space Environ Med* 65:323-326.

Kundrát, J., and Hlavacka, F., 1989, Program for estimating stabilometric data by microcomputer, *Cs Physiol* (in Slovak) 38:123-129.

Sázel, M., 1992, Galvanic vestibular stimulation of a pilot in a flight simulator, In: H Krejcová, J., Jerábek, eds., 1993, Proceedings of XVIIth Bárány Society Meeting, Castle Dobris, Czechoslovakia, Conexim, Prague, pp. 120-121.

Sázel, M., 1993, Parameters of stabilometry with galvanic vestibular stimulation in pilots (Abstract). *Physiol Res* 42:12.

45

EFFECTS OF ORBITAL SPACE FLIGHT ON VESTIBULAR REFLEXES AND PERCEPTION

Laurence R. Young

Man-Vehicle Laboratory
Massachusetts Institute of Technology
Cambridge, Massachusetts 02139

INTRODUCTION

We have considered the complex adaptive process of human spatial orientation in terms of a time varying optimal estimator (Borah et al., 1978,1988, Merfeld et al., 1993). According to this view, the brain is constantly attempting to make the best possible guess about where we are in the presence of parallel, at times conflicting, cues from the various sensory organs and from efferent copy of our own motor commands. Central to the optimal estimator is the Kalman filter, which is capable of determining mathematically the gains to be applied to each measurement signal in order to produce a "state estimate" which is optimal in the least squares error sense, for some given error cost function. At the heart of the optimal estimator is the "internal model", which includes knowledge of the dynamic characteristics of the body, of the sense organs and of the random motions to which the body is subjected. This internal model appears, in part, under many different names in the evolving literature on oculomotor and postural control, including "perceptual feedback", "corollary discharge", "velocity storage", "second integrator" and "body schema". In its full implementation the mathematics for solving the Ricatti equations for the optimal gains is daunting - but the idea is actually quite simple.

The optimal estimator constantly updates its estimate of spatial orientation - which we consider to consist of a twelve dimensional vector - three angles describing our orientation, three Cartesian coordinate numbers describing the position of the center of the body reference frame, and the velocity, or first derivative of each of these. New information from the sense organs is weighed according to the model of the frequency response of each sensor, as well as the estimate of the inherent noise in each channel. For example, mid frequency signals from the semicircular canals, in the region of 0.1 to 1 Hz, are given a high gain in the representation of angular velocity, whereas low frequency semicircular canal signals, say in the region of 0.01 Hz, are generally disregarded because of the knowledge that the canals are unreliable at frequencies well below that determined by the cupula/endolymph long time constant. Similarly, visual orientation cues dominate at the lowest frequencies and localized tactile cues are weighed most heavily at the high frequencies, where they represent a shift in body position. The inherent noise in each channel may be thought of mechanistically in terms of the variance in the resting discharge of afferent signals, or as represented by motion detection thresholds measured psychophysically (Ormsby and Young, 1977). A channel with high noise power in a given frequency region is weighed less heavily. Furthermore, the optimal estimator requires some idea of the expected patterns of input or disturbance motions, even if these motions are random. For the usual case random Gaussian noise distributions are assumed for simplicity, but even predictable discrete motions patterns which could be learned and anticipated may be incorporated into the internal model. Therefore, the knowledge that a certain motion pattern is expected will increase the weighting of sensors in the frequency range appropriate to that pattern.

Beyond the representation of the sensor dynamics and input characteristics, the internal model must characterize the body dynamics, including efferent copy of motor commands. Some of the characterization is simply a representation of Newton's laws of motion. Thus the internal model "knows" that the body's linear and angular momentum will remain unchanged in the absence of any external force. It "knows" that a jump initiated by motor commands to the leg extensors will produce a certain upward acceleration, followed by downward acceleration at 1-g until contact with the floor is made once again. Of course the internal model must be adaptive to keep up with the current actual body dynamics. Thus the first jump made when wearing a heavy backpack is likely to produce an "expected response" from the internal model which is higher than that registered by the visual and vestibular sensors, but in time the model is adaptively updated to represent the current body parameters. (The same idea applies when motor control is applied to some mechanical system, like riding a bicycle or operating an elevator, so that in time each new situation is recognized by the internal model and the correct characterization of body dynamics is used.)

Finally, the internal model must have an accurate representation of the external environment acting upon the body and the sensory system. The considerable literature concerning plasticity of the vestibulo-ocular system under conditions of reversed or magnified vision presents ample evidence of the adaptive process, and the extent to which a novel environment is "learned" with repeated exposure. The external environment change with which we have been concerned over the past decade and a half is gravity, and the adaptation of the internal model to the novel conditions of "weightlessness" achieved in orbital space flight. Let us then put ourselves in the position of an astronaut, and see what alterations to the internal model are required.

Obviously gravity is not eliminated when we go into orbit - at least not in the low earth orbit flights in which our experiments are conducted. At orbital altitudes of only 200-300 km gravity is virtually the same as on the surface - but it feels as though we were in a gravity free environment. (Of course we and the spacecraft are both falling toward the center of the earth at an acceleration of 9.8 m/sec^2, The only reason we don't impact the earth is that we are moving ahead so rapidly, at orbital velocity, that the falling path of our spacecraft just parallels that of the earth, at least for a circular orbit.) The familiar visual environment of the spacecraft, with the floor, the ceiling and the walls with rack mounted equipment just where it had been for our many months of training, certainly doesn't offer any clues that we are in "free fall". Furthermore, we don't feel as though we are falling. But as soon as we start making head movements and responding to body position with normal, 1-g postural reactions we are immediately aware of the dramatic change in the gravito-inertial environment. As a first time flier we would be readily identifiable by the 'jumping jack" motion as we inappropriately extend our legs to meet the floor - which sends us bobbing up to the ceiling. (Our more experienced crew mates would remain stationary effortlessly by use of a foothold or handhold, having learned on a previous flight to inhibit the 1-g postural reflexes.) Head movements in pitch or roll produce yet another surprise. Although the semicircular canals and the visual input confirm the intended head movement, the otolith organs fail to confirm the change in head angular orientation relative to the floor. Only gradually, over a period of hours to days, is the internal model of the environment updated so that the expected response to motor commands matches reality in weightlessness, and so that the gravito-inertial signals detected by the otolith organs are correctly identified with linear acceleration and not with head orientation relative to the vertical. Gradually the space motion sickness which accompanies the sensory-motor conflict subsides (Oman and Shubentsov, 1992; Oman, 1990) and the oculomotor, postural and spatial orientation responses reach a new steady state. With full adaptation to weightlessness the optimal estimator once again mixes external cues detected by the sense organs, with internal model representations of body dynamics - but the details are now appropriate to free-fall. Not surprisingly, postural imbalance is often observed shortly after return to the surface of the earth, but persists only long enough for the familiar 1-g internal model to be called upon and appropriate interpretations of graviceptor signals to be restored. (Kenyon and Young, 1986).

Fig. 1 is a schematic representation of postural reflexes and perception of orientation, emphasizing the three components of the "internal model" contained within the central estimator. Acceleration from external fields and body motions are detected by the various sensors, each with its own dynamic response and noise characteristics. The Central Estimator internal model includes specific models of the sensors and of the body dynamics, and keeps track of the motor commands. It also includes an internal "force field model" which describes the forces on the body associated with gravity, for example, or with the Coriolis forces in a rotating environment. In the course of adapting to weightlessness this force field model is modified appropriately.

In order to explore the way this putative internal model is updated we performed a series of "rotating dome" space experiments in which the perception of self motion was measured with varying combinations of visual, vestibular and haptic cues. Ground tests of roll vection with the head in different orientations had convinced us that the magnitude of visually induced roll rate was limited by the signals from the utricular macula, which fail to confirm the tilt relative to the vertical. (Dichgans et al., 1972; Young et al., 1973). Stated in terms of the internal model and optimal estimator, the otolith signals are given a significant weight at low frequencies as indicators of head tilt on earth. The optimal control model of spatial orientation predicted how semicircular canal signals, weighted mostly at higher frequencies, would delay the onset of roll vection and how otolith signals would limit its steady state magnitude. The five Spacelab experiments, beginning in 1983 and continuing

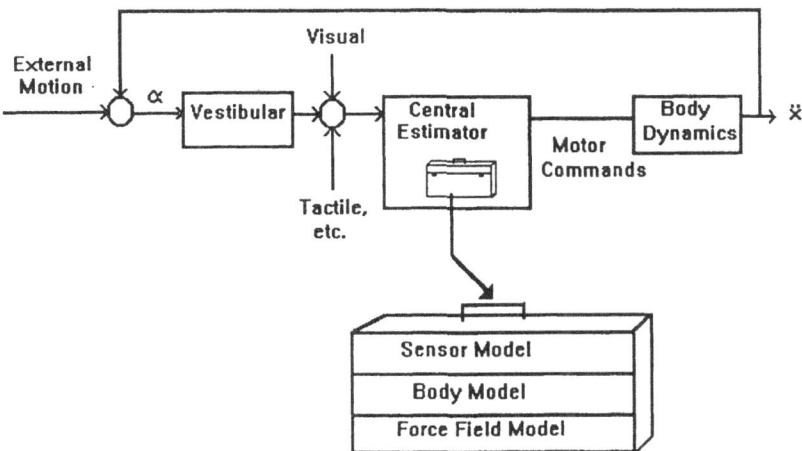

Figure 1. Block diagram of internal model, posture control representation.

through Spacelab Life Sciences-2 in 1993, demonstrated that in the absence of dependable otolith tilt information in space all subjects rearranged their multi-sensor spatial orientation estimator, with most placing increased weighting on visual cues at lower frequencies. The latest experiments explored the role of localized tactile cues (pressure on the feet) and non-localized tactile cues (from a bite board) on the visually induced tilt. The mere presence of any tactile cue, even if it provided no information about the presence or absence of body tilt or direction of sway, served to inhibit the dependence on visual field motion in determining perceived self motion in space. Surprisingly, the adaptive process continued over the course of the 14 day mission, and subjects generally became increasingly reliant upon their internal body reference cues (down is where the feet are) and were progressively less influenced by any external cues. Detailed results of the rotating dome experiments are given in the series of papers for each mission (Young et al., 1986; Young and Shelhamer, 1990; Watt et al., 1992; Young et al 1992; Young et al., 1994).

REFERENCES

Borah, J., Young, L.R., and Curry, R.E., 1978, Sensory mechanism modeling", AFHRL-TR-78-83, 1978.

Borah, J., Young, L.R., and Curry, R.E., 1988, Optimal estimator model for human spatial orientation", in Representation of Three-Dimensional Space in the Vestibular, Oculomotor, and Visual Systems: A Symposium of the Barany Society, Cohen, B. and Henn, V., eds., Annals of the New York Academy of Sciences, 545:51-73, New York.

Dichgans, J., Held, R., Young, L.R., and Brandt, Th., 1972, Moving visual scenes influence the apparent direction of gravity", *Science* 178:1217-1219.

Kenyon, R.V., and Young, L.R, 1986, MIT/Canadian vestibular experiments on the Spacelab-1 mission: 5. Postural responses following exposure to weightlessness", *Experimental Brain Research* 64:335-346.

Merfeld, D.M., Young, L.R., Oman, C.M., and Shelhamer, M.J., 1993, A multidimensional model of the effect of gravity on the spatial orientation of the monkey", *Journal of Vestibular Research*, 3(2):141-161.

Oman, C.M., 1990, Motion sickness: a synthesis and evaluation of the sensory conflict theory", *Canadian Journal of Physiology and Pharmacology*, 68:294-303.

Oman, C.M., and Shubentsov, I., 1992, Space sickness symptom severity correlates with average head acceleration", in Mechanisms and Control of Emesis, eds., Bianchi, A.L., Grelot, L., Miller, A.D., and King, G.L., Colloque INSERM/John Libbey Eurotext Ltd. 233:185-194.

Ormsby, C.C., and Young, L.R., 1977, Integration of semicircular canal and otolith information for multisensory orientation stimuli", *Mathematical Biosciences* 34:1-21.

Watt, D.G.D., Landolt, J.P., and Young, L.R., 1992, Effects of long-term weightlessness on roll circularvection", Proceedings of the Seventh CASI Conference on Astronautics, Ottawa.

Young, L.R., Jackson, D.K., Groleau, N., and Modestino, S.A., 1991, Multisensory integration in microgravity", in Sensing And Controlling Motion: Vestibular and Sensorimotor Function, Cohen, B., Tomko, D.L., and Guedry F., eds., *Annals of the New York Academy of Sciences* 656:340-353. Originally presented at the NYAS conference, Sensing and Controlling Motion, Palo Alto.

Young, L.R., Jackson, D.K., Groleau, N., and Modestino, S.A., 1992, Multisensory integration in microgravity, in Sensing And Controlling Motion: Vestibular and Sensorimotor Function, Cohen, B., Tomko, D.L. and Guedry F., eds., *Annals of the New York Academy of Sciences* 656:340-353.

Young, L.R., Mendoza, J.C., Groleau, N., and Wojcik, P.W., 1994, Tactile Influences on Astronaut Visual Spatial Orientation: Human Neurovestibular Experiments on Spacelab Life Sciences 2", (submitted to *J Applied Physiol.*).

Young, L.R., Oman, C.M., Curry, R.E., and Dichgans, J.M., 1973, A descriptive model of multi-sensor human spatial orientation with applications to visually induced sensations of motion, AIAA Paper No. 73-915, AIAA Visual and Motion Simulation Conference, Palo.

Young, L.R., Shelhamer, M., 1990, Microgravity enhances the relative contribution of visually-induced motion sensation, *Aviation Space Environmental Medicine* 61:525-530.

Young, L.R., Shelhamer, M., and Modestino, S., 1986, MIT/Canadian vestibular experiments on the Spacelab-1 mission: 2. Visual vestibular tilt interaction in weightlessness ", *Experimental Brain Research* 64:299-307.

INDEX

The manufacturer's authorised representative in the EU is Springer
Nature Customer Service Centre GmbH, Europaplatz 3, 69115 Heidelberg,
Germany. If you have any concerns regarding our products, please
contact ProductSafety@springernature.com

Printed and bound by CPI Group (UK) Ltd, Croydon, CR0 4YY
23/04/2026
02095628-0013